T0257480

Transgenic Plants: Developments and Concerns

Transgenic Plants:
Developments and Concerns

Edited by **Harvey Parker**

New York

Published by Callisto Reference,
106 Park Avenue, Suite 200,
New York, NY 10016, USA
www.callistoreference.com

Transgenic Plants: Developments and Concerns
Edited by Harvey Parker

International Standard Book Number: 978-1-63239-598-6 (Hardback)

Printed in the United States of America.

Contents

Preface

The world is advancing at a fast pace like never before. Therefore, the need is to keep up with the latest developments. This book was an idea that came to fruition when the specialists in the area realized the need to coordinate together and document essential themes in the subject. That's when I was requested to be the editor. Editing this book has been an honour as it brings together diverse authors researching on different streams of the field. The book collates essential materials contributed by veterans in the area which can be utilized by students and researchers alike.

This book on transgenic plants is an essential contribution to the field as it includes information by veteran researchers from across the world. Progress of effective transformation protocols is becoming an integral strategy to traditional breeding methods for the enhancement of crops. This book contains the novel developments carried regarding the optimization of transformation techniques. It is structured into two sections namely, Application and Crop Improvement. This book will serve as a valuable information guide to students, researchers and botanists.

Each chapter is a sole-standing publication that reflects each author's interpretation. Thus, the book displays a multi-facetted picture of our current understanding of application, resources and aspects of the field. I would like to thank the contributors of this book and my family for their endless support.

Editor

Part 1

Application

Recent Advances in Fruit Species Transformation

Hülya Akdemir[1], Jorge Gago[2],
Pedro Pablo Gallego[2] and Yelda Ozden Çiftçi[1,*]
*[1]Gebze Institute of Technology, Department of Molecular Biology and Genetics,
Plant Biotechnology Laboratory, Kocaeli,
[2]Applied Plant and Soil Biology, Faculty of Biology,
University of Vigo, Vigo,
[1]Turkey
[2]Spain*

1. Introduction

Rapid increase of human population together with global climate variability resulted in increased demand of plant based food and energy sources (Varshney et al., 2011). Fruits and nuts have essential role to enhance quality of humankind life since a diet based on cereal grains, root and tuber crops, and legumes is generally lacked a wide range of products such as fiber, vitamin, provitamins or other micronutrients and compounds exist in fruit and nut species (Heslop-Harrison, 2005). According to last FAOSTAT statistics, totally about 594.5 million t fruit crops (except melons) were produced in the world in 2009 (http://faostat.fao.org). Because an increase demand exists in global food production, many economically important fruit crops production need to be improved, however, conventional breeding is still limited due to genetic restrictions (high heterozygosity and polyploidy), long juvenile periods, self-incompatibility, resources restricted to parental genome and exposed to sexual combination (Akhond & Machray, 2009; Malnoy et al., 2010; Petri et al., 2011). Thus, there is an urgent need for the biotechnology-assisted crop improvement, which ultimately aimed to obtain novel plant traits (Petri & Burgos, 2005).

Plant genetic engineering has opened new avenues to modify crops, and provided new solutions to solve specific needs (Rao et al., 2009). Contrary to conventional plant breeding, this technology can integrate foreign DNA into different plant cells to produce transgenic plants with new desirable traits (Chilton et al., 1977; Newell, 2000). These biotechnological approaches are a great option to improve fruit genotypes with significant commercial properties such as increased biotic (resistance to disease of virus, fungi, pests and bacteria) (Ghorbel et al., 2001; Fagoaga et al., 2001; Fagoaga et al., 2006; Fagoaga et al., 2007) or abiotic (temperature, salinity, light, drought) stress tolerances (Fu et al., 2011); nutrition; yield and quality (delayed fruit ripening and longer shelf life) and to use as bioreactor to produce proteins, edible vaccines and biodegradable plastics (Khandelwal et al., 2011).

---

*Corresponding Author

Currently, public concerns and reduced market acceptance of transgenic crops have promoted the development of alternative marker free system technology as a research priority, to avoid the use of genes without any purpose after the transformation protocol as selectable and reporter marker genes. Typically, it is employed for the selection strategy that confers resistance to antibiotics and to herbicides (Miki & McHugh, 2004; Manimaran et al., 2011). A large proportion of European consumers considered genetically modified crops as highly potential risks for human health and the environment. European laws are restrictive and do not allow the deliberate release of plant modified organism (Directive 2001/18/EEC of the European Parliament and the Council of the European Union). Under these premises, great efforts have also been realized to develop alternative marker free technologies in fruit species. Recently, it was demonstrated in apple and in plum, that transgenic plants without marker genes can be recovered and confirmed its stability by molecular analysis (Malnoy et al., 2010; Petri et al., 2011). In 2011, for first time it was described authentically "cisgenic" plants in apple cv. Gala (Schouten et al., 2006a,b; Vanblaere et al., 2011).

Efficient regeneration systems for the generation of transgenic tissues still appear as an important bottleneck for most of the species and cultivars. In the literature, different protocols were described to transform fruit cells using various DNA delivery techniques, however the attempts generally focused on transformation via *Agrobacterium* or microprojectile bombardment. In this chapter, a detailed application of these techniques in fruit transformation is summarized together with usage of proper marker and selection systems and *in vitro* culture techniques for regeneration of the transgenic plants.

2. Techniques used to transform fruit species

Improvement of the plant characteristics by transfer of selected genes into fruit plant cells is possible mainly through two principal methods: *Agrobacterium*-mediated transformation and microprojectile bombardment (also called "biolistic" or "bioballistic"). Soil-borne Gram negative bacteria of the genus *Agrobacterium* infect a wound surface of the plants via a plasmid called Ti-plasmid containing three genetically important elements; *Agrobacterium* chromosomal virulence genes (*chv*), T-DNA (transfer DNA) and Ti plasmid virulence genes (*vir*) that constitute the T-DNA transfer machinery. Since Ti plasmid encodes mechanisms of integration of T-DNA into the host genome, it is used as a vector to transform plants.

Since direct gene transfer procedures involve intact cells and tissues as targets, in some species breaching of the cell wall is needed in order to enable entrance of DNA to cell (Petolino, 2002). This is accomplished by making some degree of cell injury or totally enzymatic degradation of the cell wall. Advantages of microprojectile bombardment can be summarized as i) transfer of multiple DNA fragments and plasmids with co-bombardment, ii) unnecessity pathogen (such as *Agrobacterium*) infection and usage of specialized vectors for DNA transfer (Veluthambi et al., 2003). Although microprojectile bombardment eliminates species-dependent and complex interaction between bacterium and host genome, stable integration is lower in this technique in comparison to *Agrobacterium*-mediated transformation (Christou, 1992). Moreover, the existence of truncated and rearranged transgene DNA can also lead transgene silencing in the transgenic plants (Pawlowski & Somers, 1996; Klein & Jones, 1999; Paszkowski & Witham, 2001). On the other hand, other important requirement for this technique is that the explants or target cells have to be physically available for the bombardment (Hensel et al., 2011). Also, it was described that transgenic explants regenerated can be chimeric (Sanford et al., 1990). Nevertheless, application of both of the techniques for the transfer of foreign DNA results in "transient" or "stable" expression of the DNA fragment. In the following sections, recent advances in genetic transformation of fruit species via *Agrobacterium*-mediated and direct gene transfer techniques are presented.

2.1. *Agrobacterium*-mediated gene transfer

2.1.1 A complex relationship

In the decade of the 80´s, the first reports were published related to the introduction of foreign DNA in plant genome thanks to the Ti plasmid of *Agrobacterium tumefaciens* (De Block et al., 1984; Horsch et al., 1985). After more than 25 years, *Agrobacterium*-mediated gene transfer is still the most used method for fruit species transformation including apple, almond, banana, orange, grapevine, melon etc. (Table 1; Rao et al., 2009).

Plant transformation by using *Agrobacterium* has some advantages since the technique is relatively simple; transfer and integration of foreign DNA sequences with defined ends (left and right borders of T-DNA) is precise; stable transformation is high; transgene silencing is typically low and long T-DNA sequences (>150 kb) can be transferred (Veluthambi et al., 2003). However, it is still far from to be a routine transformation application in plants because of its host-range restrictions (Gelvin, 1998).

The initial drawback of *Agrobacterium*-mediated transformation method is the host-range restrictions. However, the bacterium and the target tissue can be manipulated to overcome this obstacle (Trick & Finer, 1997). These authors proposed a new approach to facilitate *Agrobacterium* penetration into plant tissues, the sonication assisted *Agrobacterium*-mediated transformation (SAAT) method. This method consists the use of ultrasounds to produce cavitations on and below the plant surfaces and into the membrane cells, wounding the tissues to enhance *Agrobacterium* infection (Trick & Finer, 1997, 1998).. Also, this method can be combined with vacuum infiltration to promote bacteria agglutination around the tissues to increase the *Agrobacterium* infection as it was demonstrated in kidney bean (Liu et al., 2005) and in woody plants as *Eucalyptus* (Villar et al., 1999; Gallego et al., 2002; Gallego et al., 2009). Today, application of these modifications solely or in combination with other approaches has made it possible to transfer foreign DNA via *Agrobacterium* even to monocots (Hiei et al., 1994; Ishida et al., 1996; Hensel et al., 2011) which were initially transformed with direct gene transfer methods since *Agrobacterium* is not a natural host. Following its first successful usage in soybean and Ohio buckeye (Trick & Finer, 1997), SAAT was also applied recently to fruit species including orange (Oliveira et al., 2009); banana (Subramanyam et al., 2011) and grapevine (Gago et al., 2011). In the last paper the developed efficient methodology that combined SAAT with vacuum infiltration allowed to obtain reporter gene expression in different newly formed organs such as stems, petioles and leaves. Expression was related to vascular tissues due to the *EgCCR* promoter of *Eucalyptus gunnii* and demonstrated that its activity is conserved and fully functional in grapevine as it was shown by *uidA* (GUS) and *gfp* reporter marker genes. Transgenic grapevine lines were verified by Southern blot analysis in five randomly chosen transgenic lines showing simple integration patterns in four lines with different band length indicating independent transformation events into the grapevine genome. We also applied the optimized protocol to pistachio nodes to reveal out if this method of transformation and vascular-specific promoter of eucalyptus, also works in this species. Histological observations of GFP activity presence in vascular bundles and leaves (Fig. 1) together with PCR amplification of 858 bp fragment of *nptII* and 326 bp of *uidA* (Fig. 2) genes confirmed not only gene integration but also showed that SAAT in combination with vacuum infiltration and vascular specific promoter could also be used for pistachio transformation. With PCR amplification, four out of five GFP+ putative transgenic shoots showed the amplified bands of *nptII* and *uidA* genes (Fig 2).

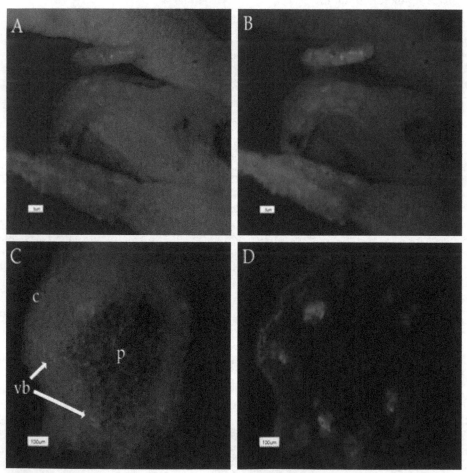

Fig. 1. Expression of GFP in pistachio transformed with *EgCCR-GFP-GUS* construct. Fluorescent images of different tissues and organs were taken 6 months after *Agrobacterium*-mediated transformation. GFP fluorescence in shoot apex (A-B, bars represent 5μm) and transverse section of the transgenic microshoot (C-D, bars represents 100μm) were carried out using a 480/40 nm exciter filters, and two-barrier filter >510 nm (wide range) and 535/550 nm (specific filter for GFP fluorescence). (Abbreviations: vascular bundles (vb), pith, p; cortex, c).

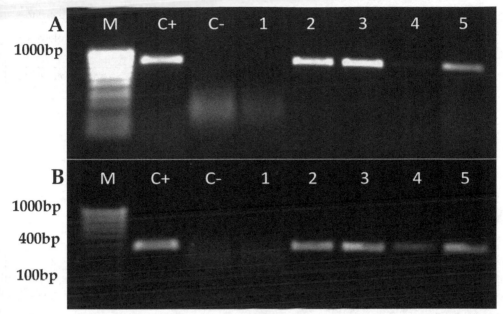

Fig. 2. Analysis of putative transgenic pistachio plants by PCR amplification using primers designed for 858 bp fragment of the *nptII* gene (A) and for a fragment of 326 bp of *uid*A gene (B). (M: DNA 1000 bp ladder, C+: positive control, C-: untransformed plant, T1-T5 putative transgenic shoots lines).

Agrobacterium-mediated transformation is highly genotype dependent for many plants (Pérez-Piñeiro, 2012) but also for fruit species. Different reports described that some cultivars were completely found to be as highly recalcitrant for transformation process whereas others are completely successful (Galun & Breiman, 1998; Petri & Burgos, 2005). This problem is widely described in different fruit species as apricot, grapevine and others (López-Pérez et al., 2008; Wang, 2011). López-Pérez and collaborators (2009) described that grapevine cultivars "Crimson Seedless" and "Sugraone" obtained different transformation efficiencies depending on the optical densities tested. Transformation of hypocotyls obtained from germination of mature seeds and nodal explants of apricot cultivars Dorada, Moniquí, Helena, Canino, Rojo Pasión and Lorna resulted in different transformation efficiencies (Wang, 2011). Some authors pointed out that one of the main goals of plant genetic engineering must be the development of genotype-independent transformation procedures, however due to this highly complex plant-pathogen interaction it will be very difficult to achieve this with the currently available technologies (Petri & Burgos, 2005).

Species	Aim	Plasmid	Transgenes	References
Apple				
Malus x domestica	Method optimisation	pBIN6	*nptII, nos*	James et al., 1989
M. x domestica	Investigation of early events in transformation	pDM96.0501	*sgfp, gusA,nptII*	Maximova et al., 1998

Species	Aim	Plasmid	Transgenes	References
M. x domestica	Investigation of influence of rolA gene on shoot growth	pMRK10	*rolA, nptII*	Holefors et al., 1998
M. x domestica	Scab resistance	p35S-ThEn42, pBIN19ESR	*ech42, nptII*	Bolar et al., 2000
M. x domestica	Resistance to fireblight	pLDB15	*attE, nptII, gusA*	Ko et al., 2000
M. x domestica	Improve rooting ability	pCMB-B	*rolB, nptII, gusA*	Zhu et al., 2001
M. x domestica	Scab resistance	pBIN(Endo+Nag)	*ech42, nag70, nptII*	Faize et al., 2003
M. x domestica	Self-fertility	pGPTV-KAN	*S₃RNase, nptII*	Broothaerts et al., 2004
M. x domestica	Method optimisation Enhance rooting	pCMB-B	*rolB, nptII, gusA*	Radchuk & Korkhovoy, 2005
M. x domestica	Method optimisation	pNOV2819	*pmi, nptII, gusA*	Degenhardt et al., 2006
M. x domestica	Investigation of function of ARRO-1 in adventitious rooting	pK7GWIWG2 (II)	*ARRO-1, nptII*	Smolka et al., 2009
M. x domestica	Stability of scab resistance	pMOG402.hth.gus.intron	*Hth, nptII, gusA*	Krens et al., 2011
M. x domestica	Development of selection system	pCAMBIAVr-ERE-GUS	*VrERE, gusA*	Chevreau et al., 2011
M. x domestica	Transformation without selectable marker gene	pPin2Att.35SGUSint+.n pPin2MpNPR1.GUS−.n ptII		Malnoy et al., 2010
Almond				
Prunus dulcis	Method optimisation	pBI121mgfp-5-ER pNOV2819	*nptII pmi*	Ramesh et al., 2006
Avocado				
Persea americana Mill.	Method optimisation	pMON9749, pTiT37-SE	*nptII, gusA*	Cruz-Hernandez et al., 1998
Banana				
Musa spp.	Method optimisation (Agro + SAAT+ Vacuum infiltration)	pCAMBIA1301	*hptII, gusA*	Subramanyam et al., 2011
Musa spp.	Resistance to Fusarium wilt	pBI121-PFLP	*pflp, nptII*	Yip et al., 2011
Blueberry				
Vaccinium spp.	Method optimisation	p35SGUS-int	*gusA*	Cao et al., 1998

Species	Aim	Plasmid	Transgenes	References
Blueberry				
V. corymbosu m L.	Method optimisation	pBISN1	*nptII, gusA*	Song & Sink, 2004
Grapevine				
Vitis vinifera	Method optimisation	Nr	*gusA, nptII*	Nakano et al., 1994; Gago et al., 2011
V. rootstocks	Resistance to viruses and crown gall	pBIN19 pGA482G	*mutant virE2, nptII GLRaV-3cp*	Xue et al., 1999
V. vinifera	Resistance to fungal pathogens	pBI121	*nptII rice chitinase gene*	Yamamoto et al., 2000
V. vinifera	Fungal resistance	pGJ42	*chitinase, rip, nptII*	Bornhoff et al., 2005
V. vinifera	Method optimisation	pGA643	*nptII, GFLVCP*	Maghuly et al., 2006
V. vinifera	Method optimisation	Nr	*egfp, nptII*	Dutt et al., 2007
V. vinifera	Resistance to powdery mildew	pGL2	*ricechitinase gene, hgt*	Nirala et al., 2010
V. vinifera	Method optimisation	pBin19-sgfp	*nptII, sgfp*	Pérez-López et al., 2008
V. vinifera	Method optimisation	pSGN	*nptII, egfp*	Li et al., 2006
V. vinifera	Method optimisation	pCAMBIA2301	*nptII, gusA*	Wang et al., 2005
Grapefruit				
Citrus paradisi	Resistance to Citrus tristeza virus	pGA482GG	*CP, RdRp, gusA, nptII*	Febres et al., 2003
C. paradisi	Resistance to Citrus tristeza virus	pGA482GG	*CP, gusA, nptII*	Febres et al., 2008
Kiwifruit				
Actinidia spp.	Hairy root induction	A722,C58, ICMP8302, ICMP8326, ID1576, LBA 4404, A4T	*gusA, nptII*	Atkinson et al., 1990
Actinidia spp.	Method optimisation	pLAN411, pLAN421	*gusA, nptII*	Uematsu et al., 1991
A.deliciosa	Improved rooting	pBIN19	*nptII, rol A,B,C*	Rugini et al., 1991
A. eriantha	Method optimisation	pART27-10	*gusA, nptII*	Wang et al., 2006
A. deliciosa	Manipulation of plant architecture	pBI121	*ipt*	Honda et al., 2011

Species	Aim	Plasmid	Transgenes	References
Mango				
Magnifera indica L.	Method optimisation	pTiT37-SE::pMON9749	*nptII, gusA*	Mathews et al., 1992
M.indica L.	Methodoptimisation	pGV3850::1103	*nptII*	Mathews et al., 1993
M.indica L.	Mediate ethylene biosynthesis	pBI121	*nptII, gusA antisense ACC oxidase, antisense ACC synthase*	Cruz Hernandez et al., 1997
M. indica L.	Rooting enhancement	Nr	*rol B*	Chavarri et al., 2010
Melon				
Cucumis melo	Resistance to ZYMV, TEV, PVY	FLCP core AS	*nptII, ZYMV coatpr.*	Fang & Grumet, 1993
C. melo	Salt resistance	pRS655	*nptII, gusA, hal1*	Bordas et al., 1997
C. melo	Resistance to ZYMV	pBI-ZCP3'UTR	*nptII, ZYMV coatpr.*	Wu et al., 2009
Nectarberry				
Rubus arcticus	Method optimisation	pFAJ3001	*gusA*	Kokko & Kärenlampi, 1998
Orange				
Citrus sinensis	Method optimisation (Agro + SAAT+ vacuum infiltration)	pGA482GG	*gusA, nptII*	Oliveira et al., 2009
C. sinensis	Research on expression of Mt-GFP	pBI. mgfp4.coxIV	*Mt-gfp*	Xu et al., 2011
C. sinensis	Influence of methylation on gene expression	pBIN.mgfp5-ER	*gfp, nptII*	Fan et al., 2011
C. sinensis	Modification of gibberellin levels	pBinJIT-CcGA20ox1-sense pBinJITCcGA20ox1-antisense	*nptII, CcGA20ox1 nptII, CcGA20ox1*	Fagoaga et al., 2007
C. sinensis	Resistance to fungi	pBI121.*P23*	*nptII, PR-5*	Fagoaga et al., 2001
C. aurantifolia	Resistance to virus	pBin19-sgfp	*nptII, sgfp, p23*	Fagoaga et al., 2006
Poncirus trifoliate	Enhanced salt tolerance	pBin438	*nptII, AhBADH*	Fu et al., 2011
Papaya				
Carica papaya	Resistance to PRSV	pRPTW	*PRSV replicase gene, neo*	Chen et al., 2001

Species	Aim	Plasmid	Transgenes	References
Pear				
Pyrus communis	Alter growth habit	pGA-GUSGF	*rolC, gusA, nptII*	Bell et al., 1999
P. communis	Method optimisation	pPZP pME504	*gusA, nptII*	Yancheva et al., 2006
P. communis	Method optimisation	PBISPG	*nptII, gusA*	Sun et al., 2011
Peanut				
Arachis hypogaea	Production of edible vaccines for *Helicobacter pylori*	pBI121.Oleosin-UreB	*ureB, nptII*	Yang et al., 2011
A. hypogaea	Improvement of salt and drought resistance	pGNFA-(pAHC17)	*AtNHX1*	Asif et al., 2011
A. hypogaea	Production of vaccines for Peste des petits ruminants (PPR)	pBI121	*Hn*	Khandelwal et al., 2011
Plum				
Prunus armeniaca	Method optimisation	pBIN19-sgfp	*nptII, gfp*	Petri et al., 2004
P. armeniaca	Method optimisation	pBIN19-sgfp, p35SGUSint	*nptII, gfp/nptII, gusA*	Petri et al., 2008
P. domestica	Transformation of marker free plants	pCAMBIAgfp94(35S) / pGA482GGi ihpRNAE10´	*nptII, gfp, gusA, ppv-cp*	Petri et al., 2011
P. domestica	New selection system with hygromycin	pC1381, pC1301, pC2301	*gusA, hpt, nptII*	Tian et al., 2009
P. domestica	Control of PPV infection	pGA482GG	*nptII, gusA, PRVcp*	Scorza et al., 1995
P. salicina	Method optimisation	pCAMBIA2202	*nptII, gfp*	Urtubia et al., 2008
Pomegranate				
Punica granatum	Method optimisation	pBIN19-sgfp	*nptII, gfp*	Terakami et al., 2007
Strawberry				
Fragaria spp.	Method optimisation	pBI121	*nptII, gusA*	Barcelo et al., 1998
Fragaria x ananassa Duch.	Modulation of fruit softening	pBI121	antisense of *endo-β-1,4-glucanase*	Lee & Kim, 2011
White mulberry				
Morus alba	Method optimisation	pBI121	*nptII, gusA*	Agarwal & Kanwar, 2007

Table 1. Some important reports on genetic transformation of fruit species via *A. tumefaciens* or *A. rhizogenes*.

Some abbreviations: *AtNHX1*: a vacuolar type Na+/H+ antiporter gene; *gfp*: green fluorescent protein coding gene; *hal1*: yeast salt tolerance gene; *hpt*: hygromycin phosphotransferase coding gene; *ipt*: isopentyl transferase gene; *neo*: neomycin phosphorate transferase coding gene; *nos*, nopaline synthase coding gene; *nptII*, neomycin phosphtransferase II coding gene; *pmi*: phosphomannose isomerase coding gene; *ppv*: Plum pox virus; *prsv*: papaya ringspot virus; *pvy*: potato virus Y; *tev*: tobacco etch virus; *gusA (uidA)*: β-glucuronidase coding gene; *UreB*: antigen gene; *zymv*: zucchini yellow mosaic virus.

2.2 Direct gene transfer

Direct gene transfer techniques include microprojectile bombardment, microinjection, electroporation, and usage of whiskers. Among them, microprojectile bombardment is an alternative technique of *Agrobacterium*-mediated transformation since its physical nature overcomes biological barriers and enables naked DNA delivery directly into host genome or alternatively into mitochondria and chloroplasts. In this technique, plasmid or linearized DNA-coated metal microparticles (gold or tungsten) at high velocity is bombarded to intact cells or tissues (Sanford et al. 1987; Klein et al. 1987; Sanford, 1988). Furthermore, biological projectiles such as bacteria (i.e., *E. coli, Agrobacterium*), yeast and phage associated with tungsten can also be used in microprojectile bombardment (Bidney, 1999; Kikkert et al. 1999).

Microprojectile bombardment was developed in the 1980s for transformation of plants which were recalcitrant to *Agrobacterium*–mediated transformation (Paszkowski et al., 1984) such as agronomically important cereals. Following the development of the first particle delivery system (Sanford et al. 1987; Sanford 1988), different effective devices such as PDS-1000/He, Biolistic® particle delivery system; particle inflow gun; electrical discharge particle acceleration; ACCELL™ technology and microtargeting bombardment device were also evolved to improve transformation capacity. Among them, PDS-1000/He, Biolistic® particle delivery system (BIO-RAD), which is a modified version of Sanford's system, is the most used system for biolistic transformation due to its efficient and relatively simple application and acquisition of reproducible results between laboratories (Taylor & Fauquet, 2002). Particle inflow gun can be an alternative to other biolistic systems due to its very low cost and it was used successfully in banana transformation (Becker et al., 2000). Electrical discharge particle acceleration, ACCELL™ technology uses high voltage electrical discharge into a droplet water to generate shock waves and project microprojectiles to different cell layers of target tissues (McCabe & Christou, 1993). Microtargeting bombardment device was designed for shoot meristem transformations (Sautter, 1993) but it is not widely used for plant transformation. All of the microprojectile bombardment systems are not depend on any plant cell type but target cells which will be bombarded need to be physically accessible (Hensel et al., 2011).

Particle bombardment were carried out not only to optimize plant transformation but also to transfer gene constructs encoding for various antimicrobial peptides or proteins for fungal resistance against to *Fusarium oxysporum* f. sp. cubense and *Mycospaerella fijiensis* or preharvest and postharvest diseases *Verticillium theobromae* or *Trachysphaera fructigen* (i.e., Remy et al., 2000; Sagi et al., 1998; Tripathi, 2003), virus (i.e., Fitch et al., 1992; Tennant et al., 1994; Gonsalves et al., 1994; Scorza et al., 1996), pest (i.e., Serres et al., 1992) and herbicide tolerance (i.e., Zeldin et al., 2002). This technique has been applied to transformation of various fruit species including banana, cranberries, citrus, grapevine, melon, papaya and peanut (Table 2).

Species	Aim	Transfer system	Plasmid	Transgenes	References
Apple					
Malus x domestica	Method optimisation	PEG-mediated	pKR10	*Gfp*	Maddumage et al., 2002
Banana					
Musa spp.	Method optimisation	Particle bombardment	pUbi-BtintORF1 pBT6.3-Ubi-NPT pUbi-BTutORF5 pBT6.3-Ubi-NPT pUGR73 pDHkan	*nptII, gusA, BBTV*	Becker et al., 2000
Musa spp.	Tolerance to Sigatoka leaf spot	Particle bombardment	pYC39	*ThEn-42, StSy, Cu, Zn-SOD*	Vishnevetsky et al., 2011
Musa spp.	Resistance to virus	Particle bombardment	pAB6, pAHC17,pH1	*gusA, bar,ubi, BBTV-G-cp*	Ismail et al., 2011
Cranberry					
Vaccinim macrocarpon	Method optimisation & Pest control	Particle bombardment	pTvBTGUS	*nptII, gusA, Bt*	Serres et al., 1992
V. macrocarpon	Herbicide resistance	Particle bombardment	pUC19	*bar, aphII*	Zeldin et al., 2002
Grapevine					
Vitis vinifera	Method optimisation	Biolistic	pBI426	*nptII, gusA*	Hebert et al., 1993
V. vinifera	Method optimisation	Particle bombardment & Agro	pGA482GG	*nptII, gusA TomRSV-CP*	Scorza et al., 1996
V. vinifera	Method optimisation	Biolistic	pSAN237	*nptII, magainin,PGL*	Vidal et al., 2003
V. vinifera	Comparison of minimal cassette with standard circular plasmids	Biolistic	pSAN168, pSAN237	*Magainin, nptII*	Vidal et al., 2006
Kiwifruit					
Actinidia spp.	Method optimisation	PEG 4000	pDW2	*Cat*	Oliveira et al., 1991
Actinidia spp.	Method optimisation	Electroporation	pBI121, pTi35SGUS	*gusA, nptII*	Oliveira et. al., 1994
A. deliciosa	Method optimisation	PEG 4000	p35SGUS	*gusA*	Raquel & Oliveira, 1996

Species	Aim	Transfer system	Plasmid	Transgenes	References
Melon					
Cucumis melo	Protection against	Particle bombardment İnfection & Agro	pGA4822GG/ CP	nptII, gusA, CMV-WLCP	Gonsalves et al., 1994
Papaya					
Carica papaya	PRV resistance	Particle bombardment	pGA482GG	PRV, nptII	Fitch et al., 1992
C. papaya	PRV resistance	Particle bombardment	pGA482GG	nptII, gusA, cpPRVHA	Tennant et al., 1994
C. papaya	Control of PRSV	Particle bombardment	pGA482GG	cpPRSV-pHA5, nptII, gusA	Cai et al., 1999
C. papaya	Method optimisation	Particle bombardment	pCAMBIA130 3 pML202	hpt, nptII, mgfp5'	Zhu et al., 2004
C. papaya	Use of PMI/Man	Particle bombardment	pNOV3610	Pmi	Zhu et al., 2005

Table 2. Some important reports on genetic transformation of fruit species via direct gene transfer.

A successful protocol was studied very recently in banana cv. Williams apical meristems with microprojectile bombardment of a new construct pRHA2 plasmid containing *bar* and coat protein of banana bunchy top nanovirus *(BBTV-cp)* genes that encoded the viral coat protein by using Biolistic™ PDS-1000/He system, 650 psi helium pressures and 5 μg DNA/shot for acquisition of virus resistance (Ismail et al., 2011). After bombardment, 62% of apical meristems were survived on the selective medium and 80% of explants produced shoots in the following first subculture and all shoots were rooted (Ismail et al., 2011). In addition to those disease-based studies, others were also carried out in order to develop efficient transformation protocols via biolistic transformation (Sagi et al., 1995; Becker et al., 2000). Among them, Sagi and co-workers (1995) reported the transformation of embryogenic cell suspensions of cooking banana 'Bluggoe' (ABB genome) and plantain 'Three Hand Planty' (AAB genome) via particle bombardment. Then, Cavendish banana cv. Grand Nain embryogenic suspension cells were co-bombarded with the plasmid containing *nptII* selectable marker gene under the control of *BBTV* promoter or the cauliflower mozaic virus *(CaMV)* 35S promoter, the β-glucuronidase *(gusA)* reporter gene and *BBTV* genes under the control of the maize polyubiquitin promoter by using particle inflow gun and stably integration was obtained in all of the tested transformed plants (Becker et al., 2000). Very recently, microprojectile bombardment was also applied to induce tolerance to Sigatoka leaf spot caused by *Mycosphaerella fijiensis* in banana by transferring endochitinase gene of *ThEn-42* from *Trichoderma harzianum* together with the grape stilbene synthase gene *(StSy)* under the control of 35 S promoter and the inducible PR-10 promoter, respectively (Vishenevetsky et al., 2011). Moreover, in order to improve scavenging of free radicals generated during fungal attack, the superoxide dismutase gene *(Cu, Zn-SOD)* of tomato was also included to this gene cassette under the control of ubiquitin promoter. Both PCR and Southern blot analysis confirmed the stable integration of the transgenes and 4-year field trial showed that several transgenic banana lines had improved tolerance not only to Sigatoka but also other fungus such as *Botrytis cinerea*. Gene transfer via microprojectile bombardment was also carried out in American cranberry (*Vaccinium macrocarpon*) firstly to increase productivity by transferring *Bacillus thuringiensis* subsp. Kurstaki crystal protein gene (*Bt*) for pest resistance (Serres et al., 1992), and latter on, by *bar* gene to confer tolerance to the phosphinothricin-based herbicide glufosinade (Zeldin et al., 2002). Although preliminary bioassays for efficiency of the *Bt* gene against an important lepidopteran demonstrated no consistently effective control in former, stable transmission and expression of herbicide tolerance was observed in both inbred and outcrossed progeny of cranberry trans clone in latter.

In tangelo (*Citrus reticulata* Blanco × *C. paradisi* Macf.) cv. Page embryogenic suspension cells were bombarded with tungsten coated plasmid containing *gusA* and *nptII* genes (Yao et al., 1996). Following to bombardment, 600 transient and 15 stable transformants were obtained and integration of the interest genes confirmed by PCR and Southern blot analyses. A large of kanamycin-resistant embryogenic calli showed also GUS activity. In another study, Kayim and associates (1996) bombarded tungsten-coated plasmid (pBI221.2) containing the *gusA* gene into lemon cv. Kütdiken nucellar cells by biolistic device and expression of the *gusA* gene was histochemically confirmed.

Feronia limonia L. is important fruit tree because of its edible fruits. It is suitable for cultivation in semi-arid tropics and also can be used for reforestation and wasteland reclamation projects (Sing et al., 1992; Purohit et al., 2007). *Feronia limonia* L. hypocotyl segments were also bombarded with tungsten-coated plasmid pBI121 having *gusA* reporter gene driven by *CaMV35S* promoter and *nptII* as a selective marker under control of *nos* promoter using Biolistic™ PDS-1000/He particle delivery system at different rupture disc pressures (1100 and 1350 psi) and target distances (6 and 9 cm) (Purohit et al., 2007). This study revealed that 1100 psi/6 cm and 1350 psi/9 cm were the optimal bombardment condition with supplying a maximum 90% of GUS transient expression.

In grapevine, the initial transformation studies via microprojectile bombardment were performed for method optimization with transferring *nptII* and *gus* genes as selective and reporter marker genes, respectively (Hebert et al., 1993; Kikkert et al., 1996; Scorza et al., 1996). Later, Vidal and co-workers (2003) studied the efficiency of biolistic cotransformation in grapevine for multiple gene transfer of *nptII* and antimicrobial genes (*magainin* and *peptidyl-glycine-leucine*). The stable transformation was confirmed by *gus* gene expression, followed by PCR and Southern blot analyses of *nptII* and antimicrobial genes showed. Three years later, same research group (Vidal et al., 2006) reported the efficient biolistic transformation of grapevine by using minimal gene cassettes, which are linear DNA fragments lacking the vector backbone sequence.

Papaya is economically important and preferred another fruit species because of its nutritional and medicinal properties grown in tropical and subtropical regions (Tripathi et al., 2011). Papaya ringspot virus (PRSV) is major limiting factor in papaya production in Hawaii (Gonsalves, 1998; Fuchs & Gonsalves, 2007). First PRSV resistant papaya plants (cv. SunUp) were obtained by PDS/1000-He particle bombardment device of cv. Sunset with the transformation vector pGA482GG/cpPRV4 containing the *prsv* coat protein (*CP*) gene (Fitch et al., 1992). The PRSV resistant papaya has been commercialized, reached to end user and improved papaya is now under production in Hawaii (Tripathi et al., 2008). This study was followed by other reports mainly on improvement of PRSV tolerance in papaya via microprojectile bombardment-based transformation (Tennant et al., 1994; Cai et al., 1999; Guzman-Gonzalez et al., 2006). The deployment of transgenic papayas has showed that virus CP protein supplies durable and stable resistance to homologous strains of PRSV (Fermin et al., 2010). Moreover, no ecological influence of transgenic papayas on adjacent non-transgenic papaya trees, microbial flora and beneficial insects was evident (Sakuanrungsirikul et al., 2005). However, political and social factors have negatively affected the technology in Thailand (Davidson, 2008).

Although there are various wild peanut species having disease resistance traits, hybridization between wild and cultivars is difficult due to self-incompatibility, low frequency of hybrid seed production and linkage drag (Stalker & Simpson, 1995) and because of that genetic transformation is a practical tool to improve disease resistant cultivars. Singsit and associates (1997) transformed peanut somatic embryos with gold-coated plasmid constructs containing both *Bacillus thuringiensis* cryIA(c) and *hph* genes driven by *CaMV35S* promoters by PDS 1000 He biolistic device for resistance lepidopteran insect larvae of lesser cornstalk borer. The embryogenic cell lines showed hygromycin resistance and integration of *hph* and *Bt* genes were confirmed by PCR and/or Southern blot analyses in regenerated plants and a progeny. 18% CryIA(c) protein of total soluble protein was detected by ELISA immunoassay in the hygromycin resistant plants. Production of peanut stripe virus (PStV) resistant peanut is another attempt for biotechnologists since the virus negatively affects seed quality and yield in Asia and China (Higgins et al., 1999). Somatic embryos of peanut cv. Gajah and cv. NC-7 were transformed by co-bombardment of *hph* gene and one of two forms of the *PStV* coat protein genes and both of the transgenic plants showed high level resistance to the homologous virus isolate (Higgins et al., 2004). Transfer of anti-apoptotic genes originated from mammals, nematods or virus into plants is another approach for enhancement of plant resistance against to biotic and abiotic stresses (Chu et al., 2008a). With this aim, peanut cv. Georgia Green embryogenic callus was bombarded with anti-apoptotic *Bcl-xL* gene by microprojectile bombardment. Although Bcl-xL protein was detected in four transgenic lines, just one transgenic line (25-4-2a-19) had stable protein expression and showed tolerance to 5μM paraquat (commercial herbicide) (Chu et al., 2008a). Around 0.6% of total population in USA is affected of IgE-mediated allergic reaction following to peanut consumption (Sicherer et al., 2003). To produce hypoallergenic peanut, peanut cv. Georgia Green embryogenic cultures were also transformed via microprojectile bombardment and silenced peanut allergens (Ara h 2 and Ara h 6) by RNA interference. Expression of these allergens was not decreased effectively but, binding of IgE to the two allergens, significantly declined (Chu et al., 2008b).

Apart from microprojectile bombardment, electroporation (Oliviera et al., 1994) and PEG-mediated transformation were also carried out in apple (Maddumage et al., 2002) and kiwifruit (Raquel & Oliveira, 1996) in order to optimize transformation protocol by transferring *gusA*, *gfp* and/or *nptII*.

3. Markers and selection of transformants

3.1 Reporter genes

Reporter genes or non-selectable marker genes are commonly used components of the plasmid constructs allowing the verification of transformation and the detection of the putative transformed cells. In many fruit transformation studies, histochemical analyses of transformed cells are visualized by using β-glucuronidase (GUS) expression as a reporter gene (Jefferson, 1987; Table 1). This enzyme is encoded by E. coli uidA (gusA) gene and histochemical localization of the gene expression is detected in subcellular levels (Daniell et al., 1991). High levels of GUS is not toxic for plant and the enzyme is very stable in cells, however, the assay is destructive to plants (Miki & McHugh, 2004). gusA generally co-transformed with other selective marker genes to enable the selection of transformants. The gene gfp encodes for the protein green fluorescent protein (GFP) (Chalfie et al., 1994). This is one of the mostly used reporter marker gene in fruit transformation protocols for monitoring transformed cells in vivo and in real time just by application of UV-light for the excitation of the fluorescent protein. GFP has not any cytotoxic effect on transformed plant cells (Stewart, 2001; Manimaran et al., 2011). In vivo detection may permit the manual selection of transformed tissues with focusing in the areas where the signal is more brightly. Fusion of GFP with other proteins of interest provides precise visualizing of intracellular localization and transport in transformed plant (Miki & McHugh, 2004; Manimaran et al., 2011). In some fruit species, it is reported that chlorophyll red autofluorescence can mask GFP expression making the detection really difficult or even impossible in species as apricot, peach and plum (Billinton & Knight, 2001; Padilla et al., 2006; Petri et al., 2008; Petri et al., 2011). However, it was described as an efficient reporter gene in other woody fruit plants, such as citrus (Ghorbel et al., 1999) and peach (Pérez-Clemente et al., 2004). These contrary results confirm the highly variability of the reporter gfp gene which is described by Hraška and co-workers (2008). Other reporter gene, luciferase (luc) (Gould & Subramani, 1988) also let the monitorization of the transgene putative cells in living tissues, however, it is not so widely employed as the gfp (van Leeuwen et al., 2000; Miki & McHugh, 2004).

3.2 Selection systems, a critical step

Selection of transformed regenerants is a critical step in any transformation procedure (Burgos & Petri, 2005). Selection systems can be classified as positive or negative, and conditional or non-conditional. Positive selection systems are those that promote the growth of transformed cells and tissues, by the contrary, negative selection systems are those that promote the death of the transgenic cells. Both systems can be conditioned by an external substrate to perform their activity. Currently, negative selection systems are used in combination with positive selection systems to eliminate transformed cells with incorrect molecular programmed excision of the T-DNA (Schaart et al., 2004; Vamblaere et al., 2011). Typically, in positive conditional selection systems the selectable marker gene encodes for an enzyme conferring resistance to some specific toxic substrate that enable the growth of the transformed cell, tissues and inhibiting or killing non transformed tissues (more information in the comprehensive review of Miki & McHugh, 2004). In the literature approximately 50 selection marker genes are described for genetic plant transformation, however, just only three genes of positive conditional selection system (nptII and hpt, resistance to the antibiotics kanamycin and hygromycin, respectively, and bar gene encoding resistance to herbicide phosphinothricin) are commonly employed in more than 90% of the papers (Miki & McHugh, 2004). These three selectable genes are also the most used ones to transform fruit species as it can be seen in Table 1 and Table 2. Escherichia coli nptII gene (also known neo) encoded protein (neomycin phosphotransferase, NPTII) inactives aminoglycoside antibiotics such as kanamycin, neomycin, geneticin (G418), and paramomycin that inhibit protein translation in the transformed cells (Padilla & Burgos, 2010). Hygromycin B is another aminoglycoside antibiotic that inhibits protein synthesis with a broad spectrum activity against prokaryotes and eukaryotes and especially it is very toxic in plants. *Escherichia coli hpt* (*aphIV*, *hph*) gene codes for the hygromycin phosphotransferase to detoxify hygromycin B by phosphorylation via an ATP-dependent phosphorylation of a 7''-hydroxyl group and it is generally used as another selection marker gene when *nptII* was not effective in plant transformation studies (Twyman et al., 2002; Miki & McHugh, 2004).

Similar to antibiotics, herbicides have different specific target sites in plants. The resistance can be achieved by various mechanisms such as usage of natural isozyme or generation of enzyme mutagenesis or detoxification of the herbicides by metabolic processes. Phosphinothricin (PPT; ammonium glufosinate) is an active component of commercial herbicides formulations and analogous to glutamate, the substrate of glutamate synthase. In plants, this enzyme catalyzes the conversion of glutamate to glutamine by removing ammonia assimilation from the cell. Inhibition of the enzyme results in ammonia accumulation and disruption of chloroplast and finally cell death due to photosynthesis inhibition (Lindsey, 1992; OECD, 1999). In plant transformation studies, as herbicide resistance selection marker gene, *pat* from *S. viridochromogenes* (Wohlleben et al., 1988) and *bar* gene from *S. hygroscopicus* (bialophos resistance; Thompson et al., 1987) encoding the enzyme phosphinothricin N-acetyltransferase (PAT) are extensively used for resistance to PPT. PAT converts PPT to a non-herbicidal acetylated form by transferring the acetyl group from acetyl CoA to the free amino group of PPT (Miki & McHugh, 2004).

Currently, an alternative to these highly employed "toxic" approaches conditional positive selection markers based on the promotion of a metabolic advantage to transformed cells are used. Some authors mentioned that this kind of selection can improve considerably the selection of the transformants, since others such as antibiotics generally cause considerable necrosis (produced by the death of non-transformed cells) that often inhibits regeneration from adjacent tissues (Petri & Burgos, 2005). Previously, results obtained with this approach demonstrated higher yields than when the toxic selective agents were employed, and seems to be broadly applicable to crop plants (Miki & McHugh, 2004). Some of the most widely used are the *AtTPS1*/glucose (Leyman et al., 2006); *galT*/galactose (Joersbo et al., 2003); xylose isomerase (Haldrup et al., 1998); D-aminoacid/*dao1* (Alonso et al., 1998) and the *pmi*/mannose (Joersbo et al., 1998). Probably, one of the most used one in fruit species is the gene *pmi* that encodes the enzyme phosphomannose isomerase (EC 5.3.1.8) that catalyzes the reversible interconversion of mannose 6-phosphate and fructose 6-phosphate. This enzyme is present in bacteria as *E. coli* and also, in humans, however it is not present in plants, as exception of soybean and other legumes. Using a media with mannose as the unique carbon source, only transformed cells can grow and develop. Glycolysis is inhibited due to the accumulation of mannose-6-phosphate converted from mannose by hexokinase with preventing cell growth and development in non-transformed cells (Miki & McHugh, 2004). Sensitivity to the toxic effect of mannose-6-phosphate is different between species, and can be avoided by combining with other sugars such as sucrose, maltose and fructose (Joersbo et al., 1999). Diverse fruit trees were selected with this system, alone or in combination with sucrose, i.e., 12 g/L mannose and 5 g/L sucrose in orange (Ballester et al., 2008); 30 g/L mannose without any sugar more in papaya (Zhu et al., 2005); 2,5 g/L mannose and 5 g/L sucrose in almond (Ramesh et al., 2006) or 1-10 g/L mannose and 5-30 g/L in apple (Degenhardt et al., 2006). In *Citrus sinensis*, the best results were obtained when 13 g/L mannose as unique source of carbon was added into the selection media. Mannose combined with other sugars promoted reduction in transformation efficiencies and escapes (Boscariol et al., 2003). Apricot cv. Helena and Canino required the lower combination of mannose with sucrose (1,25 g/L mannose and 20 g/L sucrose) in comparison with other woody fruit trees to obtain the most effective selection procedure. Moreover, safety assessments were revealed that there is no any adverse effect of the enzyme on mammalian allergenicity and toxicity (Reed et al., 2001).

Other selective strategies were developed as positive non-conditional systems, or in other words, using selectable marker genes that "promote" plant regeneration. Currently, there is more information about the genetic and biochemical control of organogenesis than embryogenesis for plant regeneration. Because of this, commonly genes related with cytokinins synthesis are employed for shoot organogenesis. More efforts are required to discover molecular mechanisms of embryogenesis to use these strategies in species highly dependent on embryogenesis regeneration to develop transgenic plants. Genes as *cki1* or the most employed isopentenyl transferase *ipt* gene encoding the enzyme IPT, catalyze the synthesis of isopentyl-adenosine-5-monophosphate, which is the first step in cytokinin biosynthesis (Miki & McHugh, 2004). This gene modify the endogenous balance between cytokinins and auxins, stimulating cell division and differentiation of the cells that promote an altered morphology, development and physiology of transgenic plants (Sundar & Sakthivel, 2008). Some authors observed that the *ipt* gene improved transformation efficiency in apricot leaf explants in comparison with the selection through *nptII* (López-Noguera et al., 2009).

3.3 A differential transgene expression: Constitutive versus specific promoters

Currently, an important debate is carrying out about the risks of the "unpredictable" behavior and recombinogenic potential of constitutive promoters (Gittins et al., 2003) and to avoid the public concerns about the risks of ubiquitous transgene expression in crops.

Commonly, most of the fruit species have been transformed with plasmidic constructions harbouring constitutive or ubiquitous promoters, as the Cauliflower Mosaic virus 35S (*CaMV35S*). In this sense, different authors described that constitutive expression may be harmful for the host plant, causing sterility, retarded development, abnormal morphology, yield penalty, altered grain composition or transgene silencing (Cai et al., 2007 and references therein) and its expression level is dependent on the cell type, the developmental stage and on the perception of environmental triggers (Hensel et al., 2011). Moreover, under constitutive promoters reporter and selectable marker, and genes of interest are expressed continuously in all tissues without any temporal control. In this sense, specific-promoters appear as an alternative approach to avoid the undesirable side effects of constitutive promoters and to target transgene expression in a spatial or temporal specific way (Gago et al., 2011; Hensel et al., 2011).

Recently, vascular specific promoter *EgCCR* from *Eucaliptus gunnii* was checked in pistachio in this study as mentioned above as well as other fruit species such as kiwifruit and grapevine (Paradela et al., 2006; Gago et al., 2011) and results demonstrated that this promoter is conserved and fully functional in these species. Vascular promoters can drive resistance to biotic or abiotic stresses related with vascular tissues. Specific promoters could be useful to synchronize transgene activity spatially and/or temporally to control with more accuracy the pathogenic process (Gago et al., 2011).

3.4 Alternative transformation systems: Transgenics without marker genes

A highly desirable approach to promote public acceptation for future commercialization of transgenic plants and products is focused on the elimination of marker genes from transformed plants or the direct production of marker-free transgenics (Kraus, 2010). These newly and promising approaches are highly dependent on previously established highly efficient regeneration protocols that may be based on organogenesis or embryogenesis (Petri et al., 2011). There are various technologies such as homologous recombination, co-transformation, site-specific recombination (Cre/loxP site specific recombination system, R/RS system, FLP/FRT system etc) or marker elimination by transposons to remove selective marker genes (Hao et al., 2011; Manimaran et al., 2011). However, there are still few marker-free fruit species transformation protocols.

Strawberry leaf explants were transformed with site-specific recombinase for the precise elimination of undesired DNA sequences and a bifunctional selectable marker gene used for the initial positive selection of transgenic tissue and subsequent negative selection for fully marker-free plants (Schaart et al., 2004).

MAT (multi-auto-transformation) (Ebinuma et al., 1997) combined with the *Agrobacterium* oncogene *ipt* gene, for positive selection with the recombinase system R/RS for removal of marker genes acting as "molecular scissors" after transformation were used as alternative approach in citrus plants (Ballester et al., 2007; 2008). Also, in apricot (López-Noguera et al., 2009) a similar strategy was used. Regeneration of apricot transgenic shoots was significantly improved to non-transformed plants (regenerated in non-selective media). Moreover, it was significantly higher in comparison with previous published data using resistance to kanamycin mediated by *nptII* gene. The lack of *ipt* differential phenotype promoted difficulties to assess the excision of the marker genes, that require periodic assays. Complete excision of marker genes ranged from 5 to 12 months, however, only 41% of the regenerated transgenic shoots R-mediated recombination occurs correctly. In *Citrus sp.*, it was also reported that anomalous excision of marker genes promoting failures in the expression of the reporter genes (Ballester et al., 2007, 2008).

Apple (Malnoy et al., 2010) and pineapple sweet orange (Ballester et al., 2010) transformation using "clean" binary vector including only the transgene of interest were carried out to create marker-free transformants. Very recently, melon (*C. melo* L. cv Hetao) was transformed with a marker-free and vector-free antisense 1-aminocyclopropane-1-carboxylic acid oxidase construct via the pollen-tube pathway and transgenic lines are choosen by PCR without using any selectable marker agent (Hao et al., 2011).

In plum (*Prunus domestica*), transformation was carried out without reporter or selectable marker genes using a high-throughput transformation system (Petri et al., 2011). Previously, authors checked the efficiency of the regeneration of transformed shoots using conventional constructs harbouring reporter marker such as *gusA* and *gfp*, and *nptII* gene. Transformation efficiency varied from 5.7-17.7%. Using a marker free construct, the intron-hairpin-RNA (ihpRNA) harbouring the Plum Pox Virus coat protein (*ppv-cp*) gene, these authors regenerated five transgenic lines confirmed by Southern blot. It is important to take into account that this kind of free marker strategy is widely dependent on highly yields in regeneration systems.

3.5 Cisgenesis, the P-DNA technology and multigene transformation

Other relevant advance in fruit species transformation was the proposal made by Schouten and coworkers (2006), the "cisgenesis". This term means the use of recombinant DNA technology to introduce genes from crossable donors plants, isolated from within the existing genome or sexually compatible relative species for centuries therefore, unlikely to alter the gene pool of the recipient species. Cisgenesis includes all the genetic events of the T-DNA as introns, flanking regions, promoters, and terminators (Vanblaere et al., 2011).

This methodology proposes to transfer the own plant DNAs, the P-DNAs. The use of this technology requires the construction of whole plant derived vector from the target species. Within the target species genome, it must be a DNA fragment with two T-DNA border-like sequences oriented as direct repeats ideally about 1-2 kb apart with suitable restriction sites for cloning of a desirable gene.

In the last years, different works were considered to step towards introducing regulatory elements and genes of interest from crossable donor plants, however with some foreign elements as marker genes in species as melon and apples (Benjamin et al., 2009; Joshi, 2010; Szankowski et al., 2009). Up to 2011 there is no any report of real "cisgenesis" plantlets, in agreement with Schouten et al. (2006) definition of the topic. In 2011, Vanblaere and coworkers developed apple cv. Gala cisgenic plants by expressing the apple scab resistance gene *HcrVf2* encoding resistance to apple scab. Marker-free system was employed for the development of three cisgenic lines containing one insert of the P-DNA after removing by recombination with using chemical induction. These lines were not observed different from non-transformed cv. Gala plants.

Cisgenic plants are essentially the same as the traditionally bred varieties, and they might be easier to commercialise than the "problematic" transgenic plants (Schouten et al., 2006; Rommens et al., 2007). Critical opinions to these proposals also were clearly exposed, the uncontrolled P-DNA integration into the plant target genome can cause mutations or affect to the expression of other native genes, altering the behaviour of that cisgenic plants in an unpredictable manner (Schubert & Willims, 2006; Akhond & Machray, 2009). Recently, interesting approaches are being proposed for genome editing using ZFNs (Zinc finger nucleases) that can promote induction of double-strand breaks at specific genomic sites and promote the replacement of native DNA with foreign T-DNA (Weinthal et al., 2010).

The multigen transfer (MGT) methodology consist in introducing more than one gene at once. Commonly, most of the transgenic plants are generated by introducing just one single gene of interest, but now MGT are being developed to obtain more ambitious phenotypes as the complete import of metabolic pathways, whole protein complex and the development of transgenic fruit species with various new traits simultaneously transferred (Naqvi et al., 2009). In this sense, this technology would be highly desirable for commercial fruit species cultivars to obtain new traits related with large fruit size, high coloration of the fruit epidermis, flesh firmness and virus resistance (Petri et al., 2011) at the same time without the need of several rounds of introgressive backcrossing.

4. *In vitro* culture techniques for the recovery of transgenic plants

Plants are complex, diverse organisms and have adapted evolutionarily to almost every ecological niche on the planet. Development of successful transformation protocol depends on a reliable and highly efficient regeneration system. Explant types are highly variable since it depends on the selected organogenetic process optimized for each species. Commonly, the genetic transformation protocols of fruit species employed explants such as ovules, anthers, seedlings, zygotic and somatic embryos, cotyledons, epicotyles, hypocotyles, leaf pieces, roots, meristems (Fagoaga et al., 2007; Lopez- Perez et al., 2008; Petri et al., 2008; Husaini, 2010; Malnoy et al., 2010; Bosselut et al., 2011; Petri et al., 2011; Gago et al., 2011). Typically, it is recommended that those tissues have high and active cell division to enhance the regeneration of the transgenic lines (Mante et al., 1991; Schuerman & Dandekar, 1993; Wang, 2011). Ideally, fruit species transformation must be done with somatic tissues such as leaves and roots to transform varieties already well known and accepted in the market by the consumers. Recently, some authors also proposed the possibility of the use of transgenic seedlings to develop new fruit varieties through subsequent cross-breeding. These transgenic seedlings can add new traits impossible to obtain in the species genome-pool (Petri et al., 2011).

Organogenesis was the strategy selected in different species to develop most of the known and efficient regeneration protocols for fruit species, concretely for fruit trees (Petri et al., 2011). Almond (Costa et al., 2006); apple (Smolka et al. 2009; Lau & Korban, 2010; Vanblaere et al., 2011); banana (Subramanyam et al., 2011); fig (Yancheva et al., 2005); kiwifruit (Tian et al., 2011); peach (Padilla et al., 2006); strawberry (Mercado et al., 2010); peanut (Asif et al., 2011); watermelon (Huang et al., 2011) and pear (Sun et al., 2011) are some examples of transformed cultivars for some fruit species that the transformed tissues were regenerated via organogenesis. Since organogenesis protocols are developed for many different fruit species, it is easier to adapt the regeneration system into genetic transformation methods (Frary & Eck, 2005). However, some risks also are assumed in using this regeneration system. Generally, it is considered that the origin of the new adventitious shoots is based on the involvement of few cells (George et al., 2008), enhancing the risks of chimera development.

Somatic embryogenesis that leads the formation of an embryo from somatic cells is another procedure to regenerate fruit transformants such as banana (Vishnevetsky et al., 2011); papaya (Zhu et al., 2001); grapevine (Nirala et al., 2010) and mango (Chavarri et al., 2010). Regeneration from transformed embryos can be achieved via direct germination or shoot organogenesis and the method is useful for large-scale and rapid propagation of transformants. In grapevine most of the approaches are being performed by using embryogenic cultures from different tissues such as zygotic embryos, leaves, ovaries and anther filaments to provide cells amenable to gene insertion and regeneration (Mezzetti et al., 2002; Dutt et al., 2007; López-Noguera et al., 2009). However, these techniques are highly genotype dependent and for many cultivars they have been difficult to obtain successful results (Dutt et al., 2007). Moreover, it is considered that anther filaments, as commonly employed in grapevine for embryogenic calli, are laborious, cultivar-dependent, depend on availability of immature flowers and may affect strongly the phenotype of the regenerated plantlets (Mezzetti et al., 2002). However, it is really interesting to take into account that regeneration from somatic embryos and secondary somatic embryos are currently assumed that they are derivatives of single cell origin.

In the decade of the 90´s some unsuccessful efforts were reported to transform meristems from micropropagated shoot tips due to high explant mortality and uncontrolled *Agrobacterium* overgrowth after coculture stages (Ye et al., 1994; Druart et al., 1998; Scorza et al., 1995). Mezzetti and co-workers (2002) described in grapevine the development of meristematic bulk tissues, a highly aggregate of meristematic cells produced after three months in increased concentrations of BA (N^6-benzyladenine) and the removal of the apical meristem. After 90 days, under the previous conditions, these highly regenerative tissues produced easily adventitious shoots and can be transformed by *Agrobacterium*, being able to regenerate several transgenic lines. Other interesting approach was the genetic transformation of shoot apical meristems, previously subjected to a dark growth stage after wounding for transformation. Authors reported that 1% of shoot tips produced stable transgenic lines after weeks (Dutt et al., 2007). Ismail and co-workers (2011) transformed successfully banana apical meristems via microprojectile bombardment and regenerated 80.3% percent of the transformed meristematic tissues.

4.1 The chimeric question: Are my transgenic plants genetically uniform?

This is one of the most exciting questions that plant biotechnology researchers ask to themselves after all the long extensive, intensive and difficult labour needed to transform most of the fruit species. Some of the transformed regenerants can be chimeras, a mix of transformed and non transformed cells in the tissues, in other words, non genetically uniform organisms (Hanke et al., 2007). Recently, Petri and collaborators (2011) described that most of the known and efficient regeneration methods for fruit trees are based on organogenesis, where new adventitious shoot formation is originated from a determined number of cells. So, it comes hard to detect non chimeric and stable transgenic lines without using a selectable marker gene. Very recently, different authors using marker free technology as alternative systems or with genetically programmed marker excision reported the appearance of chimeric transformants in apple, strawberry, lime, citrus or plum (Domínguez et al., 2004; Schaart et al., 2004; Ballester et al., 2007; Malnoy et al., 2011; Petri et al., 2011).

Strawberry is highly sensitive to kanamycin selection, and it was described that selection of transgenic regenerants in these sensitive tissues can be associated with chimeric shoots containing transgenic and non-transgenic sections (Husaini, 2010). It was observed that increasing antibiotic concentration gradually avoid chimerisms in strawberry (Mathews et al., 1998; Husaini et al., 2010). Even under this strictly methodology some authors pointed out the inactivation events on the selection agent must be performed through the transformed cells, so, non transformed cells can develop and grown (Petri & Burgos, 2005; Wang, 2011). A useful methodology was also proposed for the quick and low-cost identification of chimeras by Faize and collaborators (2010) in tobacco and in apricot based in quantitative real-time PCR even in early developmental stages, and also let to monitor their dissociation.

5. Future perspectives and concluding remarks

Currently, most of the fruit genetic transformation protocols integrated the new genes randomly and in unpredictable copy numbers influencing negatively its expression. Also public concerns and reduced market acceptance of transgenic crops have promoted the development of alternative marker free technologies in fruit species. For those reasons development of protocols to obtain transgenic fruits without marker genes and the use of the own plant DNA resources, such as "cisgenic" fruit plants, are the big challenges. ZFNs have also been succesfuly used to drive the replacement of native DNA sequences with foreign DNA molecules and to mediate the integration of the targeted transgene into native genome sequences.

Most of the fruit transgenic plants are generated by introducing just one single new character (gene of interest), however, some authors proposed that multigene transfer technology (MGT) needs to be developed to obtain new traits related at the same time. The combination of multiple traits can be a highly interesting approach as it could be applied to achieve resistance to several biotic or abiotic stresses and traits related to fruit quality such as large fruit size, high coloration of the fruit epidermis, increase flesh firmness to improve ripening without the need of several rounds of introgressive backcrossing.

The future of fruit genetic transformation is required of genotype-independent protocols, accuracy molecular tools to drive the T-DNA insertion and its expression, and efficiency and highly-yield selection and regeneration in vitro culture methodology. But *Agrobacterium*-mediated transformation procedure is a high non linear complex biological process, and its complexity can be understood with the composition of many different and interacting elements governed by non-deterministic rules and influenced by external factors. In this sense, the emergent technology dedicated to meta-analysis can be really useful to increase our understanding of fruit genetic transformation, making possible to identify relationships among several factors and extracting useful information generating understable and reusable knowledge (Gago et al., 2011; Gallego et al., 2011; Perez-Pineiro et al., 2012) Under these perspectives, modeling any fruit transformation procedure (*Agrobacterium*-mediated, biolistics, electroporation etc.) including the genetic engineering, *in vitro* plant tissue culture and regeneration stages will be improved for the next years.

6. Acknowledgements

The authors wish to thank to Dr. Mariana Landín and to Dr. María José Clemente for critical reading of this work. This work was supported by Regional Government of Xunta de Galicia: exp.2007/097 and PGIDIT02BTF30102PR. This study was also partially supported by # TBAG- 209T030 from TUBITAK—The Scientific and Technical Research Council of Turkey. PG thanks to Minister of Education of Spain for funding the sabbatical year at Faculty of Science, University of Utrecht, Netherlands.

7. References

Agarwal, S. & Kanwar, K. (2007). Comparison of genetic transformation in *Morus alba* L. via different regeneration systems. *Plant Cell Rep* 26, pp. 177–185

Akhond, MAY. & Machray, GC. (2009). Biotech crops: technologies, achievements and prospects. *Euphytica* 166, pp. 47-59.

Alonso, J.; Barredo, JL.; Diez, B.; Mellado, E.; Salto, F.; Garcia, JL. & Cortes, E. (1998). D-Amino-acid oxidase gene from *Rhodotorula gracilis* (*Rhodosporidium toruloides*) ATCC 26217. *Microbiology*, 144, pp. 1095–1101

Asif, MA.; Zafar, Y, Iqbal, J.; Iqbal, MM.; Rashid, U.; Ali, GM.; Arif, A. & Nazir, F. (2011). Enhanced expression of AtNHX1, in transgenic groundnut (*Arachis hypogaea* L.) Improves Salt and Drought Tolerence. *Molecular Biotechnology* 49, pp. 250-256.

Atkinson, RG.; Candy, CJ. & Gardner, RC. (1990). *Agrobacterium* infection of five New Zealand fruit crops. *Zealand of Crop and Horticultural Science,* 18, pp. 153-156

Ballester, A.; Cervera, M. & Peña, L. (2007). Efficient production of transgenic *Citrus* plants using isopentenyl transferase positive selection and removal of the marker gene by site-specific recombination. *Plant Cell Reports,* 26, pp. 39-45

Ballester, A. Cervera, M. & Peña, L. (2008). Evaluation of selection strategies alternative to nptII in genetic transformation of citrus. *Plant Cell Reports* 27, pp. 1005-1015

Ballester, A. Cervera, M. & Pena, L. (2010). Selectable marker-free transgenic orange plants recovered under non-selective conditions and through PCR analysis of all regenerants. *Plant Cell Tissue and Organ Culture,* 102, pp. 329-336

Barceló, M.; El-Mansouri, I.; Mercado, JA.; Quesada, MA. & Alfaro, FP. (1998). Regeneration and transformation via *Agrobacterium tumefaciens* of the strawberry cultivar Chandler. *Plant Cell Tissue and Organ Culture,* 54, pp. 29-36

Becker, D.; Dugdale, B, Smith, MK.; Harding, RMJ. & Dale, JL. (2000). Genetic transformation of Cavendish banana (*Musa* spp. AAA group) cv. 'Grand Nain' via microprojectile bombardment. *Plant Cell Reports,* 19, pp. 229-234

Bell, RL.; Scorza, R.; Srinivasan, C. & Webb, K. (1999). Transformation of 'Beurre Bosc' pear with the *rolC* Gene. *J. Amer. Soc. Hort. Sci.* 124(6), pp. 570-574

Benjamin, I. Kenigsbuch, D. Galperin, M. Abrameto, J.A. & Cohen, Y. (2009). Cisgenic melons over expressing glyoxylate-aminotransferase are resistant to downy mildew. *European Journal of Plant Pathology,* 125, pp. 355-365

BIORAD (2002). http://www.bio-rad.com.

Bidney, D. (1999). Plant transformation method using *Agrobacterium* species adhered to microprojectiles. *United States Patent,* 5, 932, 782

Billinton, N. & Knight, AW. (2001). Seeing the wood through the trees: a review of techniques for distinguishing green fluorescent protein from endogenous autofluorescence. *Analytical Biochemistry,* 291, pp. 175-197

Bolar, JP.; Norelli, JL.; Wong, KW.; Hayes, CK.; Harman, GE. & Aldwinckle, HS. (2000). Expression of endochitinase from *Trichoderma harzianum* in transgenic apple increases resistance to apple scab and reduces vigor. *Phytopathology,* 90, pp. 72-77

Bordas, M.; Montesinos, C.; Dabauza, M.; Salvador, A.; Roig, LA.; Serrano, R. & Moren, V. (1997). Transfer of the yeast salt tolerance gene HAL1 to *Cucumis melo* L. cultivars and in vitro evaluation of salt Tolerance. *Transgenic Research,* 6, pp. 41-50

Bornhoff B.A.; Harst M.; Zyprian E. & Töpfer R. (2005). Transgenic plants of *Vitis vinifera* cv. Seyval blanc. *Plant Cell Reports,* 24, pp. 433-438.

Boscariol, RL.; Almeida, WAB.; Derbyshire, MTVC.; Mourâo-Filho, FAA. & Mendes, BMJ. (2003). The use of the PMI/mannose selection system to recover transgenic sweet orange plants (*Citrus sinensis* L. Osbeck). *Plant Cell Reports,* 22, pp. 122-128

Bosselut, N.; Ghelder, CV.; Claverie, M.; Voisin, R.; Onesto, JP.; Rosso, MN. & Esmenjaud, D. (2011). *Agrobacterium rhizogenes*-mediated transformation of *Prunus* as an alternative for gene functional analysis in hairy-roots and composite plants. *Plant Cell Reports,* 30(7), pp. 1313-1326

Broothaerts, W.; Keulemans, J. & Van Nerum, I. (2004). Self-fertile apple resulting from S-RNase gene silencing. *Plant Cell Reports*, 22, pp. 497–501

Cai, M.; Wei, J.; Xianghua, L.; Caiguo, X. & Shiping, W. (2007). A rice promoter containing both novel positive and negative cis-elements for regulation of green tissue specific gene expression in transgenic plants, *Plant Biotechnology Journal*, 5, pp. 664-674

Cai, WQ.; Gonsalves, C.; Tennant, P.; Fermin, G.; Souza, M.; Sarindu, N.; Jan F.J.; Zhu H.Y. & Gonsalves, D. (1999). A protocol for efficient transformation and regeneration of *Carica papaya* L. *In Vitro Cellular and Developmental Biology-Plant*, 35(1), 61-69

Cao, X.; Liu, Q.; Rowland, LJ. & Hammerschlag, FA. (1998). GUS expression in blueberry (*Vaccinium spp.*): factors influencing Agrobacterium-mediated gene transfer efficiency. *Plant Cell Reports*, 18, pp. 266–270

Chalfie, M.; Tu, Y.; Euskirchen, G.; Ward, W. & Prasher, D. (1994). Green fluorescent protein as a marker for gene expression. *Science*, 263, 802–805

Chavarri, M.; García, AV.; Zambrano, AY.; Gutiérrez, Z. & Demey, JR. (2010). Insertion of *Agrobacterium rhizogenes rolB* gene in Mango. *Interciencia*, 35 (7), pp. 521-525.

Chen, G.; Ye, CM.; Huang, JC.; Yu, M. & Li, BJ. (2001). Cloning of the papaya ringspot virus (PRSV) replicase gene and generation of PRSV-resistant papayas through the introduction of the PRSV replicase gen. *Plant Cell Reports*, 20, pp. 272–277

Chevreau, E.; Dupuis, F.; Taglioni, JP.; Sourice, S.; Cournol, R.; Deswartes, C.; Bersegeay, A.; Descombin, J.; Siegwart, M. & Loridon, K. (2011). Effect of ectopic expression of the eutypine detoxifying gene Vr-ERE in transgenic apple plants. *Plant Cell Tissue and Organ Culture*, 106, pp. 161–168

Chilton, MD.; Drummond, MH.; Merio, DJ.; Sciaky, D.; Montoya, AL.; Gordon, MP. Nester, EW. (1977). Stable incorporation of plasmid DNA into higher plant cells: the molecular basis of crown gall tumorigenesis. *Cell*, 11, pp. 263-271

Christou, P. (1992) Genetic transformation of crop plants using microprojectile bombardment. *Plant Journal*, 2, pp. 275-281

Chu, Y.; Faustinelli, P.; Ramos, ML.; Hajduch, M.; Stevenson, S.; Thelen, JJ.; Maleki, SJ.;Cheng, H. & Ozias-Akins, P. (2008b). Reduction of IgE binding and nonpromotion of *Aspergillus flavus* fungal growth by simultaneously silencing Ara h 2 and Ara h 6 in peanut. *J. Agric. Food Chem.* 56, pp. 11225–11233

Chu, Y. & Deng, XY.; Faustinelli, P.; Ozias-Akins, P. (2008a). Bcl-xL transformed peanut (*Arachis hypogaea* L.) exhibits paraquat tolerance. *Plant Cell Reports*, 27, pp. 85–92

Costa, MS.; Miguel, CM. & Oliveira, MM. (2006). An improved selection strategy and the use of acetosyringone in shoot induction medium increase almond transformation efficiency by 100-fold. *Plant Cell, Tissue and Organ Culture*, 85, pp. 205-209

Cruz-Hernández, A.; Witjaksono Litz, RE. & Gomez Lim, M. (1998). *Agrobacterium tumefaciens*-mediated transformation of embryogenic avocado cultures and regeneration of somatic embryos. *Plant Cell Reports*, 17, pp. 497–503

Cruz-Hernandez, A.; Gomez Lim, MA. & Litz, RE. (1997).Transformation of mango somatic embryos. *Acta Horticulturae*, 455, pp. 292–298

Daniell, H.; Krishnan, M. & McFadden, B.F. (1991). Transient expression of β-glucuronidase in different cellular compartments following biolistic delivery of foreign DNA into wheat leaves and calli. *Plant Cell Reports*, 9, pp. 615–619

Davidson, SN. (2008). Forbidden Fruit: Transgenic Papaya in Thailand. *Plant Physiology*, 147, pp. 487-493

De Block, M.; Herrera-Estrella, L.; Van Montagu, M.; Schell, J. & Zambriski, P. (1984). Expression of foreign genes in regenerated plants and their progeny. *EMBO Journal*, 3, pp. 1681-1689

Degenhardt, J.; Poppe, A.; Montag, J. & Szankowski, I. (2006). The use of the phosphomannose-isomerase/mannose selection system to recover transgenic apple plants. *Plant Cell Reports*, 25, pp. 1149–1156

Dominguez, A.; Guerri, J.; Cambra, M.; Navarro, L.; Moreno, P. & Pena, L. (2000). Efficient production of transgenic citrus plants expressing the coat protein gene of citrus tristeza virus. *Plant Cell Reports*, 19, pp. 427–433

Druart, P. Delporte, F.; Brazda, M.; Ugarte-Ballon, C.; da Câmara Machado, A.; Laimer da Câmara Machado, M.; Jacquemin, J. & Watillon, B. (1998). Genetic transformation of cherry trees. *Acta Horticulturae*, 468, pp. 71–76

Dutt, M.; Li, ZT.; Dhekney, SA. & Gray, DJ. (2007). Transgenic plants from shoot apical meristems of *Vitis vinifera* Thompson Seedless via *Agrobacterium*-mediated transformation. *Plant Cell Reports*, 26, pp. 2101–2110

Ebinuma, H.; Sugita, K.; Matsunaga, E.; Endo, S.; Yamada, K. (1997). Selection of marker-free transgenic plants using the isopentyl transferase gene. *Proceedings of National Academy of Science USA*, 94, pp. 2117–2121

Fagoaga, C.; Tadeo, FR.; Iglesias, DJ.; Huerta, L.; Lliso, I.; Vidal, AM.; Talon, M.; Navarro, L.; Garcia-Martinez, JL.; Peña, L. (2007). Engineering of gibberellin levels in citrus by sense and antisense overexpression of a GA 20-oxidase gene modifies plant architecture. *Journal of Experimental Botany*, 58(6), pp.1407–1420

Fagoaga, C.; López, C.; Hermoso de Mendoza, A.; Moreno, P.; Navarro, L.; Flores, R. & Peña, L. (2006). Post-transcriptional gene silencing of the p23 silencing supperssor of Citrus tristeza virus confers resistance to the virus in transgenic Mexican lime. *Plant Molecular Biology*, 60, pp. 153-165

Fagoaga, C.; Rodrigo, I.; Conejero, V.; Hinarejos, C.; Tuset, JJ.; Arnau, J.; Pina, JA.; Navarro, L. & Peña, L. (2001). Increased tolerance to *Phytophthora citrophthora* in transgenic orange plants constitutively expressing a tomato pathogenesis related PR-5. *Molecular Breeding*, 7, pp. 175-185

Faize, M.; Malnoy, M.; Dupuis, F.; Chevalier, M.; Parisi, L. & Chevreau, E. (2003). Chitinases of *Trichoderma atroviride* induce scab resistance and some metabolic changes in two cultivars of apple. *Phytopathology*, 93, pp. 1496-1504

Faize, M.; Faize, L. & Burgos, L. (2010). Using quantitative real-time PCR to detect chimeras in transgenic tobacco and apricot and to monitor their dissociation. *BMC Biotechnology*, 10, p.53

Fan, J.; Liu, X.; Xu, SX.; Xu, Q. & Guo, WW. (2011). T-DNA direct repeat and 35S promoter methylation affect transgene expression but do not cause silencing in transgenic sweet orange. *Plant Cell Tissuse and Organ Culture*, 107, pp. 225-232

Fang, G. & Grumet, R. (1993). Genetic engineerring of potyvirus resistance using constructs derived from the zucchini yellow mosaic virus coat protein gene. *Molecular Plant-Microbe Interactions* 6, pp. 358–367

Febres, VJ.; Lee RF. & Moore GA. (2008). Transgenic Resistance to Citrus Tristeza Virus in Grapefruit. *Plant Cell Reports*, 27(1), pp. 93-104

Febres VJ.; Niblett, CL.; Lee, RF. & Moore, GA. (2003). Characterization of grapefruit plants (*Citrus paradisi* Macf.) transformed with Citrus Tristeza Closterovirus genes. *Plant Cell Reports*, 21, pp. 421-428

Fitch, MM.; Manshardt, RM.; Gonsalves, D.; Slightom, JL. & Sanford, JC. (1992). Virus resistant papaya derived from tissues bombarded with the coat protein gene of papaya ringspot virus. *Bio/Technology*, 10, pp. 1466-1472

Frary, A. & Eck, JV. (2005). Organogenesis From Transformed Tomato Explants. In: *Methods in molecular biology, vol. 286: Transgenic plants: methods and protocols*. L., Pena; NJ., Totowa (Eds.), 141-151, Humana Press Inc.

Fu, X.; Khan, EU.; Hu, SS.; Fan, QJ. & Liu, JH. (2011). Overexpression of the betaine aldehyde dehydrogenase gene from *Atriplex hortensis* enhances salt tolerance in the transgenic trifoliate orange (*Poncirus trifoliate* L. Raf). *Environmental and Experimental Botany* (in press)

Fuchs, M. & Gonsalves, G. (2007). Safety of virus-resistant transgenic plants two decades after their introduction: lesson from realistic field risk assessment studies. *Annual Review of Phytopathology*, 47, pp. 173-202

Gago, J. (2009). *Biotechnology of Vitis vinifera L.: Modellization through Artificial Intelligence*. Doctoral Thesis. University of Vigo (Spain) (in Spanish).

Gago, J.; Grima-Pettenati, J. & Gallego, PP. (2011). Vascular-specific expression of GUS and GFP reporter genes in transgenic grapevine (*Vitis vinifera* L. cv. Albariño) conferred by the EgCCR promoter of *Eucalyptus gunnii*. *Plant Physiology and Biochemistry*, 49, pp. 413-419

Gallego, PP.; Rodriguez, R.; de la Torre, F. & Villar, B. (2002). Genetic transformation of *Eucalyptus globulus*. In: *Sustainable Forestry Wood Products and Biotechnology*. S. Espinel, Y. Barredo, & E. Ritter (Eds.), 163-170, DFA-AFA Press, Vitoria-Gasteiz, Spain

Gallego, PP.; Landín, M. & Gago, J. (2011). Artiticial neural networks technology to model and predict plant biology process. In: *Artificial Neural Networks- Methodological Advances and Biomedical Applications*. K. Suzuki (Ed.), 197-216, Intech Open Access Publisher: Croatia

Gallego, PP.; Rodríguez, R.; de la Torre, F. & Villar, B. (2009). *Procedimiento para transformar material vegetal procedente de árboles adultos*. Patent number: ES2299285 (A1), Universidade de Vigo, Spain.

Galun, E. & Breiman, A. (1998). *Transgenic Plants*. Imperial College Press, London, UK.

Gelvin, SB. (1998). The introduction and expression of transgenes in plants. *Current Opinion in Biotechnology*, 9, pp. 227-232

George, EF.; Hall, MA. & De Klerk, GJ. (2008). Adventitious regeneration. In: *Plant Propagation by Tissue Culture*, 3rd edn. E.F. George, M.A. Hall & G-J, De Klerk. (Eds.), 355-401, Dordrecht, Netherlands: Springer

Ghorbel, R.; Juarez, J.; Navarro, L. & Peña, L. (1999). Green fluorescent protein as a screenable marker to increase the efficiency of generating transgenic woody fruit plants. *Theoretical and Applied Genetics*, 99, pp. 350-358

Ghorbel, R.; López, C.; Fagoaga, C.; Moreno, P.; Navarro, L.; Flores, R. & Peña, L. (2001). Transgenic Citrus plants expressing the citrus tristeza virus p23 protein exhibit viral-like symptoms. *Molecular Plant Pathology* 2, pp. 27-36

Gittins, JR.; Pellny, TK.; Biricolti, S.; Hiles ER.; Passey, AJ. & James DJ. (2003). Transgene expression in the vegetative tissues of apple driven by the vascular-specific rolC and CoYMV promoters. *Transgenic Research,* 12, pp. 391–402

Gonsalves C.; Xue B.; Yepes M.; Fuchs M.; Ling K. & Namba S. (1994). Transferring cucumber mosaic virus-white leaf strain coat protein gene into *Cucumis melo* L. and evaluating transgenic plants for protection against infections, *Journal of American Society for Horticultural Science,* 119, pp. 345-355.

Gonsalves, D. (1998). Control of papaya ringspot virus in papaya: a case study. *Annual Review of Phytopathology,* 36, pp. 415–437

Gould, SJ. & Subramani, S. (1988). Firefly luciferase as a tool in molecular and cell biology. *Analytical Biochemistry,* 175, pp. 5–13

Guzmán-González, S.; Valádez-Ramírez, P.; Robles-Berber, RE.; Silva-Rosales, L. & Cabrera-Ponce, JL. (2006). Biolistic genetic transformation of *Carica papaya* L. using Helios™ gene gun. *HortScience,* 41, pp. 1053-1056

Haldrup, A.; Petersen, SG. & Okkels, FT. (1998). Positive selection: a plant selection principle based on xylose isomerase, an enzyme used in the food industry. *Plant Cell Reports,* 18, pp. 76–81

Hanke, MV.; Reidel, M.; Reim, S. & Flachowsky, H. (2007). Analysis of tissue uniformity in transgenic apple plants. *Acta Horticulturae,* 738, pp. 301–306

Hao, J.; Niu, Y.; Yang, B.; Gao, F.; Zhang, L.; Wang, J. & Hasi, A. (2011). Transformation of a marker-free and vector-free antisense ACC oxidase gene cassette into melon via the pollen-tube pathway. *Biotechnology Letters,* 33, pp. 55–61

Hebert, D.; Kikkert, JR.; Smith, FD. & Reisch, BI. (1993). Optimization of biolistic transformation of embryogenic grape cell suspensions. *Plant Cell Reports,* 12, pp. 585–589

Hensel, G.; Himmelbach, A.; Chen, W.; Douchkov, DK. & Kumlehn, J. (2011). Transgene expression systems in the *Triticeae* cereals. *Journal of Plant Physiology,* 168, pp. 30–44

Heslop-Harrison, JS. (2005). Introduction. In: *Biotechnology of Fruit and Nut Crops* (Ed. RE Litz). CAB International, London, UK. p. xix.

Hiei, Y.; Ohta, S.; Komari, T. & Kumashiro, T. (1994). Efficient transformation of rice (*Oryza sativa* L.) mediated by Agrobacterium and sequence analysis of the boundaries of the T-DNA. *The Plant Journal,* 6, pp. 271-282

Higgins, CM.; Dietzgen, RG.; Akin M.; Sudarsono, H.; Chen, K. & Xu, Z. (1999). Biological and molecular variability of peanut stripe potyvirus. *Curr Topics Virol,* 1, pp. 1–26

Higgins, CM.; Hall, RM.; Mitter, N.; Cruickshank, A. & Dietzgen, RG. (2004). Peanut stripe potyvirus resistance in peanut (*Arachis hypogaea* L.) plants carrying viral coat protein gene sequences *Transgenic Research,* 13, pp. 59–67

Holefors, A.; Xue ZT. & Welander, M. (1998). Transformation of the apple rootstock M26 with the *rol*A gene and its influence on growth. *Plant Science,* 136, pp. 69-78

Honda, C.; Kusaba, S.; Nishijima, T. & Moriguchi, T. (2011). Transformation of kiwifruit using *ipt* gene alters tree architecture. *Plant Cell Tissue and Organ Culture* 107, pp. 45–50

Horsch, RB.; Fry, JE.; Hoffman, NL.; Eichholtz, D.; Rogers, D. & Fraley, RT. (1985). A simple and general method for transferring genes into plants. *Science,* 227, pp. 1229-1231

Hraška, M.; Heřmanová, V.; Rakouský, S. & Ĉurn, V. (2008). Sample topography and position within plant body influence the detection of the intensity of green fluorescent protein fluorescence in the leaves of transgenic tobacco plants. *Plant Cell Reports*, 27, pp. 67–77

Huang, YC.; Chiang, CH.; Li, CM. & Yu, TA. (2011). Transgenic watermelon lines expressing the nucleocapsid gene of Watermelon silver mottle virus and the role of thiamine in reducing hyperhydricity in regenerated shoots. *Plant Cell Tissuse and Organ Culture* 106, pp. 21–29

Husaini, AM. (2010). Pre- and post-agroinfection strategies for efficient leaf disk transformation and regeneration of transgenic strawberry plants. *Plant Cell Reports*, 29, pp. 97-110

Ishida, Y.; Satio, H.; Ohta, S.; Hiei, Y.; Komari, T. & Kumashiro, T. (1996). High efficiency transformation of maize (*Zea mays* L.) mediated by *Agrobacterium tumefaciens*. *Nature Biotechnology*, 14, pp. 745–750

Ismail, RM.; El-Domyati, FM.; Wagih, EE.; Sadik, AS. & Abdelsalam, AZE. (2011). Construction of banana bunchy top nanovirus-DNA-3 encoding the coat protein gene and its introducing into banana plants cv. Williams. Journal of Genetic *Engineering and Biotechnology*, 9 (1), pp. 35-41

James, JD.; Passey, AJ.; Barbara, DJ. & Bevan, M. (1989). Genetic transformation of apple (*Malus pumila* Mill) using a disarmed Ti-binary vector. *Plant Cell Reports*, 7, pp. 658-661

Jefferson, RA. (1987). Assaying chimeric genes in plants: the *gus* gene fusion system. *Plant Molecular Biology Reporter*, 5, pp. 387–405

Joersbo, M.; Donaldson, I.; Kreiberg, J.; Guldager Petersen, S.; Brunstedt, J. & Okkels, FT. (1998). Analysis of mannose selection used for transformation of sugar beet. *Molecular Breeding*, 4, pp. 111-117

Joersbo, M.; Joregensen, K. & Brunstedt, J. (2003). A selection system for transgenic plants based on galactose as selective agent and a UDP-glucose: galactose-1-phosphate uridyltransferase gene as selective gene. *Molecular Breeding*, 11, pp. 315-323

Joersbo, M.; Petersen, SG. & Okkels, FT. (1999). Parameters interacting with mannose selection employed for the production of transgenic sugar beet. *Physiologia Plantarum*, 105, pp. 109-115

Joshi, SG. (2010). *Towards durable resistance to apple scab using cisgenes*. Wageningen, The Netherlands: Wageningen University, PhD Thesis

Kayim, M.; Koc, NM. & Tor, M. (1996). Gene Transfer into citrus (*Citrus limon* L.) nucellar cells by particle bombardment and expression of *gus* activity. *Turkish Journal of Agriculture and Forestry*, 20, pp. 349-352

Khandelwal, A.; Renukaradhya, GJ.; Rajasekhar, M.; Sita, GL. & Shaila, MS. (2011). Immune responses to hemagglutinin-neuraminidase protein of peste des petits ruminants virus expressed in transgenic peanut plants in sheep. *Veterinary Immunology and Immunopathology*, 140, pp. 291-296

Kikkert, JR.; Hebert-Soule, D.; Wallace, PG.; Striem, MJ. & Reisch, BI. (1996). Transgenic plantlets of 'Chancellor' grapevine (*Vitis* sp.) from biolistic transformation of embryogenic cell suspensions. *Plant Cell Reports*, 15, pp. 311–316

Kikkert, JR.; Humiston, GA.; Roy, MK. & Sanford, JC. (1999). Biological projectiles (phage, yeast, bacteria) for genetic transformation of plants. *In Vitro Cellular & Developmental Biology –Plant*, 35, pp. 43–50

Klein, TM.; Wolf, ED.; Wu, R. et al. (1987). High velocity microprojectiles for delivering nucleic acids into living cells. *Nature*, 327, pp. 70-73

Klein, TM. & Jones, TJ. (1999). Methods of genetic transformation: the gene gun. In: *Molecular Improvement of Cereal Crops: Advances in Cellular and Molecular Biology of Plants*, Vol. 5, I.K. Vasil, (Ed.), 21-42, Dordrecht, The Netherlands: Kluwer Academic Publishers

Ko, K.; Norelli, JL.; Reynoird, JP.; Boresjza-Wysocka, E.; Brown, SK. & Aldwinckle, HS. (2000). Effect of untranslated leader sequence of AMV RNA 4 and signal peptide of pathogenesis-related protein 1b on attacin gene expression, and resistance to fire blight in transgenic apple. *Biotechnology Letters*, 22, pp. 373-381

Kokko HI. & Kärenlampi, SO. (1998). Transformation of arctic bramble (*Rubus arcticus* L.) by Agrobacterium tumefaciens. *Plant Cell Reports*, 17, pp. 822–826

Kraus J. (2010). Concepts of Marker Genes for Plants. In: *Genetic Modification of Plants*. F., Kempken, C., Jung (Eds.), 39-60, Biotechnology in Agriculture and Forestry, Volume 64, 1, Springer-Verlag, Berlin-Heidelberg

Krens, FA.; Schaart, JG.; Groenwold, R.; Walraven, AEJ.; Hesselink, T. & Thissen, JTNM. (2011). Performance and long-term stability of the barley hordothionin gene in multiple transgenic apple lines. *Transgenic Research*, 20, pp. 1113-1123

Lau, JM. & Korban, SS. (2010). Transgenic apple expressing an antigenic protein of the human respiratory syncytial virus. *Journal of Plant Physiology*, 167, pp. 920-927

Lee, YK. & Kim, IJ. (2011). Modulation of fruit softening by antisense suppression of endo-b-1,4-glucanase in strawberry. *Molecular Breeding*, 27, pp. 375–383

Leyman, B.; Avonce, N.; Ramon, M.; Van Dijck, P.; Iturriaga, G. & Thevelein, JM. (2006). Trehalose-6-phosphate synthase as an intrinsic selection marker for plant transformation. *Journal of Biotechnology*, 121, pp. 309–317

Li, ZT.; Dhekney, S.; Dutt, M.; Van Aman, M.; Tattersall, J.; Kelley, KT. & Gray, DJ. (2006). Optimizing *Agrobacterium* -mediated transformation of grapevine. *In Vitro Cellular and Developmental Biology –Plant*, 42, pp. 220–227

Lindsey, K. (1992). Genetic manipulation of crop plants. *Journal of Biotechnology* 26, pp. 1–28

Liu, Z.; Park, BJ.; Kanno, A. & Kameya, T. (2005). The novel use of a combination of sonication and vacuum infiltration in *Agrobacterium* -mediated transformation of kidney bean (*Phaseolus vulgaris* L.) with leagene. *Molecular Breeding*, 16, pp. 189–197

López-Noguera, S.; Petri, C. & Burgos, L. (2009). Combining a regeneration-promoting ipt gene and site-specific recombination allows a more efficient apricot transformation and the elimination of marker genes. *Plant Cell Reports*, 28, pp. 1781-1790

Lopez-Perez, AJ.; Velasco, L.; Pazos-Navarro, M. & Dabauza, M. (2008). Development of highly efficient genetic transformation protocols for table grape Sugraone and Crimson Seedless at low agrobacterium density. *Plant Cell Tissue and Organ Culture*, 94, pp. 189–199

Maddumage, R.; Fung, RMW.; Ding, H.; Simons, JL. & Allan, AC. (2002). Efficient transformation of suspension culturederived apple protoplasts. *Plant Cell Tissue and Organ Culture*, 70, pp. 77– 82

Maghuly, F.; Leopold, S.; Machado, A.; Borroto Fernández, E.; Khan, MA.; Gambino, G.; Gribanno, I.; Schartl, A. & Laimer, M. (2006) Molecular characterization of grapevine plants transformed with GFLV resistance genes: II. *Plant Cell Reports*, 25, pp. 546–553

Malnoy, M.; Boresjza, EE.; Norelli, JL.; Flaishman, MA.; Gidoni, D. & Aldwinckle, HS. (2010). Genetic transformation of apple (Malus x domestica) without use of a selectable marker gene. *Tree Genetics and Genome*, 6, pp. 423-433

Manimaran, P.; Ramkumar, G.; Sakthivel, K.; Sundaram, RM.; Madhav, MS. & Balachandran, SM. (2011). Suitability of non-lethal marker and marker-free systems for development of transgenic crop plants: Present status and future prospects. *Biotechnology Advances*, 29, pp. 703-714.

Mante, S.; Morgens, PH.; Scorza, R.; Cordts, JM. & Callahan, AM. (1991). *Agrobacterium*-mediated transformation of plum (*Prunus domestica* L) hypocotyls slices and regeneration of transgenic plants. *Bio-Technology*, 9, pp. 853-857

Mathews, H.; Dewey, V.; Wagner, W. & Bestwick, RK. (1998). Molecular and cellular evidence of chimaeric tissues in primary transgenics and elimination of chimaerism through improved selection protocols. *Transgenic Research*, 7, pp. 123-129

Mathews, H.; Litz, RE.; Wilde, DH.; Merkel, S. & Wetzstein, HY. (1992). Stable integration and expression of β-glucuronidase and NPT II genes in mango somatic embryos. *In Vitro Cellullar and Developmental Biology – Plant*, 28, pp. 172-178

Mathews, H.; Litz, RE.; Wilde, HD. & Wetzstein, HY. (1993). Genetic transformation of mango. *Acta Horticulturae*, 341, pp. 93-97

Maximova, SN.; Dandekar, AM. & Guiltinan, MJ. (1998). Investigation of Agrobacterium-mediated transformation of apple using gren fluorescent protein: high transient expression and low stable transformation suggest that factors other than T-DNA transfer are rate-limiting. *Plant Molecular Biology*, 37, pp. 549-559

McCabe, D. & Christou, P. (1993). Direct DNA transfer using electrical discharge particle acceleration (ACCELL™ technology). *Plant Cell Tissue and Organ Culture*, 33, pp. 227-236

Mercado, JA.; Trainotti, L.; Jiménez-Bermúdez, L.; Santiago-Doménech, N.; Posé, S.; Donolli, R.; Barceló, M.; Casadoro, G.; Pliego-Alfaro, F. & Quesada, MA. (2010). Evaluation of the role of the endo-β-(1,4)-glucanase gene FaEG3 in strawberry fruit softening . *Postharvest Biology and Technology*, 55, pp. 8-14

Mezzetti, B.; Pandolfini, T.; Navacchi, O. & Landi, L. (2002). Genetic transformation of Vitis vinifera via organogenesis. *BMC Biotechnology*, 2, pp. 18-27

Miki, B. & McHugh, S. (2004.) Selectable marker genes in transgenic plants: applications, alternatives and biosafety. *Journal of Biotechnology*, 107, pp. 193-232

Nakano, M.; Hoshino, Y. & Mii, M. (1994). Regeneration of transgenic plants of grapevine (Vitis vinifera L.) via *Agrobacterium rhizogenes*- mediated transformation of embryogenic calli. *Journal of Experimental Botany*, 45, pp. 649-656

Naqvi, S.; Farré, G.; Sanahuja, G.; Capell, T.; Zhu, C. & Christou, P. (2009). When more is better: multigene engineering in plants. *Trends in Plant Science*, 15, pp. 48-56

Newell, CA. (2000). Plant transformation technology; developments and applications. *Molecular Biotechnology*, 16, pp. 53-65

Nirala, NK.; Das, DK.; Srivastava, PS.; Sopory, SK. & Upadhyaya, KC. (2010). Expression of a rice chitinase gene enhances antifungal potential in transgenic grapevine (*Vitis vinifera* L.). *Vitis*, 49, pp. 181-187

OECD. (1999). Concensus document on general information concerning the genes and their enzymes that confer tolerance to phosphinothricin herbicide. *Series on Harmonization of Regulatory Oversight in Biotechnology*, No. 11

Oliveira, MLP.; Febres, VJ.; Costa, MGC.; Moore, GA. & Otoni, WC. (2009). High-efficiency *Agrobacterium* –mediated transformation of citrus via sonication and vacuum infiltration. *Plant Cell Reports,* 28, pp. 387–395

Oliveira, MM.; Borrosa, JG.; Martins, M. & Pais, MS. (1994). In Y. P. S. Bajaj (Ed.), *Biotechnology in Agriculture and Forestry,* 29 pp. 189–210. Berlin: Springer

Oliveira, MM.; Barroso, J. & Pais, MS. (1991). Direct gene transfer into kiwifruit protoplasts: analysis of transient expression of the CAT gene using TLC autoradiography and a GC-MS based method. *Plant Molecular Biology,* 17, pp. 235–242

Padilla IMG. & Burgos L. (2010). Aminoglycoside antibiotics: structure, functions and effects on in vitro plant culture and genetic transformation protocols. *Plant Cell Reports,* 29, pp. 1203–1213

Padilla, IMG.; Golis, A.; Gentile, A.; Damiano, C. & Scorza, R. (2006). Evaluation of transformation in peach *Prunus persica* explants using green fluorescent protein (GFP) and beta-glucuronidase (GUS) reporter genes. *Plant Cell Tissue and Organ Culture,* 84, pp. 309–314

Paradela, MR.; De Paz, P. & Gallego, PP. (2006) Comparation of two kiwifruit transformation methods, *Proceedings of 11th International Association for Plant Tissue Culture & Biotechnology Congress. Biotechnology and Sustainable Agriculture 2006 and Beyond,* pp 76 (P-1121), Beijing China, September 27-30, 2006

Paszkowski, J.; Shillito, RD.; Saul, M.; Mandak, V.; Hohn, T.; Hohn, B. et al. (1984). Direct gene transfer to plants. *EMBO Journal,* 3, pp. 2717–2722

Paszkowski, J. & Witham, SA. (2001). Gene silencing and methylation processes. *Current Opinion in Plant Biology,* 4, pp. 123–129

Pawlowski, WP. & Somers, DA. (1996). Transgene inheritance in plants genetically engineered by microprojectile bombardment. *Molecular Biotechnology,* 6, pp. 17–30

Pérez-Clemente, RM.; Pérez-Sanjuán, A.; García-Férriz, L.; Beltrán, JP. & Cañas, LA. (2004) Transgenic peach plants (*Prunus persica* L.) produced by genetic transformation of embryo sections using the green fluorescent protein (GFP) as an in vivo marker. *Molecular Breeding,* 14, pp. 419–427

Pérez-Piñeiro, P.; Gago, J.; Landín, P. & Gallego, PP. (2012). *Agrobacterium*-mediated transformation of wheat: general overview and new approaches to model and identify the key factors involved. In: *Transgenic Plants.* Y., Ozden Çiftçi (Ed.), XX_XX, Intech Open Access Publisher: Croatia (in press)

Petolino, J. (2002). Direct DNA delivery into intact cells and tissues. In: *Transgenic Plants and Crops.* G. Khachatourians, A. McHughen, R. Scorza, W-K. Nip &Y. Hui, (Eds.), 137–141, Mercel Dekker Inc.; New York

Petri, C. & Burgos, L. (2005). Transformation of fruit trees. Useful breeding tool or continued future prospect? *Transgenic Research,* 14, pp. 15-26

Petri, C.; Alburquerque, N.; García-Castillo, S.; Egea, J. & Burgos, L. (2004) Factors affecting gene transfer efficiency to apricot leaves during early *Agrobacterium*-mediated transformation steps. *Journal of Horticultural Science and Biotechnology,* 79, pp. 704-712

Petri, C.; Hily, J.M.; Vann, C.; Dardick, C. & Scorza, R. (2011) A high-throughput transformation system allows the regeneration of marker-free plum plants (*Prunus domestica*). *Annals of Applied Biology,* 159, pp. 302-315

Petri, C.; Wang, H.; Alburquerque, N.; Faize, M. & Burgos, L. (2008). Agrobacterium-mediated transformation of apricot (*Prunus armeniaca* L.) leaf explants. *Plant Cell Reports*, 27, pp. 1317–1324

Purohit, SD.; Raghuvanshi, S. & Tyagi AK. (2007). Biolistic-mediated DNA delivery and transient expression of GUS in hypocotyls of *Feronia limonia* L.- A fruit tree. *Indian Journal of Biotechnology*, pp. 504-507

Radchuk, VV. & Korkhovoy, VI. (2005). The *rolB* gene promotes rooting in vitro and increases fresh root weight in vivo of transformed apple scion cultivar 'Florina'. *Plant Cell Tissue and Organ Culture*, 81, pp. 203–212

Ramesh, S.; Kaiser, BN.; Franks, TK.; Collins, GG. & Sedgley, M. (2006). Improved methods in *Agrobacterium*-mediated transformation of almond using positive (mannose/pmi) or negative (kanamycin resistance) selection-based protocols. *Plant Cell Reports*, 25, pp. 821–828

Rao, AQ.; Bakhsh, A.; Kiani, S.; Shahzad, K.; Shahid, AA.; Husnain, T. & Riazuddin, S. (2009). The myth of plant transformation. *Biotechnology Advances*, 27, pp. 753–763

Raquel, MH. & Oliveira, MM. (1996) - Kiwifruit leaf protoplasts competent for plant regeneration and direct DNA transfer. *Plant Science*, 121, pp. 107-114

Reed, J.; Privalle, L.; Powell, M.L.; Meghji, M.; Dawson, J.; Dunder, E.; Suttie, J.; Wenck, A.; Launis, K.; Kramer, C.; Chang, YF.; Hansen, G. & Wright, M. (2001). Phosphomannose isomerase: an efficient selectable marker for plant transformation *In Vitro Cellular and Developmental Biology -Plant*, 37, pp. 127–132

Remy S.; Buyens, A.; Cammue, BPA.; Swennen, R. & Sagi, L. (2000). Production of transgenic banana plants expressing antifungal proteins. International Symposium on Banana in the Subtropics. *Acta Horticulturae*, 490, pp. 219–277

Rommens, CM.; Haring, MA.; Swords, K.; Davies, HV. & Belknap, WR. (2007). The intragenic approach as a new extension to traditional plant breeding. *Trends in Plant Science*, 12, pp. 397–403

Rugini, E.; Pellegrineschi A.; Mecuccini M. & Mariotti, D. (1991). Increase of rooting ability in the woody species kiwi (Actinidia deliciosa A. Chev.) by transformation with *Agrobacterium rhizogenes* rol genes. *Plant Cell Reports*, 6, pp. 291–5

Sagi, L.; Panis, B.; Remy, S.; Schoofs, H.; Smet, K.; Swennen, R. & Gammue, BPA. (1995). Genetic transformation of banana and plantain (*Musa* spp.) via particle bombardment. *Biotechnology*, 13, pp. 481–485

Sagi, L.; Remy, S. & Swennen, R. (1998). Genetic transformation for the improvement of bananas – a critical assessment. In: INIBAP Annual Report. pp. 33–36. Montpellier: FRA.

Sakuanrungsirikul, S.; Sarindu, N.; Prasartsee, V.; Chaikiatiyos, S.; Siriyan, R.; Sriwatanakul, M.; Lekananon, P.; Kitprasert, C.; Boonsong, P.; Kosiyachinda, P. et al (2005). Update on the development of virus-resistant papaya: virus-resistant transgenic papaya for people in rural communities of Thailand. *Food Nutr Bull*, 26, pp. 422–426

Sanford JC, Klein TM, Wolf ED, Allen N. (1987). Delivery of substances into cells and tissues using a particle bombardment process. *Journal of Paniculate Science and Technology*, 6, pp. 559-563

Sanford JC. (1988). The Biolistic Process. *Trends in Biotechnology* 6, pp. 299-302

Sanford, JC. (1990). Biolistic plant transformation. *Physiologia Plantarum* 79, pp. 206-209

Sautter C. (1993). Development of a microtargeting device for particle bombardment of plant meristems. *Plant Cell Tissue and Organ Culture*, 33, pp. 251–257

Schaart, JG.;, Krens, FA.; Pelgrom, KT.; Mendes, O. & Rouwendal, GJ. (2004). Effective production of marker-free transgenic strawberry plants using inducible site-specific recombination and a bifunctional selectable marker gene. *Plant Biotechnology Journal*, 2, pp. 233–240

Schouten, HJ.; Krens, FA. & Jacobsen, E. (2006a). Do cisgenic plants warrant less stringent oversight? *Nature Biotechnology*, 24, p. 753

Schouten, HJ.; Krens, F.A. & Jacobsen, E. (2006b) Cisgenic plants are similar to traditionally bred plants. *EMBO Reports*, 7, pp. 750–753

Schubert, D. & Willims, D. (2006).'Cisgenic' as a product designation. *Nature Biotechnology*, 24, pp. 1327–1329

Schuerman, PL. & Dandekar, AM. (1993). Transformation of temperate woody crops - Progress and Potentials. *Scientia Horticulturae*, 55, pp. 101-124

Scorza, R.; Cordts, JM.; Gray, DJ.; Gonsalves, D.; Emershad, RI. & Ramming, DW. (1996). Producing transgenic 'Thompson Seedless' grape (*Vitis vinifera* L.) plants. *Journal of American Society for Horticultural Science*, 121, pp. 616–619

Scorza, R.; Levy, L.; Damsteegt, V.; Yepes, L.M.; Cordts, J.; Hadidi, A.; Slightom, J. & Gonsalves, D. (1995). Transformation of plum with the papaya ringspot virus coat protein gene and reaction of transgenic plants to plum pox virus. *Journal of American Society for Horticultural Science*, 120, pp. 943-952

Serres, R.; Stang, E.; McCabe, D.; Russell, D.; Mahr, D. & McCown, B. (1992) Gene transfer using electric discharge particle bombardment and recovery of transformed cranberry plants. *Journal of American Society for Horticultural Science*, 117, pp. 174 – 180

Sicherer, SH.; Munoz-Furlong, A. & Sampson, HA. (2003). Prevalence of peanut and tree nut allergy in the United States determined by means of a random digit dial telephone survey: a 5-year followup study. *Journal of Allergy and Clinical Immunology*, 112, pp. 1203–1207

Singh, S. (1992). Fruit crops of wasteland. Scientific Publishers, Jodhpur, India.

Singsit, C.; Adang, MJ.; Lynch, R.; Anderson, WF.; Wang, A.; Cardineau, G. & Ozias-Akins, P. (1997). Expression of a *Bacillus thuringiensis* cryIA(c) gene in transgenic peanut plants and its efficacy against lesser cornstalk borer. *Transgenic Research*, 6, pp. 169–176

Smolka, A.; Welander, M.; Olsson, P.; Holefors, A. & Zhu, LH. (2009). Involvement of the ARRO-1 gene in adventitious root formation in apple. *Plant Science*, 177, pp. 710–715

Song, GQ. & Sink, KC. (2004). Agrobacterium tumefaciens-mediated transformation of blueberry (*Vaccinium corymbosum* L.). *Plant Cell Reports*, 23, pp. 475–484

Stalker, HT. & Simpson, CE. 1995: Genetic resources in *Arachis*. In: *Advances in Peanut Science*, Pattee HE., & Stalker HT, 14–53. American Peanut Research and Educational Society, Stillwater, OK, USA

Stewart, CN. (2001). The utility of green fluorescent protein in transgenic plants. *Plant Cell Reports*, 20, pp. 376–382

Subramanyam, K.; Subramanyam, K.; Sailaja, KV.; Srinivasulu, M. & Lakshmidevi K. (2011). Highly efficient *Agrobacterium* -mediated transformation of banana cv. Rasthali (AAB) via sonication and vacuum infiltration. *Plant Cell Reports*, 30, pp. 425–436

Sun, Q.; Zhao, Y.; Sun, H.; Hammond, RW.; Davis, RE. & Xin, L. (2011). High-efficiency and stable genetic transformation of pear (*Pyrus communis* L.) leaf segments and regeneration of transgenic plants. *Acta Physiologiae Plantarum*, 33, pp. 383–390

Sundar, IK. & Sakthivel, N. (2008). Advances in selectable marker genes for plant transformation. *Journal of Plant Physiology*, 165, pp. 1698–1716

Szankowski, I.; Waidmann, S.; Degenhardt, J.; Patocchi, A.; Paris, R.; Silfverberg-Dilworth, E.; Broggini, G. & Gessler, C. (2009). Highly scab-resistant transgenic apple lines achieved by introgression of HcrVf2 controlled by different native promoter lengths. *Tree Genetics and Genomes*, 5, pp. 349–358

Taylor, NJ. & Fauquet, CM. (2002). Microparticle Bombardment as a Tool in Plant Science and Agricultural Biotechnology. *DNA and Cell Biology*, 21, pp. 963–977

Tennant, PF.; Gonsalves, C.; Ling, KS.; Fitch, MM.; Manshardt, R.; Slightom, LJ. & Gonsalves, D. (1994). Differential protection against papaya ringspot virus isolates in coat protein gene transgenic papaya and classically cross-protected papaya. *Phytopathology*, 84, pp. 1359-1366

Terakami, S.; Matsuda, N.; Yamamoto, T.; Sugaya, S.; Gemma, H. & Soejima, J. (2007). *Agrobacterium* -mediated transformation of the dwarf pomegranate (*Punica granatum* L. var. nana). *Plant Cell Reports*, 26, pp.1243-1251

Thompson, CJ.; Movva, NR.; Tizard, R.; Crameri, R.; Davies, JE.; Lauwereys, M. & Botterman, J. (1987). Characterization of the herbicide-resistance gene bar from *Streptomyces hygroscopicus*. *EMBO Journal*, 6, pp. 2519–2523

Tian, N.; Wang, J. & Xu Z.Q. (2011). Overexpression of Na+/H+ antiporter gene AtNHX1 from *Arabidopsis thaliana* improves the salt tolerance of kiwifruit (*Actinidia deliciosa*). *South African Journal of Botany*, 77, pp. 160–169

Tian, L.; Canli, F. A.; Wang, X. & Sibbald, S. 2009. Genetic transformation of *Prunus domestica* L. using the *hpt* gene coding for hygromycin resistance as the selectable marker. *Scientia Horticulturae*, 119, pp. 339-343

Trick, HN. & Finer, JJ. (1998). Sonication-assisted *Agrobacterium*-mediated transformation of soybean *Glycine max* L. Merrill embryogenic suspension culture tissue. *Plant Cell Reports*, 17, pp. 482–488

Trick, HN. & Finer, JJ. (1997). SAAT: sonication-assisted *Agrobacterium* mediated transformation. *Transgenic Research*, 6, pp. 329-336

Tripathi, S.; Suzuki, JY.; Carr, JB.; McQuate, GT.; Ferreira, SA.; Manshardt, RM.; Pitz, KY.; Wall, MM. & Gonsalves D. (2011). Nutritional composition of Rainbow papaya, the first commercialized transgenic fruit crop. *Journal of Food Composition and Analysis*, 24, pp. 140–147

Tripathi, L. (2003). Genetic engineering for improvement of *Musa* production in Africa. *African Journal of Biotechnology*, 2, pp. 503–508

Tripathi, S.; Suzuki, J.Y.; Ferreira, S.A. & Gonsalves, D. (2008). Papaya ringspot virus-P: characteristics, pathogenicity, sequence variability and control. *Molecular Plant Pathology*, 9, pp. 269–280

Twyman, R.M.; Stöger, E.; Kohli, A.; Capell, T. & Christou, P. (2002). Selectable and screenable markers for rice transformation. *Molecular Methods of Plant Analysis*, 22, pp. 1–17

Uematsu, C.; Murase, M.; Ichikawa, H. & Imamura, J. (1991). *Agrobacterium* -mediated transformation and regeneration of kiwi fruit. *Plant Cell Reports*, 10, pp. 286–290

Urtubia C.; Devia J.; Castro A.; Zamora P.; Aguirre C.; Tapia E.; Barba P.; Dell'Orto P.; Moynihan M.R.; Petri C.; Scorza R. & Prieto H. (2008). *Agrobacterium*-mediated genetic transformation of *Prunus salicina*. Plant Cell Reports, 27, pp. 1333–1340

van Leeuwen, W.; Hagendoorn, MJM.; Ruttink, T.; van Poecke, R.; van der Plas, LHW. & van der Krol, AR. (2000). The use of the luciferase reporter system for *in planta* gene expression studies. *Plant Molecular Biology Reports*, 18, pp. 143a–143t

Vanblaere, T.; Szankowski, I.; Schaart, J.; Schouten, H.; Flachowsky, H.; Broggini, GAL. & Gessler, C. (2011). The development of a cisgenic apple plant. *Journal of Biotechnology*, 154, pp. 304– 311

Varshney, RK.; Kailash, C.; Bansal, KC.; Aggarwal, PK.; Datta, SK. & Craufurd, PQ. (2011). Agricultural biotechnology for crop improvement in a variable climate: hope or hype? *Trends in Plant Science*, 16, pp. 363-371

Veluthambi, K.; Gupta, A.K. & Sharma, A. (2003). The current status of plant transformation technologies. *Current Science*, 84, pp. 368-380

Vidal, JR.; Kikkert, JR.;Wallace, PG. & Reisch, BI. (2003). High-efficiency biolistic co-transformation and regeneration of 'Chardonnay' (*Vitis vinifera* L.) containing *npt-II* and antimicrobial peptide genes. *Plant Cell Reports*, 22, pp. 252–260

Vidal, JR.; Kikkert, JR.; Malnoy, MA.; Wallace, PG.; Barnard, J. & Reisch, BI. (2006). Evaluation of transgenic Chardonnay (*Vitis vinifera*) containing magainin genes for resistance to crown gall and powdery mildew. *Transgenic Research*, 15, pp. 69-82

Villar, B.; Oller, JJ.; Teulieres, C.; Boudet ,AM. & Gallego, PP. (1999). *In planta* transformation of adult clones of *Eucalyptus globulus* sp using an hypervirulent *Agrobacterium tumefaciens* strain. In: *Application of Biotechnology to Forest Genetic*. Espinel, S. & Ritter, E. (Eds.), pp. 22-25, DFA-AFA Press, Vitoria-Gasteiz, Spain.

Vishnevetsky, J.; White, Jr. TL.; Palmateer, AJ.; Flaishman, M.; Cohen, Y.; Elad, Y.; Velcheva, M.; Hanania, U.; Sahar, N.; Dgani, O. & Perl A. (2011). Improved tolerance toward fungal diseases in transgenic Cavendish banana (*Musa* spp. AAA group) cv. Grand Nain. *Transgenic Research*, 20, pp. 61–72

Wang, T.; Ran, Y.; Atkinson, RG.; Gleave, AP. & Cohen D. (2006). Transformation of *Actinidia eriantha*: A potential species for functional genomics studies in *Actinidia*. *Plant Cell Reports*, 25, pp. 425–431

Wang, H. (2011). *Development of a genetic transformation protocol genotype-independent in apricot*. Doctoral Thesis. CEBAS-CSIC, University of Murcia (Spain) (in Spanish).

Wang, Q.; Li, P.; Hanania, U.; Sahar, N.; Mawassi, M.; Gafny, R.; Sela, I.; Tanne, E. & Perl, A. (2005). Improvement of *Agrobacterium*-mediated transformation efficiency and transgenic plant regeneration of *Vitis vinifera* L. by optimizing selection regimes and utilizing cryopreserved cell suspensions. *Plant Science*, 168, pp. 565-571

Weinthal, D.; Tovkach, A.; Zeevi, V. & Tzfira, T. (2010). Genome editing in plant cells by zinc finger nucleases. *Trends in Plant Science*, 15, pp. 308-321

Wohlleben, W.; Arnold, W.; Broer, I.; Hilleman, D.; Strauch, E. & Puhler, A. (1988). Nucleotide sequence of the phosphinothricin N-acetyltransferase gene from Streptomyces viridochromogenes Tü494 and its expression in *Nicotiana tabacum*. *Gene* ,70, pp. 25-37

Wu, HW.; Yu, TA.; Raja, JAJ.; Wang, HC. & Yeh, SD. (2009). Generation of transgenic oriental melon resistant to Zucchini yellow mosaic virus by an improved cotyledon-cutting method. *Plant Cell Reports*, 28, pp. 1053–1064

Xu, SX.; Cai, XD.; Tan, B. & Guo, WW. (2011). Comparison of expression of three different sub-cellular targeted GFPs in transgenic Valencia sweet orange by confocal laser scanning microscopy. *Plant Cell Tissue and Organ Culture,* 104, pp. 199-207

Xue, B.; Ling, KS.; Reid, CL.; Krastanova, S.; Sekiya, M.; Momol, EA.; Sule, S.; Mozsar, J.; Gonsalves, D. & Burr, TJ. (1999). Transformation of five grape rootstocks with plant virus genes and a *vir*E2 gene from *Agrobacterium tumefaciens*. *In Vitro Cellular and Developmental Biology – Plant,* 35, pp. 226-231

Yamamoto, T.; Iketani, H.; Ieki, H.; Nishizawa, Y.; Notsuka, K.; Hibi T, Hayashi T, Matsuta N (2000) Transgenic grapevine plants expressing a rice chitinase with enhanced resistance to fungal pathogens. *Plant Cell Reports,* 19, pp. 639-646

Yancheva, SD.; Shlizerman, LA.; Golubowicz, S.; Yabloviz, Z.; Perl, A.; Hanania, U. & Flaishman, MA. (2006). The use of green fluorescent protein (GFP) improves *Agrobacterium* -mediated transformation of 'Spadona' pear (*Pyrus communis* L.). *Plant Cell Reports,* 25, 183-189

Yancheva, SD.; Golubowicz, S.; Yablowicz, Z.; Perl, A. & Flaishman, MA. (2005). Efficient Agrobacterium-mediated transformation and recovery of transgenic fig (*Ficus carica* L.) plants. *Plant Science,* 168, pp. 1433-1441

Yang, C.; Chen, S. & Duan, G. (2011). Transgenic peanut (*Arachis hypogaea* L.) expressing the urease subunit B gene of *Helicobacter pylori*. *Curr Microbiol,* 63, pp. 387-391

Yao, JL.; Wu JH.; Gleave, AP. & Morris, BAM. (1996). Transformation of citrus embryogenic cells using particle bombardment and production of transgenic embryos. *Plant Science,* 113, pp. 175-183

Ye, XJ.; Brown, SK.; Scorza, R.; Cordts, JM. & Sanford, JC. (1994) Genetic transformation of peach tissues by particle bombardment. *Journal of the American Society of Horticultural Science,* 119, pp. 367-373

Yip, MK.; Lee SW.; Su, KC.; Lin, YH.; Chen, TY. & Feng, TY. (2011). An easy and efficient protocol in the production of pflp transgenic banana against Fusarium wilt. *Plant Biotechnology Reports,* 5, pp. 245-254

Zeldin, E.;, Jury, TP.; Serres, RA. & McCown, BH. (2002). Tolerance to the herbicide glufosinate in transgenic cranberry (*Vaccinium macrocarpon* Ait.) and enhancement of tolerance in progeny. *Journal of American Society for Horticultural Science,* 127, pp. 502-507

Zhu Y.J.; Agbayani R.; McCafferty H.; Albert HH. & Moore PH. (2005). Effective selection of transgenic papaya plants with the PMI/Man selection system. *Plant Cell Reports,* 24, pp. 426-432

Zhu YJ.; Agbayani R. & Moore PH. (2004). Green fluorescent protein as a visual selection marker for papaya (*Carica papaya* L.) transformation *Plant Cell Reports,* 22, pp. 660-667

Zhu, LH.; Holefors, A.; Ahlman, A.; Xue, ZT. & Welander, M. (2001). Transformation of the apple rootstock M.9/29 with the *rol*B gene and its influence on rooting and growth. *Plant Science,* 160, pp. 433-439

Green Way of Biomedicine – How to Force Plants to Produce New Important Proteins

Aneta Wiktorek-Smagur[1], Katarzyna Hnatuszko-Konka[2], Aneta Gerszberg[2],
Tomasz Kowalczyk[2], Piotr Luchniak[2] and Andrzej K. Kononowicz[2]
[1]Nofer Institute of Occupational Medicine
[2]Department of Genetics Plant Molecular Biology and
Biotechnology University of Lodz
Poland

1. Introduction

Recombinant proteins can be expressed in transformed cell cultures of bacteria, yeasts, molds, mammals, plants, insects, or via transgenic plants and animals. Numerous factors influence quality, functionality, yield and protein production rate, so the choice of appropriate expression system is of primary importance. During last few years, plants have become an increasingly promising and attractive platform for recombinant protein production (Basaran & Rodriguez–Cerezo, 2008). Progress in recombinant DNA technology, plant transformation and *in vitro* regeneration techniques are major reasons why plants have emerged as efficient expression systems. Plant expression systems offer significant advantages over the other expression systems (Table 1). First of all, plants have a higher eukaryote protein synthesis pathway very similar to animal cells with only minor differences in protein glycosylation. Therefore, plant biosynthesis pathway ensures correct structure even in the case of highly complex proteins. In contrast to plants, bacteria are not able to carry out most of posttranslational modifications essential for eukaryotic proteins activity. There is no risk of contamination of recombinant proteins with human or animal pathogens (HIV, hepatitis viruses, prions), bacteria endotoxins or oncogenic DNA sequences (Sharma & Sharma, 2009).

Other advantages of the plant–based expression systems include: high scalability (in the case of field cultivation), low production cost of biomass (agriculture), in some cases low upstream costs (edible vaccines, purification process can be omitted), and what is most important - the ability to produce target proteins with desired structures and biological functions (Boehm, 2007). Recombinant proteins expressed in plants can be accumulated to a high level in seed endosperm, fruit or storage organs (e.g. tubers, roots) or secreted directly to the culture media. Because plant culture media contain no exogenous proteins, the recovery of recombinant proteins from a medium is expected to be much simpler and less expensive than the recovery from homogenized biomass (Cox et al., 2009).

Features	Transgenic plants	Plants viruses	Yeast	Bacteria	Mammalian cell culture	Transgenic animals
Cost/storage	Cheap	Cheap	Cheap	Cheap	Expensive	Expensive
Distribution	Easy	Easy	Feasible	Feasible	Difficult	Difficult
Gene size	Not limited	Limited	Unknown	Unknown	Limited	Limited
Glycosylation	Correct	Correct	Incorrect	Absent	Correct	Correct
Production costs	Low	Low	Medium	Medium	High	High
Production scale	Worldwide	Worldwide	Limited	Limited	Limited	Limited
Propagation	Easy	Feasible	Easy	Easy	Hard	Feasible
Protein folding accuracy	High	High	Medium	Low	High	High
Protein homogeneity	High	Medium	Medium	Low	Medium	Low
Protein yield	High	Very high	High	Medium	Medium-high	High
Safety	High	High	Unknown	Low	Medium	High
Scale up costs	Low	Low	High	High	High	High
Therapeutic risk	Unknown	Unknown	Unknown	Yes	Yes	Yes
Time required	Medium	Low	Medium	Low	High	High

Table 1. Comparison of features of recombinant protein production in existing systems (according to Fischer and Emans 2004; worked out / modified on the basis of Demain and Vaishnav 2009).

The usage of aquatic plants e.g. *Lemnaceae* seems to be a good solution. For example Rival et al. (2008) made studies on obtaining aprotinin from *Spirodela oligorrhiza* (duckweed). Their experiments show that significant amounts of recombinant aprotinin can be produced using *Spirodela* as a plant host. Whereas Cox and co-workers (2009) expressed human monoclonal antibody (mAbs) in *Lemna minor*. The micro-alga *Chlamydomonas reinhardtii* has recently been shown as a promising platform for foreign protein production (Muto et al., 2009). This photosynthetic single-celled plant possesses several interesting features in comparison to the majority of plants as it has a rapid doubling time (ca. 10 h); its homogenous culture is easily scaled up; it has a rapid sexual cycle (ca. 2 weeks) with stable and viable haploids. All these attributes make the time of petting a final product on a large-scale much shorter in comparison to higher plants (months or years). Growth in containment bioreactors allows to control conditions of farming as well as reduces the risk of contamination and loss of algae due to pathogens. It is worth mentioning that all three genomes of *C. reinhardtii* have been fully sequenced affording strong foundation for targeted genetic manipulation (Specht et al., 2010).

Feasible storage of recombinant proteins in desiccated plant parts excludes the requirement for its immediate isolation and lowers the risk of the loss of biological function during prolonged freezing of preparations. For example, antibodies or vaccines expressed in cereal seeds remain stable at ambient temperatures for years (Stoger et al., 2002). Until recently, low accumulation levels have been the major bottleneck for plant-made recombinant protein production. However, several breakthroughs have been done during past few years allowing for high accumulation levels. Mainly through chloroplast, vacuole, ER lumen transient expression, coupled with subcellular targeting and protein fusions (Sharma and Sharma, 2009). Viral transfection and agroinfiltration are promising alternative strategies ensuring increase in yields and speeding up the development of an expression platform (Gleba et al., 2005). On the other hand, plant–based expression systems are different from the mammalian host pattern of glycosylation. The occurrence has raised concerns regarding the potential immunogenicity of plant-specific complex N-glycans (α 1,3-fucose and β 1,2-xylose residue), which are present in the heavy chains of plant-derived antibodies (Gomord and Faye 2004). The above mentioned residues have been confirmed not only to induce immune response but also to make foreign proteins undergo a conformational change making them different from the native ones which results in decrease in their biological activity. However, some achievements in humanized glycosylation or removal of enzymatic pathway generating immunogenic residues on glycoproteins have been reported. Recently it has been shown that glycoengineered moss (*Physcomitrella patens*) can synthesize proteins carrying a humanized glycosylation pattern (Decker and Reski, 2008). A few years ago *Physcomitrella patens* platform was developed and commercialized as a contained tissue culture system for recombinant protein production in photo-bioreactors [Biotech GmbH (© greenovation)]. *P. patens* has some characteristic features which make it a suitable system for foreign protein production. Firstly, it grows rapidly under photoautotrophic conditions and secondly the moss protonema can release the desired protein into the medium. The moss remains productive in the system for a period of six months, in contrast to animal cell cultures (20 days) (Decker and Reski, 2008).

Other approaches to overcome undesirable glycosylation accommodate export of foreign proteins into subcellular compartments: ER lumen, where glycosylation characteristic of plants does not take place; cytosol, where glycosylation process is not found; or recombinant protein expression export into plastids (proteins do not undergo glycosylation there). According to several studies ER targeting gives higher yield of biologically active protein than cytosol targeting (referred by Boehm, 2007).

Potential disadvantages of transgenic plants include possible contamination with pesticides, herbicides, and toxic plant metabolites. Proteolytic degradation, post/transcriptional gene silencing, position effect and transgenic recombination are other obstacles affecting stability or expression level of transgenic plants (Basaran and Rodriguez–Cerezo, 2008).

The public concern about health and environmental risk associated with transgenic plants is being considered at different levels: inherent risk of transgene leakage into non-transgene crops or naturally occurring wild type species (transgene escape through pollen); transgene spread by seed or fruit dispersal; horizontal gene transfer by asexual means; unintentional exposure of non-targeted organisms (e.g. birds, insects or soil microorganism); elicitation of allergic response/reaction in people (Basaran and Rodriguez–Cerezo, 2008). There are some strategies which allow to alleviate these problems including usage of closed culture facilities, such as greenhouses, hydroponic or suspension bioreactors or plastid transformation (as plastids are inherited through maternal tissues in most species and the pollen does not contain chloroplasts, hence the transgene cannot be transferred) (Basaran and Rodriguez–Cerezo, 2008).

From economical point of view, plants can be an alternative system for recombinant protein production (especially biopharmaceutical) in comparison to those exploiting mammalian or bacterial cell cultures. In this system a desired foreign protein can be produced at 2-10% of the cost of microbial fermentation system and at 0.1% of mammalian cell cultures, although it depends on the protein of interest, product field and a plant used. In general, the recombinant protein yields up to 1.5% of the total soluble protein (TSP). For example the content of antibodies does not exceed 0.35%-2% and vaccines- 0.01-0.4% of TSP (Basaran and Rodriguez–Cerezo, 2008). On the other hand, phytase from *A. niger* was obtained at the level 14% of the total tobacco soluble protein, but hirudin from *H. medicinalis* at 1% of canola seed weight and GUS from *E. coli* was produced in corn at 0.7% of TSP (Demain and Vaishnav 2009).

2. Expression strategies

Gene expression and synthesis of proteins is a complex multi-step process. For efficient expression of recombinant proteins in plants, it is essential to optimize every step of the process for the plant machinery. This includes the methods of plant transformation, the choice of a transgene promoter, improvement of transcript stability and the efficiency of its translation. After translation, the protein needs to be accumulated in plant cells or effectively secreted.

2.1 Stable nuclear transformation

The first step in plant transformation consists in the entrance of a desired genomic sequence into a plant cell. Stable nuclear transformation is caused by integration of the recombinant DNA in the nuclear genome. DNA can be transferred into the nuclear genome by either direct (e.g. biolistics) or indirect (e.g. *Agrobacterium*) methods, it depends on the plant species and the type of tissue (Thanavala et al., 2006).

In the stable nuclear transformation whole plants can be regenerated, eventually producing a seed stock or a plant tissue maintained in an aseptic culture. The advantage of this system is that the transgene is heritable, permitting the establishment of a seed stock for future use. Establishment and characterization of stable transgenic lines can be costly and time consuming. Large numbers of transgenic lines need to be screened and analyzed before a single optimal line can be selected for protein production (Ling et al., 2010). Other disadvantages are gene silencing and position effects.

Nuclear transformation has been employed and extensively studied in many plant species, however, it generally results in low expression of soluble foreign proteins (Yap & Smith, 2010).

Recombinant proteins can be targeted to different subcellular compartments in plant cells, such as cytostol, apoplast, endoplasmic reticulum, vacuole or chloroplast.

2.2 Transplastomics

Using particle bombardment or polyethylene glycol (PEG) treatment, DNA can be targeted into the chloroplast genome (Yusibov & Rabindran, 2008). Each cell contains a large number of plastids, ~100 chloroplasts per cell, and each of them contains about 100 genomes. Transplastomic lines vs. nuclear ones have significantly greater yield of foreign proteins (1-20% TSP) due to the high number of copies of the chloroplast genome and they offer major advantage in terms of transgene containment, as chloroplast genomes are predominantly maternally inherited, limiting out-crossing of the transgenic pollen. No transcriptional or post-transcriptional silencing effects have been observed in chloroplast transformation (Yap & Smith, 2010). Chloroplasts also support operon based on transgene allowing the expression of multiple proteins from a single transcript. There are two disadvantages of the chloroplast system – first: chloroplast transformation is not a standard procedure and is thus far limited to a relatively small number of crops, second: lack of some of the eukaryotic machinery for post-translational modification (Yusibov & Rabindran, 2008).

Gene integration in the plastid genome occurs by means of two homologous recombinant events mediated by a bacterial-like Rec A based system. Vectors include two 'targeting' regions flanking the selectable marker gene and a cloning site for insertion of the gene of interest. The targeting regions are between 1 and 2 kb in size and are plastid DNA sequences able to direct transgenic integration into plastome intergenic regions. Integration by homologues recombination in a preselected genome region enables insertion of only transgenic sequences and prevents uncontrollable variation in the expression of transgene. Strong promoters for plastid encoded polymerase (PEP) from the *rrn* operon and the *psbA* gene are used. Rregulatory sequences at the 5'-terminus must include a 5' untranslated region (UTR). Plastid transgene expression can be also achieved with the use of the T7 phage promoter and nuclear-encoded, plastid imported T7 RNA polymerase. In some cases protein accumulation was enhanced by translational fusion of a plastid gene N-terminal sequence with the protein of interest by including sequences downstream of the ATG start codon (downstream box) in the transgene 5'cassette that resulted in improved translation and/or protein stability. The 3'cassettes derived from 3'UTR of plastid genes generally function as inefficient terminators of transcription, but are important for plastid transcripts stability (Cardi et al., 2010).

2.3 Optimization of expression level

Increasing the transcription rate of stably transformed gene sequences is the most direct and efficient approach to increase protein expression. This is mainly achieved with the use of a strong constitutive or inducible promoter. Constitutive promoters directly drive the expression in all plant tissues and are independent of the production host developmental stage. The best known and most widely used constitutive promoter in plant biotechnology is derived from *Cauliflower Mosaic Virus (CAMV35S)*. It is more effective in dicots than monocots. Alternative constitutive promoters frequently used in plant cell transformation are the *ubiquitin* promoter, histone *H2B* promoter and the (*ocs*)*3mas* promoter (Hellwig et al., 2004). The *ubiquitin* promoter, isolated from a variety of plants including maize, *Arabidopsis*, potato, sunflower, tobacco and rice, has been frequently used to express biopharmaceuticals in plant cells. The (*ocs*)*3mas* promoter, constructed from octopine synthase (*osc*) and mannopine synthetase (*mas*) agrobacterial promoter sequences , was used for the expression of *Hepatitis B* antigen in a soybean cell culture (Smith et al., 2002). Other constitutive promoters used for expression of foreign genes in transgenic plants include: tobacco cryptic constitutive promoter (Menassa et al., 2004), Mac promoter which is a hybrid mannopine synthetase promoter and cauliflower mosaic virus 35S promoter enhancer region (Dai et al., 2000), rice actin promoter (Huang et al., 2006), banana actin promoter (Herman et al., 2001), C1 promoter of cotton leaf curl Multan virus (Xie et al., 2003), nopaline synthase promoter (Stefanov et al., 1991).

Inducible promoters allow external regulation by chemical stimuli such as alcohol, steroids, salts, sucrose or environmental factors such as temperature, light, oxidative stress and wounding. Inducible expression is advantageous as this allows protein production to be separated from cell growth. The use of chemical inducible promoters in combination with the chemical responsive transcription factor can further restrict the target transgene expression to specific organs, tissues or even cell types (Zuo & Chua, 2000). The examples of inducible promoters and synthetic transcription activators are: the rice α-amylase 3D (*RAmy3D*) promoter, which is induced by sucrose starvation; the oxidative stress-inducible a peroxidase (*SWAPA2*); an estradiol-inducible chimeric XVE transcription activator and dexamethasone-inducible pOp/4v transcription activator (Xu et al., 2011), hydroxyl-3-methylglutaryl CoA reductase 2 promoter, which is inducible by mechanical stress (Cramer et al., 1996).

Tissue-specific promoters control gene expression in a tissue or in a developmental stage specific way. The transgen driven by such a promoter is expressed in a specific tissue leaving all the other tissues unaffected. It helps to force transgene expression in storage organs like seeds, tubers or fruits. Several of such promoters were tested: tuber specific patatin promoter (Jefferson et al., 1990), fruit specific E8 promoter (Jiang et al., 2007), arcelin promoter (Osborn et al., 1988), maize globulin 1 promoter (Rusell & Fromm, 1997), 7s globulin promoter (Fogher, 2000), rice glutelin promoter (Wu et al., 1988) and soybean P-conglycinin subunit promoter (Chen et al., 1986).

The optimization of promoters activity can be further improved by means of engineered DNA elements - enhancers, activators or repressors located up or downstream of the core promoter. Enhancers are shown to increase gene expression when placed proximally to the promoter, they bind activator proteins and promote RNA polymerase II placement at the TATA box. Transcription is also enhanced with flanking the transgene by nuclear scaffold/matrix attachment regions (S/MARs) important for structural organization of eukaryotic chromatin (Halweg et al., 2005).

The translational efficiency of a transgene is determined by proper processing (capping, splicing, polyadenylation, nuclear export) and mRNA stability. The 5' and 3' untranslated region (UTR) of the plant mRNA plays crucial roles in its processing (Cowen et al., 2007). The 5'-UTR is very important for 5' capping and enables translation initiation, the 3'-UTR is indispensable in transcript polyadenylation which in turn influences the stability of mRNA (Chan and Yu, 1998). These untranslated sequences can be manipulated for the optimization of protein expression.

As the protein is synthesized, it undergoes several modifications before final delivery to its cellular destination. These modifications include enzyme involving glycosylation, phosphorylation, methylation, ADP-ribosylation, oxidation, acylation, proteolytic cleavage and non-enzymatic modifications like deamidation, glycation, racemization and spontaneous changes in protein conformation (Gomord & Faye, 2004). Post-translational proteolysis can be effectively minimized by targeting the foreign proteins to sub-cellular compartments such as the endoplasmic reticulum (ER). Proteolysis is more likely to occur in the apoplast and cytosol. ER retrieval signal (e.g. KDEL, HDEL) retains the expressed protein in the ER lumen and has been used to improve foreign protein stability. The ER contains many molecular chaperones facilitating nascent proteins folding or assembly and it is regarded as an ideal compartment for accumulating many classes of foreign proteins (Nuttal et al., 2002).

Other strategies for proteolytic degradation reduction are: co-expression of recombinant protein and protease inhibitors, co-expression of protein co-factors or subunits, knockout mutations in the genes encoding specific proteolytic enzymes.

The recent advent of highly efficient transient expression systems has completely changed the concept and revolutionized plant made pharmaceutical research. Transient transformation implies the expression of foreign DNA which cannot be inherited but is still transcribed within the host cell in a transient manner. Transient gene expression provides a rapid alternative to the time consuming stable transformation methods. This approach uses the plant hosts - *Arabidopsis thaliana, Nicotiana tabacum, Nicotiana benthamina, Lactuca sativa*. Transient expression of recombinant proteins in plants is performed by the use of engineered plant viruses and/or *Agrobacterium* mediated DNA transfer (agroinfection/agroinfiltration). Fast and high level expression is the major advantage of the transient expression systems. Full expression of a gene of interest in agroinjected leaves may be achieved in 3-4 days after infiltration with *Agrobacteria*. This system is simple and experimental procedures do not require expensive supplies and equipment. Leaves of greenhouse grown plants are infiltrated using a syringe without a needle, vacuum infiltration or the wound and agrospray inoculation method (Medrano et al., 2009). Supplementation of the infiltration media with Silwet L-77, Tween-20, or Triton X-100 improves the efficiency of transformation. In the transient expression system one can use different virus types: Tobamoviruses, Potexviruses, Potyviruses, Bromoviruses, Comoviruses and Gemniviruses. Prolific production of any given protein using the plant virus approach results from the fact that a virus can infect a plant systemically by moving in its symplast. The *Agrobacterium* based method involves the injection or vacuum infiltration of whole plants or their parts with a suspension of bacteria harboring the construct of interest (Gómez et al., 2009). *Agrobacterium* delivered plant viral vectors use the RNA polymerase II mediated nuclear export route including 5′ end capping, splicing and 3′ end formation. Plant RNA viruses replicate in the cytoplasm and are not adapted to nuclear splicing machinery which recognizes and removes cryptic introns from viral RNA leading to its degradation. The *Agrobacterium* delivered so called 'first generation' TMV and PVX vectors have low production capacity and require coinjection of a plasmid encoding gene silencing suppressor such as tombusvirus p19 or potyvirus P1/HC-Pro (Komarova et al., 2010).

A major breakthrough in viral expression strategies was facilitated by the recent advent of deconstructed virus vectors. Originally reported for the TMV-based magnICON system developed by ICON Genetics GmbH merges advantages of *Agrobacterium*-mediated DNA delivery and upgraded TMV based vectors where putative cryptic splice sites were removed and multiple plant introns inserted. Thus the basic idea is to amplify the foreign gene delivered by *Agrobacterium tumefaciens* to multiple areas of the plant allowing the virus to replicate and spread. In this process, bacteria start initial infection delivering the T-DNA encoded viral replicon to the nuclei of a large number of cells. Then, the transcripts are transported to the cytoplasm where the viral RNA amplification renders high yields of the desired protein (Gleba et al., 2005).

In conclusion, the two major strategies for expressing proteins in whole plants are transient expression with viral vectors and stable transformation where transgenes are targeted to either the nuclear or chloroplast genome. Stable transformation offers the advantage that protein production is scalable to large field production methods. However, this can be offset by low expression levels and the long time required for creating expressor lines stable across multiple generations. Today's most promising direction in the referred field is emerging from synthesis of genetically engineered agrobacteria, viruses and plants in one precisely tailored system where synthetic and system biology meet each other.

3. Overview of plant-derived medical recombinant proteins

3.1 Plant derived antibodies

Over the last few decades, medical biotechnology has led to major advances in diagnosis and therapy. At present most diseases can be detected at an early stage, and their treatment is more specific and potent. Biotechnological methods allow to identify the molecular mechanisms of a disease facilitating development of new diagnostic techniques and speeding up development of novel molecularly targeted drugs. One of the therapeutic strategies in the treatment of many diseases is the use of antibodies. Antibodies are a class of topographically homologous multidomain glycoproteins produced by the immune system and they display a remarkably diverse range of binding specificities. Since the first production of monoclonal antibodies by Kohler and Milstein in 1975 they have become an extremely important and valuable tool in medicine (Yarmush et al., 2003).

Constantly increasing demand for new and safe monoclonal antibodies forces development of high-performance production systems. Since the first report on antibody production in *N. tabacum* plants (Hiatt et al., 1989), plantibodies have been produced in various plant systems (Table 2).

Product	Disease/Pathogen	Plant	Promoter	Expression level	Organ	Reference
Human anti-rabies monoclonal antibody	Rabies	Tobacco	CaMV 35S promoter with duplicated upstream B domains	0.07% TSP	Leaves	Ko et al., 2003
Human monoclonal antibody	Hepatitis-B virus	Tobacco	CaMV 35S promoter with the omega sequence	0.2-0.6% TSP	Suspension cell cultures	Yano et al., 2004

Product	Disease/Pathogen	Plant	Promoter	Expression level	Organ	Reference
Full-length monoclonal mouse IgG1 (MGR48)	-	Tobacco	CaMV 35S, TR2' promotor	30–60 mg of fresh weight	Leaves	Stevens et al., 2000
Human-derived, monoclonal antibody	Anthrax	Tobacco	CaMV35S	-	Leaves	Hull et al., 2005
Anti-*Salmonella enterica* single-chain variable fragment (scFv) antibody	*Salmonella enterica*	Tobacco	EntCUP4, single and double-enhancer versions CaMV 35S	41.7 ug of scFv/g leaf tissue	Leaves	Makvandi-Nejad et al., 2005
Human anti-rabies virus monoclonal antibody	Rabies	Tobacco	CaMV 35S with duplicated upstream B domains (Ca2p), (Pin2p)	30 ug/g of cell dry weight	Cell suspension culture	Girard et al., 2006
BoNT antidotes	Botulinum neurotoxins (BoNTs)	Tobacco	CaMV35S	20–40 mg/kg	Leaves	Almquist et al., 2006
TheraCIM recombinant humanized antibody	Skin cancer	Tobacco	CaMV35S/ Agroinfiltration	1.2 mg/kg of leaves	Leaves	Rodríguez et al., 2005
Human monoclonal antibody 2F5	Activity against HIV-1	Tobacco	duplicated CaMV35S	2.9 ug/g fresh weight	Cell suspension	Sack et al., 2007
mAb BR55-2 (IgG2a)	Carcinomas, particularly breast and colorectal cancers	Tobacco	CaMV 35S	30 mg kg of fresh leaves	Leaves	Brodzik et al., 2006
LO-BM2, a therapeutic IgG antibody	Possible tool to prevent graft rejection	Tobacco	En2pPMA4	99 ug in the cell extract of a 100-ml culture, 12.81 ug. medium-associated antibody	Leaf and cell suspension culture	De Muynck et al., 2009
Monoclonal antibody H10 (mAb H10)	Tumour-associated antigen tenascin-C (TNC)	Tobacco	CaMV 35S with omega translational enhancer sequence from (TMV)	50–100 mg/kg fresh plant tissue	Leaves	Villani et al., 2009

Table 2. Plant derived antibodies.

3.2 Plant derived vaccines

Plants can be used to produce inexpensive and highly immunogenic vaccines. It is connected with heterologous expression of antigens. These are further purified to formulate injectable vaccine or are applied as edible vaccines. The latter idea is a very attractive alternative to injection, mostly because of low costs (no need for protein purification) and comfort of administration. However, there are some essential conditions which have to be satisfied. First of all, plants used for oral vaccine production should produce edible parts that can be consumed uncooked (antigens are often heat sensitive). Besides, these parts should be rich in protein because the antigen protein will constitute only a minor portion (0.01-0.4%) of TSP. Seeds seems to be a good choice because of antigen extended stability, even at ambient storage temperatures. As many studies revealed, vaccine antigens present in plant tissues were resistant to digestion in the gastrointestinal tract, on the other hand during this process they were release to elicite both mucosal and systemic immune responses (Sharma and Sood, 2011). Current progress in the matter is summarized in Table 3.

Vaccines	Disease	Plant	Promoter	Expression level	Organ	References
Subunit HAC1 and HAI-05	H1N1, H5N1 influenza	Tobacco	Not reported	HAC1 90 mg/ and HAI-05 50 mg/kg of plant biomass	Leaves	Shoji et al., 2011
VP1-capsid protein	FMDV (Foot and Mouth Disease Virus)	Tobacco	*psbA*	51% TSP	Leaves (Chloroplasts)	Lentz et al., 2010
TonB protein	Immunizatio n against *Helicobacter infections*	*A. thaliana*	CaMV 35S	0.05% TSP	Entirely plant	Kalbina et al., 2010
Mycobacteria l antigens Ag85B	Vaccine against tuberculosis	Tobacco	CaMV 35S	4 % TSP	Leaves	Floss et al., 2010
Surface protein 4/5 (PyMSP4/5)	Plasmodium	Tobacco	MagnICON® viral vector system	10% TSP or 1-2 mg/g of fresh weight	Leaves	Webster et al. 2009
TetC and PTX S1 antigens	DTP (diphtheria– tetanus– pertussis)	Tobacco Daucus carrota	CaMV 35S	Not reported	Leaves; Hypocotyls	Brodzik et al., 2009
HN glycoprotein	Newcastle Disease Virus (NDV)	Tobacco	P-RbcS	3µg of HN protein per mg of total leaf protein	Leaves	Gómeza et al., 2009
HBsAg	HBV (hepatitis B virus)	*Lactuca sativa*	CaMV 35S	Not reported	Shoots	Marcondes & Hansen, 2008

Vaccines	Disease	Plant	Promoter	Expression level	Organ	References
HPV-16 L1 protein	HPV (Human Papilloma Virus)	Tobacco	*psbA* promoter	24 % TSP	Leaves	Fernández-San Millán et al. 2008
16 E7 oncoprotein	HPV	Tomato; Potato	CaMV 35S	0.5 % of the cell protein-potato	Potato protoplast; leaves	Briza et al., 2007
G protein	Rabies virus	*Daucus carotta*	CaMV 35S	0.2–1.4% (TSP)	Carrot roots	Royas-Anaya et al., 2009
Capsid protein VP6	Rotavirus	Potato	P2	0.01%	Leaves, tubers	Yu & Landgridge, 2003

Table 3. Plant derived vaccines.

3.3 Plant derived biopharmaceuticals

Plants can be used to produce inexpensive biopharmaceuticals (Table 4).

Biopharmaceutical	Potential application	Plant	Promoter	Expression level	References
IL-10	Inflammatory and autoimmune diseases	Rice seeds	Glutelin B-1 promoter	2 mg pure IL-10	Fujiwara et al., 2010
Human transfferin	Receptor-mediated endocytosis pathway	Rice seeds	Glutelin 1 G-1 promoter	1% seed dry weight	Zhang et al., 2010
Glutamic acid decarboxylase (GAD65)	Autoimmune T1DM	Tobacco leaves	CaMV 35S	2.2% total soluble protein	Avesani et al., 2010
hGH, somatotropin	Growth hormone-treatment of dwarfism	*N. benthamiana*	CaMV 35S	60 mg per kilogram offresh tissue; 7%	Rabindran et. al., 2009;
Human erythropoietin (EPO)	Anemia, Renal failure	*N. tabacum*	CaMV 35S	0.05% of total soluble protein	Conley et al., 2009
Human serum albumin (HSA)	Deficiences	Tobacco, potato	Prrn; B33	11.1%TSP% (tobacco chloroplasts); 0.2%TSP (potato tuber)	Faran et al., 2002
Human lactoferrin (hLF)	Anti-inflammatory and immuno-modulation effects	Potato	Tandem promoter: P2& CaMV 35S	0.10% TSP	Chong et al., 2000
Enkephalins	Painkiller	Cress, *A. thaliana*	------------	0.10% seed protein	Daniell et al., 2001
Staphylokinase	Thrombolytic factor	*A. thaliana*	CaMV 35S	not reported	Wiktorek-Smagur et al., 2011

Table 4. Plant derived biopharmaceuticals.

3.4 Nutraceutical and non-pharmaceutical plant derived proteins

Antimicrobial nutraceutics, such as human lactoferrin and lysozymes, have now been successfully produced in several crops (Stefanova et al., 2008), and are commercially available (Table 5). Cobento Biotechnology (Denmark) has recently received approval for its *Arabidopsis* derived human intrinsic factor which is used against vitamin B12 deficiency and it is now commercially available as Coban. Other nutraceutical products are listed in Table 5.

Trypsin is a proteolytic enzyme that is used in a variety of commercial applications, including processing of some biopharmaceuticals (Sharma & Sharma, 2009). In 2004, the first plant derived recombinant protein product (bovine sequence trypsin; trade name – trypZean) developed in corn plant (Prodi Gene, USA) was commercialized. Avidin, a glycoprotein found in avian, reptilian and amphibian egg white, is primarily used as a diagnostic reagent. The plant optimized avidin coding sequence was expressed in corn and now it is available on the market. β-glucuronidase, peroxidase, laccase, cellulase, aprotinin were also developed and marketed (Basaran & Rodrigez-Cerezo, 2008).

Spider silk proteins, elastin and collagen, have been expressed in transgenic plants (Scheller et al., 2004). These are promising biomaterials for regenerative medicine.

Product name	Company name	Plant	Commercial name	Source
Avidin	Prodigene	Corn	Avidin	Obembe at al., 2011
β-glucuoronidase	Prodigene	Corn	GUS	Obembe at al., 2011
Trypsin	Prodigene	Corn	TrypZean	Obembe at al., 2011
Recombinant human lactoferrin	Meristem Therapeutic, Ventria Bioscience	Corn, Rice	Lacromin	http://www.meristemthera-peutics.com
Recombinant human lysozyme	Ventria Bioscience	Rice	Lysobac	http://www.ventria.com
Aprotinin	Prodigene	Corn, Tobacco	AproliZean	Obembe at al., 2011
Recombinant lipase	Meristem Therapeutic	Corn	Merispase	http://www.meristemthera peutics.com
Recombinant human intrinsic factor	Cobento Biotech AS	*Arabidopsis*	Coban	http://www.cobento.dk
Human growth factors	ORF Genetics	Barley	ISOkine™	http://www.orfgenetics.com
Food additive for shrimps	SemBioSys	Safflower	Immuno-spherte	http://www.sembiosys.com

Table 5. Transgenic plants based on products commercially available in the market.

4. Recombinant protein purification

4.1 Affinity chromatography

Isolation and purification of a biologically active protein from a crude lysate is often difficult and costly. Simple, cheap and more efficient strategies of its purification on the laboratory and industrial scale are thus on great demand. One of the numerous approaches in this field is an affinity tags system easily applicable for recombinant protein purification by affinity chromatography. The term 'affinity chromatography' was introduced in 1968 by Pedro Cuatrecasas, Meir Wilchek, and Christian B. Anfinsen (1968). Now it is the method of choice (Kabir et al., 2010). Affinity chromatography is based on specific interaction between two molecules in order to isolate the protein of interest from a pool of unwanted proteins and other contaminants. For this purpose a fusion protein is created. A short fragment of DNA can be ligated to the 5 ' or 3' - terminus of the target gene. This peptide or protein coding sequence (so called tag), which is translated in frame with protein of interest exhibits a characteristic property, strong and selective binding to the molecules immobilized on the solid matrices (Fong et al., 2010). Purification process is effective and simple. During passage of the cell extract containing the fusion protein and contaminants through an appropriate column the tagged protein is retained, while all the others migrate freely through the column (Fig. 1).

In the next step, the bound protein is eluted by a change in buffer composition /parameters (i.e. competitors, chelators, pH, ionic strength or temperature). Affinity tags are divided into three main classes according to their properties and the properties of molecules that interact with them: 1) tags, binding to small molecule ligands linked to a solid support (i.e. HIS-tag), 2) protein tags binding to a macromolecular partner immobilized on chromatography support (i.e. CBP-tag), 3) the protein-binding partner attached to the resin in an antibody which recognizes a specific peptide epitope in a recombinant protein (i.e. FLAG-tag) (Lichty et al., 2005, Arnau et al., 2006, Waugh et al., 2005). To date large number of gene fusion tags has been described, the most commonly used ones are presented in Table 6.

Tag	Comments	References
His-tag	Purification by interaction between immobilized metal ions and chelating amino acids	Valdez-Ortiz et al., 2005, Vaquero et al., 2002
FLAG	Purification based on binding the FLAG peptide to antibodies	Brodzik et al., 2009, Zhou and Li., 2005
Strep-tag II	Strong specific interaction between Streptag and strep-Tactin (streptavidin derivate) immobilised on resin	Witte et al., 2004

Table 6. Some examples of affinity tags commonly used for protein purification.

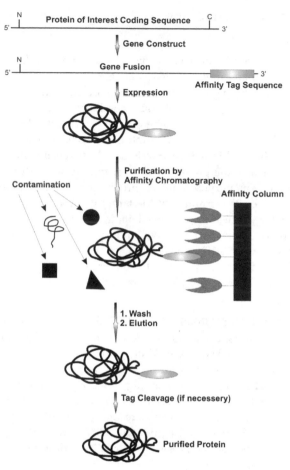

Fig. 1. Schematic representation of the recombinant protein purification process by affinity chromatography (Hearn & Acosta, 2001, modified).

4.2 Elastin-like polypeptides in recombinant protein purification

While affinity chromatography is used for purification of a broad spectrum of recombinant proteins it is not free from drawbacks. The main limitations associated with the use of this method are: 1) high cost of chromatography packing materials, 2) volume-limited sample throughput, 3) dilution of the protein product in elution buffer, 4) additional concentration step may cause loss in protein yield (Chow et al., 2008). Taking into account the above, there is a need to introduce new alternative methods for purification of recombinant proteins.

One of the possible solutions is application of non-chromatographic purification tags. Elimination of resins allows us to reduce some of the aforementioned problems.

Elastin-like polypeptides (ELP), artificial polymers containing Val-Pro-Gly-Xaa-Gly pentapeptide repeats, are an example of such tags. Such repeats occur naturally in the hydrophobic domain of human tropoelastin (soluble precursor of elastin) and they play an important role in the process of elastin formation (Mithieux & Weiss 2005, Valiaev et al., 2008). Xaa (so called guest residue) in the ELP sequence can contain any amino acid except for proline (Meyer & Chilkoti, 1999). Occurrence of proline at these positions eliminates distinctive and very useful properties of these polymers (Trabbic-Carlson et al., 2004). Literature classification of ELP is based on the type and number of amino acids present in the guest residue positions (Meyer & Chilkoti 2004).

Elastin-like polypeptides belong to one of the three classes of thermosensitive biopolymers (Mackay and Chilkoti, 2008) whose properties are changed under the influence of moderate temperature differences. Aqueous solutions of ELP exhibit lower critical solution temperature (LCST) which causes that the above phase transition temperature (T_t) ELP pass from soluble to an insoluble form (Ge et al., 2006) in a narrow temperature range (~ 2 ° C) (Ge and Filipe, 2006). This is a reversible process called coacervation. In solutions with temperature below T_t , free polymer chains remain in a disordered soluble form. The opposite occurs in solutions with temperatures above T_t, when the polymer chains have more ordered structure (called β-helix), stabilized by hydrophobic interactions (Rodriguez-Cabello et al., 2007) that increase association of polymer chains (Serrano et al., 2007). This process is reversible. The fact that ELP –protein fusions are prone to reversible transition is of great importance (Kim et al., 2004). The process of ELP-tagged protein purification involves increasing ionic strength and/or temperature of the cell lysate to induce ELP-fusion protein aggregation (Fig. 2). Next sample centrifugation/filtration separates the ELP fusion protein from contaminants. After resolubilization of an ELP fusion, another centrifugation/filtration removes denatured and aggregated biomolecules. This process called Inverse Transition Cycling (ITC) can be repeated to achieve the required purity of the product (Floss, Schallau et al., 2010).

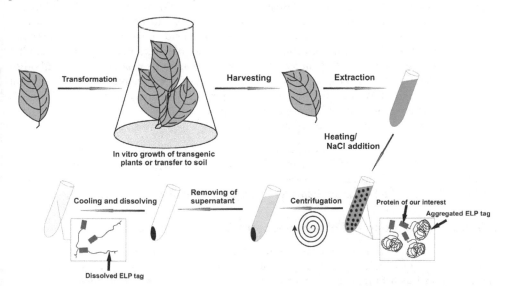

Fig. 2. Purification of ELPylated target proteins from plants using ITC (Floss et al., 2010 modified.

Purification of proteins using elastin-like polypeptides has several advantages over the traditional chromatographic methods: 1) purification of proteins with ELP tags by ITC appears to be universal for soluble recombinant proteins, 2) chromatography beads are not required, which significantly reduces the costs, 3) final concentration step is not required (Chow et al., 2008).

4.3 Application of ELP to the process of production and purification of recombinant proteins in transgenic plants

Scheller and co-workers (2004) achieved efficient and stable expression of spider's silk-ELP fusion protein in the ER of transgenic tobacco and potato. Application of ITC allowed them to obtain 80mg pure recombinant protein from 1kg tobacco leaf material. Purified biopolymer was tested as a potential component used for the cultivation of anchorage-dependent CHO-K1 cells and human chondrocytes. The most common coating substances such as collagen, fibronectin and laminin are derived from animal sources, so there is a risk of contamination of cell cultures by viruses or prions which is essentially undesirable in the case of medical applications. What is more, production of this fusion protein in plants is less costly. Lin and associates (2006) obtained active soluble glycoprotein 130 which seems to be potent drug in Crohn's disease, rheumatoid arthritis and colon cancer therapy. This work a presents creation and expression of mini-gp130-ELP. A fusion protein containing Ig-like domain and cytokine binding module of gp 130 fused to 100 repeats of ELP was expressed in tobacco leaves (ER retention). Inverse transition cycling (ITC) purification resulted in 141 μg of active mini-gp130-ELP per 1g of leaf fresh weight. Floss and co-workers (2010) demonstrated the ability of genetically engineered tobacco to produce mycobacterial antigens Ag85B and ESAT-6 as the vaccine against tuberculosis. In this work Ag85B-ELP and ESAT-6-ELP (TBAg) fusions were created, purified by inverse transition cycling and tested on animals. Production of this TBAg-ELP fusion proteins reached 4% of the tobacco leaf total soluble proteins (TSP) for the best producer plants. Further testing of the vaccine showed mycobacterium-specific immune response with no side effects in an animal model. What is more, this study also confirmed that ELP had no immunomodulating activity. Joensuu and co-workers (2009) demonstrated ELP application in production of antibodies for Foot-and-mouth disease virus (FMDV) therapy. Single chain variable antibody fragment (scFv) recognizing FMDV coat protein VP1 was expressed in transgenic tobacco plants. To recover the fusion protein in the active form the plants, ITC was performed. Finally, the authors demonstrated that scFv expressed in plants were able to bind FMDV.

It has been shown for spider silk proteins (Scheller et al., 2004), murine interleukin-4, human interleukin-10 (Patel et al., 2007) and anti-HIV type 1 antibodies (Floss et al., 2008, Floss et al., 2009) that the ELP fusion significantly enhances accumulation of recombinant proteins produced in plants. So far the mechanism of that phenomenon is not known.

5. Status of plant-derived biopharmaceuticals in clinical development

At present some non-pharmaceutical products from plants are on the market (Basaran and Rodriguez-Cerezo, 2008). Although no plant made pharmaceutical (PMP) has been commercialized as a human drug, several PMPs are at the late stage of development and some have already received regulatory approval, including a vaccine and several nutraceuticals (Table,7, 8, 9).

Antibodies	Target	Plant	Clinical trial status	Company	Source
DoxoRx	Side-effects of cancer therapy	Tobacco	Phase I	Planet Biotechnology	http://www.planet biotechnology.com
RhinoRX	Common cold	Tobacco	Phase I	Planet Biotechnology	http://www.planet biotechnology.com
IgG (ICAM1)	Common cold	Tobacco	Phase I	Planet Biotechnology	http://www.planet biotechnology.com
CaroRX	Dental caries	Tobacco	EU approved as medical advice	Planet Biotechnology,	http://www.planet biotechnology.com

Table 7. Plant derived antibodies in clinical phages of development.

Antigen or vaccine	Disease	Plant	Clinical trial status	Company	Source
Hepatitis B antigen	Hepatitis B	Lettuce	Phase I	Thomas Jefferson University	Strcatfield, 2006
Hepatitis B antigen	Hepatitis B	Potato	Phase II	Arizona State University	Streatfield, 2006
Fusion proteins	Rabies	Spinach	Phase I	Thomas Jefferson University	http://www.labome.org
Heat labile toxin B subunit of *E.coli*	Diarrhea	Potato	Phase I	ProdiGene	Tacket, 2005
Capsid protein Norwalk virus	Diarrhea	Potato	Phase I	Arizona State University	Khalsa et al., 2004
Vibrio cholerae	Cholera	Potato	Phase I	Arizona State University	Tacket, 2005
HN protein of Newcastle disease virus	Newcastle disease (Poultry)	Tobacco	USDA Approved	Dow Agro Sciences	http://www.dowagro.com
Viral vaccine mixture	Diseases of horses, dogs	Tobacco	Phase I	Dow Agro Sciences	http://www.dowagro.com
Poultry vaccine	Coccidiosis infection	Canola	Phase II	Guardian Bioscence	Basaran & Rodrigez-Cerezo, 2008
Gastroenteritis virus (TGFV) capsid protein	Piglet gastroenteritis	Maize	Phase I	ProdiGene	Basaran & Rodrigez-Cerezo, 2008
H5N1 vaccine candidate	H5N1 pandemic influenza	Tobacco	Phase I	Medicago	http://www.medicago.com

Table 8. Plant derived vaccines in clinical phages of development.

Therapeutic humans protein	Disease	Plant	Clinical trial status	Company	Source
α-Galactosidase	Fabry disease	Tobacco	Phase I	Planet Biotechnology	http://www.planet biotechnology.com
Lactoferon	Hepatitis C	Duckweed	Phase II	Biolex	http://www.biolex.com
Fibrinolytic drug	Blood clot	Duckweed	Phase I	Biolex	http://www.biolex.com
Human glucocerebrosidase	Gaucher's disease	Carrot	Waiting USDA's approval	Prostalix Biotherapeutic	http.//www.prostalix.com
Insulin	Diabetes	Safflower	Phase III	SemBioSys	http.//www.sembiosysys.com
Apolipoprotein	Cardio vascular	Safflower	Phase I	SemBioSys	http.//www.sembiosysys.com

Table 9. Plant derived pharmaceuticals in clinical phages of development.

In 2006 the world's first plant made vaccine candidate for Newcastle disease in chickens, produced in a suspension cultured tobacco cell line by Dow Agro Science, was registered and approved by the US Department of Agriculture (USDA) – the final authority for veterinary vaccines. In addition, two plant made pharmaceuticals are moving through Phase II and Phase III human clinical trials. Biolex's product candidate, Locteron®, is in Phase IIb clinical testing for the treatment of chronic hepatitis CA. This company uses two genera, *Lemna* and *Spirodela*, as a platform for production of their biopharmaceuticals. The positive outcome of Phase III trials of Protalix's glucocerebrosidase (UPLYSO®) for the treatment of Gaucher's disease which is now waiting for USDA's approval is another positive example. The successful completion of Phase III trial that concerned SemBioSys insulin bioequivalent of the commercial standard represents an important landmark in the plant made pharmaceuticals scenario and, most likely, in the next few years recombinant human insulin produced in safflower will become commercially available for diabetic people.

Medicago Inc. of Canada was invited to the sixth WHO meeting about evaluation of pandemic influenza prototype vaccines in clinical trials. One of the purposes of this meeting was to make recommendations on research activities that will contribute to the development of effective pandemic vaccines. Medicago has recently reported positive results from a Phase I human clinical trial with its H5N1 avian influenza vaccine candidate (a VLP based vaccine produced with a transient expression system). The vaccine was found to be safe, well tolerated and it also induced a solid immune response. Based on these results, Medicago will process with Phase II clinical trial with the first plant made influenza vaccine (Franconi et al., 2010). These examples will pave the way to easy public acceptance of transgenic plants as new production platforms for human therapeuticals.

6. Concluding remarks

Biopharming is still a relatively new field in plant science but in the coming years it may become the premier expression system for a wide variety of new biopharmaceuticals. The use of plants as factories for the synthesis of therapeutic protein molecules will undoubtedly develop. Since the first development of a genetically modified plant in 1984, numerous comprehensive review articles have been published demonstrating the tremendous potential of plants for pharmaceutical production. As it has been clearly shown plants are no longer considered only in terms of diet or beauty. The proteins targeted for biopharmaceutical technology form three broad categories: antibodies, vaccines, and other therapeutics. Plant bioreactors represent an attractive alternative for their synthesis requiring the lowest capital investment of all tested production systems. The events of heterologous proteins in planta production were rapidly followed with development/improvement of significant technologies (e.g. DNA delivery systems, selection methods). At present a number of promoters with tissue-specific activity or sub-cellular targeting sites that offer protein stability are known and many are still under intense study. Obviously, the construction of a transgenic plant synthesizing a functional therapeutic is a multidisciplinary process and the society of biotechnologists takes a keen interest in its success. However, over the past years various plant expression platforms have been tested and it is evident that further development and improvement are needed for more effective molecular farming. Apart from continuously increasing transgene yields efforts will need to ensure that plant-derived biopharmaceuticals would meet the same safety and efficacy standards as products of non-plant origin. There is no doubt that sooner or later the scientific limitations of molecular farming will be overcome, especially when numerous therapeutics and plant platforms are developed by many laboratories and companies. Thus, this is the regulatory requirements and public acceptance which are the greatest challenge of modern plant biotechnology. Of course, molecular farming raises less objection than technologies using genetically modified animals, but still the existing or proposed regulations remain based on public fears rather than on scientific facts.

In conclusion, "the molecular farming industry" means a natural advance in drug production technology. The dynamics of optimization and improvement of plant expression platforms illustrates its potential and tremendous scientific background. The possible success in this field will have to face the question of public acceptance. Thus, the scientists should send the clear massage to the public opinion that molecular farming is a strictly controlled technology that has strong benefits. And that probably will be more difficult than the construction of functional bioreactor itself.

7. References

Almquist, KC.; McLean, MD.; Niu, Y.; Byrne, G.; Olea-Popelka, FC.; Murrant, C.; Barclay, J. & Hall, JC. (2006). Expression of an anti-botulinum toxin A neutralizing single-chain Fv recombinant antibody in transgenic tobacco. *Vaccine*, Vol. 24, No. 12, (December, 2006), pp. 2079-2086, ISSN 0264-410X

Arnau, J.; Lauritzen, C.; Petersen, GE. & Pedersen, J. (2006). Current strategies for the use of affinity tags and tag removal for the purification of recombinant proteins. *Protein Expression and Purification*, Vol. 48, No. 1,(July 2006) pp. 1-13, ISSN 1046-5928

Avesani, A.; Vitale, A.; Pedrazzini, E.; de Virgilio, M.; Pompa, A.; Barbante, A.; Gecchele, E.; Dominici, P.; Morandini, F.; Brozzetti, A.; Falorni, A. & Mario Pezzotti, M. (2010). Recombinant human GAD65 accumulates to high levels in transgenic tobacco plants when expressed as an enzymatically inactive mutant. *Plant Biotechnology Journal*, Vol. 8, No. 8, (September 2010), pp. 862–872, ISSN 14677644

Basaran, P. & Rodriguez-Cerezo, E. (2008). Plant molecular farming: opportunities and challenges. *Critical Reviws in Biotechnology*, Vol. 28, No. 3, (March 2008), pp. 153-172, ISSN 1549-7801.

Boehm, R. (2007). Bioproduction of Therapeutic Proteins in the 21st Century and the Role of Plants and Plant Cells as Production Platforms. *Annals of the New York Academy of Sciences*, Vol. 1102, (April 2007), pp. 121–134, ISSN 1749-6632

Briza, J.; Pavingerowa, D.; Vlasak, J.; Ludvikova, V. & Niedermeierova, H. (2007). Production of human papillomavirus type 16 E7 oncoprotein fused with β-glucuronidase in transgenic tomato and potato plants. *Biologia Plantaru* ,Vol 51, No. 2, (June 2007), pp. 268-276, ISSN 0006-3134

Brodzik, R.; Glogowska, M.; Bandurska, K.; Okulicz, M.; Deka, D.; Ko, K.; van der Linden, J.; Leusen, JH.; Pogrebnyak, N.; Golovkin, M.; Steplewski, Z. & Koprowski, H. (2006). Plant-derived anti-Lewis Y mAb exhibits biological activities for efficient immunotherapy against human cancer cells. *Proceedings of the National Academy of Sciences of the United States of America*, Vol. 103, No. 23, (May 2006), pp. 8804-8809, ISSN 0027-8424

Brodzik, R.; Spitsin, S.; Pogrebnyak, N.; Bandurska, K.; Portocarrero, C.; Andryszak, K.; Koprowski, H. & Golovkin M. (2009). Generation of plant-derived recombinant DTP subunit vaccine. *Vaccine*, Vol. 27, No. 28, (June 2009), pp. 3730-3734, ISSN 0264-410X

Cardi, T.; Lenzi, P. & Maliga, P. (2010). Chloroplast as expression platforms for plant produced vaccines. *Expert Review of Vaccines*, Vol. 9, No. 8, (October 2009), pp. 893-911, ISSN 1476-0584.

Chan, MT. & Yu, SM. (1998). The 3′ untranslated region of a rice alpha amylase gene function as a sugar dependent mRNA stability determinant. *Proceedings of the National Academy of Sciences of the United States of America*, Vol. 95, No. 11, (May 1988), pp. 6543-6547, ISSN 0027-8424.

Chen, ZL.; Schuler, MA. & Beachy, RN. (1986). Functional analysis of regulatory elements in a plant embryo specific gene. *Proceedings of the National Academy of Sciences of the United States of America*, Vol. 83, No. 22, (November 1986), pp. 8560-9564, ISSN 0027-8424.

Chong, D.K.X. & Langridge, W.H.R. (2000). Production of full- length bioactive antimicrobial human lactoferrin in potato plants. *Transgenic Research*, Vol. 9, No. 1, (January 2000), pp. 71-78, ISSN 0962-8819

Chow, D.; Nunalee, ML.; Lim, DW.; Simnick, AJ. & Chilkoti., A. (2008). Peptide-based Biopolymers in Biomedicine and Biotechnology. *Materials science & engineering. R, Reports : a Review Journal*, Vol. 62, No. 4, (January 2008) pp. 125-155, ISSN 0927-796X

Conley, A.J.; Mohib, K.; Jevnikar, A.M. & Brandle, J.E. (2009). Plant recombinant erythropoietin attenuates inflammatory kidney cell injury. *Plant Biotechnology Journal*, Vol. 7, No. 2, (November 2009), pp. 183-199, ISSN 14677644

Cowen, NM.; Smith, KA. & Armstrong, K. (2007). Use of regulatory sequences in transgenic plants. Unitated States Patent, No. 7179902.

Cox, K.M.; Sterling, J.D.; Regan, J.T.; Gasdaska, J.R.; Frantz, K.K.; Peele, C.G.; Black, A.; Passmore, D.; Moldovan-Loomis, C.; Srinivasan, M.; Cuison, S.; Cardelli, P.M. & Dickey, L.F. (2006). Glycan optimization of a human monoclonal antibody in the aquatic plant Lemna minor. *Nature Biotechnology*, Vol. 24, No. 11 (November 2006) pp. 1591-1597, ISSN 1087-0156

Cramer, CL.; Weissenborn, DL.; Oishi, KK.; Grabau, EA.; Benett S.; Ponce E.; Grabowski, GA. & Radin, DN. (1996) Bioproduction of human enzymes in transgenic tobacco.

Proceedings of the National Academy of Sciences of the United States of America, Vol. 729, No. 1, (May 1996), pp. 62-71, ISSN 0077-8923.

Cuatrecasas, P.; Wilchek, M. & Anfinsen, C.B. (1968). Selective enzyme purification by affinity chromatography. *Proceedings of the National Academy of Sciences,* Vol. 61, No. 2, (October 1968) pp. 636-643, ISSN-0027-8424

Dai, Z.; Hooker, BS.; Anderson, DB. & Thomas, SR. (2000). Improved plant based production of E1 endoglucanase using potato: expression. *Mol Breeding,* Vol. 6, No. 3, (June 2000), pp. 277-285, ISSN 1380-3743.

Daniell, H.; Streatfield, S.J. & Wycoff K. (2001). Medical molecular farming: production of antibodies, biopharmaceuticals and edible vaccines in plants. *Trends in Plant Science,* Vol. 6, No. 5 (May 2001), pp. 219-226, ISSN 1360-1385

Decker, E.L. & Reski, R. (2008). Current achievements in the production of complex biopharmaceuticals with moss bioreactors. *Bioprocess Biosystems Engineering,* Vol. 31, No. 1, (January 2008), pp. 3-9, ISSN 1615-76

Demain, A.L. & Vaishnav, P. (2009). Production of recombinant proteins by microbes and higher organism. *Biotechnology Advances,* Vol. 27, No. 3, (June 2009), pp. 297-306, ISSN 0734-9750

De Muynck, B.; Navarre, C.; Nizet, Y.; Stadlmann, J. & Boutry, M. (2009). Different subcellular localization and glycosylation for a functional antibody expressed in Nicotiana tabacum plants and suspension cells. *Transgenic Research,* Vol. 18, No. 3, (January 2009), pp. 467-482, ISSN 0962-8819

Fernández-San Millán, A.; Ortigosa, S.M.; Hervás-Stubbs, S.; Corral-Martínez, P.; Seguí-Simarro, J.M.; Gaétan, J.; Coursaget, P. & Veramendi, J. (2008) Human papillomavirus L1 protein expressed in tobacco chloroplasts self-assembles into virus-like particles that are highly immunogenic. *Plant Biotechnology Journal,* Vol. 6, No. 6, (April 2008), pp. 427–441, ISSN 14677644

Fischer, R. & Emans, N. (2000). Molecular farming of pharmaceutical proteins. *Transgenic Research,* Vol. 9, No. 4-5. (August 2000), pp. 279-299, ISSN 1573-9368

Farran, I.; Sánchez-Serrano, J.J.; Medina, J.F.; Prieto, J. & Mingo-Casel, A.M. (2002). Targeted expression of human serum albumin to potato tubers. *Transgenic Research,* Vol. 11, No. 4, (August 2002), pp. 337-346, ISSN 0962-8819

Floss, D.M.; Mockey, M.; Zanello, G.; Brosson, D.; Diogon, M.; Frutos, R.; Bruel, T.; Rodrigues, V.; Garzon, E.; Chevaleyre, C.; Berri, M.; Salmon, H.; Conrad, U. & Dedieu, L. (2010). Expression and immunogenicity of the mycobacterial Ag85B/ESAT-6 antigens produced in transgenic plants by elastin-like peptide fusion strategy. *Journal of Biomedicine Biotechnology,* 2010:274346. Epub 2010 Apr 13, ISSN 1110-7243

Floss, D.M.; Sack, M.; Arcalis, E.; Stadlmann, J.; Quendler, H.; Rademacher, T.; Stoger, E.; Scheller, J.; Fischer, R. & Conrad U. (2009). Influence of elastin-like peptide fusions on the quantity and quality of a tobacco-derived human immunodeficiency virus-neutralizing antibody. *Plant Biotechnology Journal,* Vol. 7, No. 9, (December 2009), pp. 899-913, ISSN 1467-7644

Floss, D.M.; Sack, M.; Stadlmann, J.; Rademacher, T.; Scheller, J.; Stöger, E.; Fischer, R. & Conrad, U. (2008). Biochemical and functional characterization of anti-HIV antibody-ELP fusion proteins from transgenic plants. *Plant Biotechnology Journal,* Vol. 6, No. 4, (May 2008), pp. 379-391, ISSN 1467-7644

Floss, D.M.; Schallau, K.; Rose-John, S.; Conrad, U. & Scheller, J. (2010). Elastin-like polypeptides revolutionize recombinant protein expression and their biomedical application. *Trends in Biotechnology*, Vol. 28, No. 1, (January 2010), pp. 37-45, ISSN 0167-7799

Fogher, C. (2000). A synthetic polynucleotide coding for human lactoferrin, vectors, cells and transgenic plants containing it. Gene bank acc no AX006477, Patent: W00004146.

Fong, BA.; Wu, WY.; & Wood, DW. (2010). The potential role of self-cleaving purification tags in commercial-scale processes. *Trends in Biotechnology*, Vol. 28, No. 5, (May 2010) pp. 272-279, ISSN 0167-7799

Franconi, R.; Demurtas, OC. & Massa, S. (2010). Plant derived vaccines and other therapeutics produced in contained systems. *Expert Review of Vaccines*, Vol. 9, No. 8, (October 2010), pp. 877-892, ISSN 1476-0584.

Fujiwara, Y.; Aiki, Y.; Yang, L.; Takaiwa, F.; Kosaka, A.; Tsuji, N.M.; Shiraki, K. & Sekikawa, K. (2010). Extraction and purification of human interleukin-10 from transgenic rice seeds. *Protein Expression and Purification*, Vol. 72, No. 1, (February 2010), pp. 125-130, ISSN 1046-5928

Ge, X. & Filipe, C.D. (2006) Simultaneous phase transition of ELP tagged molecules and free ELP: an efficient and reversible capture system. *Biomacromolecules*, Vol. 7, No. 9, (Septrember 2006), pp. 2475-2478, ISSN 1525-7797

Ge, X.; Trabbic-Carlson, K.; Chilkoti, A. & Filipe, C.D.M. (2006). Purification of an elastin-like fusion protein by microfiltration. *Biotechnology and Bioengineering*, Vol. 95, No. 3, (October 2006), pp. 424-32 46-851, ISSN 0006-3592

Girard, L.S.; Fabis, M.J.; Bastin, M.; Courtois, D.; Pétiard, V. & Koprowski, H. (2006). Expression of a human anti-rabies virus monoclonal antibody in tobacco cell culture. *Biochemical and Biophysical Research Communications*, Vol. 345, No. 2, (January 2006) pp. 602-607, ISSN 0006-291X

Gleba, Y.; Klimyuk, V. & Marillonnet S. (2005). Magnifection – a new platform for expressing recombinant vaccines in plants. *Vaccine*, Vol. 23, No. 17-18, (March 2005), pp. 2042-2048, ISSN 0264-410X.

Gómeza, E.; Zotha, S.Ch.; Asurmendia, S.; Vázquez Roverea, C. & Berinstein, A. (2009). Expression of Hemagglutinin-Neuraminidase glycoprotein of Newcastle Disease Virus in agroinfiltrated *Nicotiana benthamiana* plants. *Journal of Biotechnology*, Vol. 144, No. 4, (September 2009), pp. 337–340, ISSN 0168-1656

Gomord, V. & Faye L. (2004). Posttranslational modification of therapeutic proteins in plants. *Current Opinion in Plant Biology*, Vol. 7, No. 2, (April 2004), pp. 171-181, ISSN 1369-5266

Halweg, C.; Thompson, WF. & Spiker, S. (2005). The rb7 matrix attachment region increase the likehood and magnitude of transgene expression in tobacco cells: a flow cytometric study. *Plant Cell*, Vol. 17, No. 2, (February 2005), pp. 418-429, ISSN 1040-4651.

Hearn, M.T. & Acosta, D. (2001). Applications of novel affinity cassette methods: use of peptide fusion handles for the purification of recombinant proteins. *Journal of Molecular Recognition*, Vol. 14, No. 6, (December 2001), pp. 323-369, ISSN 0952-3499

Hellwig, S.; Drossard, J.; Twyman, RM. & Fischer R. (2004). Plant cell cultures for the production of recombinant proteins. *Nature Biotechnology*, Vol. 22 No. 11, (November 2004), pp. 1415-1422, ISSN 1087-0156.

Herman, SR.; Harding, RM. & Dale, JL. (2001). The banana *actin 1* promoter drives near constitutive transgene expression in vegetative tissue of banana (*Musa* spp.). *Plant Cell Report*, Vol. 20, No. 6, (July 2001), pp. 525-530, ISSN 0721-7714.

Hiatt, A.; Cafferkey, R. & Bowdish, K. (1989). Production of antibodies in transgenic plants. *Nature*, Vol. 342, No. 6245, (July-August 1989), pp. 76-78, ISSN 0028-0836

Huang, Z.; Santi, L.; LePore, K.; Kilbourne, J.; Arntzen, CJ. & Mason HS. (2006). Rapid, high level production of hepatitis B core antigen in plant leaf and its immunogenicity in mice. *Vaccine*, Vol. 24, No. 14, (December 2006), pp. 2506-2513, ISSN 0264-410X.

Hull, AK.; Criscuolo, CJ.; Mett, V.; Groen, H.; Steeman, W.; Westra, H.; Chapman, G.; Legutki, B.; Baillie, L. & Yusibov, V. (2005). Human-derived, plant-produced monoclonal antibody for the treatment of anthrax. *Vaccine*, Vol. 23, No. 17-18, (March 2005), pp. 2082-2086 , ISSN 0264-410X

Jeferson, R.; Goldsbrough, A. & Bevan, M. (1990). Transcriptional regulation of a *patatin-1* gene in potato. *Plant Molecular Biology*, Vol. 14, No. 6, (February 1990), pp. 995-1006, ISSN 0735-9640.

Jiang, XL.; He, ZM.; Peng, ZQ.; Qi, Y.; Chen, Q & You, SY. (2007). Cholera toxin B protein in transgenic tomato fruit induces systemic immune response in mice. *Trangenic Research*, Vol. 16. No. 2, (April 2007), pp. 169-175, ISSN 1573-9368

Joensuu, J.J.; Brown, K.D.; Conley, A.J.; Clavijo, A.; Menassa, R. & Brandle, J.E. (2009) Expression and purification of an anti-Foot-and-mouth disease virus single chain variable antibody fragment in tobacco plants. *Transgenic Research*, Vol. 18, No. 5, (October 2009), pp. 685-696, ISSN 0962-8819

Kabir, M.E.; Krishnaswamy, S.; Miyamoto, M.; Furuichi, Y. & Komiyama, T. (2010). Purification and functional characterization of a Camelid-like single-domain antimycotic antibody by engineering in affinity tag. *Protein Expression and Purification*, Vol. 72, No. 1, (July 2010), pp. 59-65, ISSN 1046-5928

Kaiser, J. (2008). Is the drought over for pharming? *Science*, Vol. 320, No. 5875, (April 2008), pp. 473-475, ISSN 1095-9203

Kalbina, I.; Engstrand, L.; Andersson, S. & Strid, A. (2010).Expression of *Helicobacter pylori* TonB Protein in Transgenic *Arabidopsis thaliana*: Toward Production of Vaccine Antigens in Plants. *Helicobacter*, Vol. 15, No. 5, (October 2010), pp. 430-437, ISSN 1523-5378

Khalsa, G.; Mason, HS. & Arntzen CJ. (2004). Plant derived vaccines: progress and constrains, In: *Molecular farming: plant made pharmaceuticals and technical proteins*, R. Fischer, S. Schillberg, (Ed), pp. 135-158, Wiley-VCH, ISBN 9783527603633, Weinheim, Germany.

Kim, J.Y.; Mulchandani, A. & Chen, W. (2004). Temperature-triggered purification of antibodies. *Biotechnology and Bioengineering*, Vol. 90, No. 3, (May 2004), pp. 373-379, ISSN 0006-3592

Ko, K.; Tekoah, Y.; Rudd, PM.; Harvey, DJ.; Dwek, RA.; Spitsin, S.; Hanlon, CA.; Rupprecht, C.; Dietzschold, B.; Golovkin, M. & Koprowski, H. (2003). Function and glycosylation of plant-derived antiviral monoclonal antibody. *Proceedings of the National Academy of Sciences of the United States of America*, Vol. 100, No. 13, (June 2003), pp. 8013-8018, ISSN 0027-8424

Komarova, TV.; Baschieri, S.; Donini, M.; Marusic, C.; Benvenuto, E. & Dorokhov YL. (2010). Transient expression system for plant derived biopharmaceuticals. *Expert Review of Vaccines*, Vol. 9, No. 8, (October 2009), pp. 859-876, ISSN 1476-0584.

Lentz, EM.; Segretin, M.E.; Mauro M.; Wirth, SA.; Mozgovoj, MV.; Wigdorovitz, A. & Bravo-Almonacid, F.F. (2010). High expression level of a foot and mouth disease virus epitope in tobacco transplastomic plants. *Planta*, Vol. 231, No. 2, (November 2010), pp. 387-395, ISSN 0032-0935

Lichty, JJ.; Malecki, JL.; Agnew, HD.; Michelson-Horowitz, DJ. & Tan, S. (2005). Comparison of affnity tags for protein purifcation. *Protein Expression and Purification*, Vol. 41, No. 1, (May 2005), pp. 98-105, ISSN 1046-5928

Lin, M.; Rose-John, S.; Grötzinger, J.; Conrad, U. & Scheller. J. (2006). Functional expression of a biologically active fragment of soluble gp130 as an ELP-fusion protein in transgenic plants: purification via inverse transition cycling. *The Biochemical Journal*, Vol. 398, No. 3, (September 2006), pp. 577-583, ISSN 0264-6021

Ling, HY.; Pelosi, A. & Walmsley, AM. (2010). Current status of plant made vaccines for veterinary purposes. *Expert Review of Vaccines*, Vol. 9, No. 8, (October 2009), pp. 971-982, ISSN 1476-0584

Mackay, J.A. & Chilkoti, A. (2008). Temperature sensitive peptides: engineering hyperthermia-directed therapeutics. *International Journal of Hyperthermia : the Official Journal of European Society for Hyperthermic Oncology, North American Hyperthermia Group*, Vol. 24, No. 6, (September 2008), pp. 846-851, ISSN 0265-6736

Makvandi-Nejad, S.; McLean, M.D.; Hirama,T.; Almquist, K.C.; Mackenzie, C.R. & Hall, J.C. (2005). Transgenic tobacco plants expressing a dimeric single-chain variable fragment (scfv) antibody against Salmonella enterica serotype Paratyphi B. *Transgenic Research*, Vol. 14, No. 5, (October 2005), pp. 785-792, ISSN 0962-8819

Marcondes, J. & Hansen, E. (2008).Transgenic Lettuce Seedlings Carrying Hepatitis B Virus Antigen HBsAg *Brazilian Journal of Infectious Diseases*, Vol. 12, No. 6, (December 2008), pp. 469-471, ISSN 1413-8670

Medrano, G.; Reidy, MJ.; Liu J.; Ayala, J.; Dolan, M.C. & Cramer CL. (2009). Rapid system for evaluating bioproduction capa city complex pharmaceutical proteins in plants. *Method in Molecular Biology*, Vol. 483, (January 2008), pp. 51-67, ISSN 1064-3745.

Menassa, R.; Zhu, H.; Karatzs, CN.; Lazaris, A.; Richman, A. & Brandle J. (2004). Spider dragline silk proteins in transgenic tobacco leaves: accumulation and field production. *Plant Biotechnol Journal*, Vol. 2, No 5. (September 2004), pp. 431-438, ISSN 1229-2818.

Meyer, D.E. & Chilkoti, A. (1999). Purification of recombinant proteins by fusion with thermally-responsive polypeptides. *Nature Biotechnology*. Vol. 17, No.11, (January 1999) pp. 1112-1115, ISSN 1087-0156

Meyer, D.E. & Chilkoti, A. (2004). Quantification of the effect of chain length and concentration on the thermal behavior of elastin-like polypeptides. *Biomacromolecules*, Vol. 5, No. 3, (May – June 2004), pp. 846-851, ISSN 1525-7797

Mithieux, S.M. & Weiss, A.S. (2005). Elastin. *Advances in Protein Chemistry*, Vol. 70, pp. 437-461, ISSN 0065-3233

Muto, M.; Ryan, HE. & Mayfield, SP. (2009). Accumulation and processing of a recombinant protein designed as a cleavable fusion to the endogenous Rubisco LSU protein in

Chlamydomonas chloroplast. *BMC Biotechnology*, Vol. 9, (March 2009), pp. 26, doi:10.1186/1472-6750-9-26, ISSN 1472-6750

Ohana, R.,F.; Encell, L.P.; Zhao, K.; Simpson D.; Slater, M.R.; Urh, M.; & Wood, K.V. (2009). HaloTag7: A genetically engineered tag that enhances bacterial expression of soluble proteins and improves protein purification. *Protein Expression and Purification*, Vol. 68, No. 1, (November 2009), pp. 110-120, ISSN 1046-5928

Osborn, TC.; Burrow, M. & Bliss, FA. (1988). Purification and characterization of arcelin seed protein from common bean. *Plant Physiology*, Vol. 86, No. 2, (February 1988), pp. 399-405, ISSN 2345476478

Patel, J.; Zhu, H.; Menassa, R.; Gyenis, L.; Richman, A. & Brandle, J. (2007). Elastin-like polypeptide fusions enhance the accumulation of recombinant proteins in tobacco leaves. *Transgenic Research*, Vol. 16, No. 2, (April 2007) pp. 239-249, ISSN 0962-8819

Rabindran, S.; Roy, N.; Fedorkin, G. & Skarjinskaia, M. (2009). Plant-Produced Human Growth Hormone Shows Biological Activity in a Rat Model. *Biotechnology Progress*, Vol. 25, No. 2, (March 2009), pp. 530-534, ISSN 1520-6033

Rival, S.; Wisniewski, JP.; Langlais, A.; Kaplan, H.; Freyssinet, G.; Vancanneyt , G.; Vunsh, R.; Perl, A. & Edelman M. (2008). Spirodela (duckweed) as an alternative production system for pharmaceuticals: a case study, aprotinin. *Transgenic Research*, Vol. 17, No. 4, (August 2008), pp. 503-513, ISSN 0962-8819

Rodríguez, M.; Ramírez, NI.; Ayala, M.; Freyre, F.; Pérez, L.; Triguero, A.; Mateo, C.; Selman-Housein, G.; Gavilondo, JV. & Pujol, M. (2005). Transient expression in tobacco leaves of an aglycosylated recombinant antibody against the epidermal growth factor receptor. *Biotechnology and Bioengineering*, Vol. 89, No. 2, (January 2005), pp. 188-194, ISSN 0006-3592

Rodriguez-Cabello, J.C.; Prieto, S.; Reguera, J.; Arias, F.J. & Ribeiro, A. (2007). Biofunctional design of elastin-like polymers for advanced aplications in nanobiotechnology. *Journal of Biomaterials Science. Polymer Edition*, Vol. 18, No. 3, (March 2007), pp. 269-286, ISSN 0920-5063

Rojas-Anaya, E.; Loza-Rubio, E.; Olivera-Flores, M.T. & Gomez-Lim, M. (2009). Expression of rabies virus G protein in carrots (Daucus carota) *Transgenic Research*, Vol. 18, No. 6, (December 2009), pp.911-919, ISSN 0962-8819

Sack, M.; Paetz, A.; Kunert, R.; Bomble, M.; Hesse, F.; Stiegler, G.; Fischer, R.; Katinger, H.; Stoeger, E. & Rademacher, T. (2007). Functional analysis of the broadly neutralizing human anti-HIV-1 antibody 2F5 produced in transgenic BY-2 suspension cultures. he FASEB journal : official publication of the *Federation of American Societies for Experimental Biology*, Vol. 21, No. 8, (June 2007) pp. 1655-1664, ISSN 0892-6638

Scheller, J.; Hengger, D.; Viviani, A. & Conrad, U. (2004). Purification of spider silk elastin from transgenic plants and application for human chondrocyte proliferation. *Transgenic Research*, Vol. 13, No. 1, (February 2004) pp. 51-57, ISSN 1573-9368.

Serrano, V.; Liu, W. & Franzen, S. (2007). An infrared spectroscopic study of the conformational transition of elastin-like polypeptides. *Biophysical Journal*, Vol. 93, No. 7, (October 2007), pp. 2429-2435, ISSN 0006-3495

Sharma, AK. & Sharma, MK. (2009). Plants as bioreactors: Recent developments and emerging opportunities. *Biotechnology Advances*, Vol. 27, No. 6, (June 2009), pp. 811-832, ISSN 1476-0584.

Sharma, M. & Sood, B. (2011). A banana or syringe: journey to edible vaccines. *World Journal of Microbiology and Biotechnology*, Vol. 27, No. 3, (June 2011), pp. 471-477, ISSN 1573-0972

Shoji, Y.; Chichester, JA.; Jones, M.; Manceva, SD.; Damon, E.; Mett, V.; Musiychuk, KB.; H. Farrance, Ch.; Shamloul, M.; Kushnir, N.; Sharma S. & Vidadi YV. (2011) Plant-based rapid production of recombinant subunit hemagglutinin vaccines targeting H1N1 and H5N1 influenza. *Human Vaccines*, Vol. 7, Supplement, (January/February 2011), pp. 41-50, ISSN 1554-8600

Smith, ML.; Mason, HS. & Shuler, ML. (2002). Hepatis B surface antigen (HBsAg) expression in plant cell culture: kinetics of antigen accumulation in batch culture and its intracellular form. *Biotechnology Bioengineering*, Vol. 80, No. 7, (December 2002), pp. 812-822, ISSN 0006-3592.

Steen, J.; Uhlén, M.; Hober, S. & Ottosson, J. (2006). High-throughput protein purification using an automated set-up for high-yield affinity chromatography. *Protein Expression and Purification*, Vol. 46, No. 2, pp. 173-178, ISSN 1046-5928

Stefanov, I.; Illubaev, S.; Feher, A.; Margoczi, K. & Dudits D. (1991). Promoter and genotype dependent transient expression of a reporter gene in plant protoplasts. *Acta Biologica Hungarica*, Vol. 42, No. 4, (April 1991), pp. 323-330, ISSN 0236-5383.

Stefanova, G.; Vlahlova, M. & Atanassov A. (2008). Production of recombinant human lactoferrin from transgenic plants. *Biology Plantarum*, Vol. 52, No. 3, (May 2008), pp. 423-428, ISSN 1573-8264

Stevens, LH.; Stoopen, GM.; Elbers, IJ.; Molthoff, JW.; Bakker, HA.; Lommen, A.; Bosch, D. & Jordi, W. (2000). Effect of climate conditions and plant developmental stage on the stability of antibodies expressed in transgenic tobacco. *Plant Physiology*, Vol. 124, No. 1, (May 2000), pp. 173-182 , ISSN 0032-0889

Stoger, E .; Sack , M.; Perrin, Y.; Vaquero, C.; Torres, E.; Twyman, RM.; Christou, P. & Fischer, R. (2002) Practical considerations for pharmaceutical antibody production in different crop systems. *Molecular Breeding*, Vol. 9, No. 3, (September 2002), pp. 149 – 158, ISSN 1380-3743

Streatfield, SJ. (2006). Mucosal immunization using recombinant plant based oral vaccines. *Methods*, Vol. 38, No. 2, (February 2006), pp. 150-157, ISSN 1046-2023.

Tacket, CO. (2005). Plant derived vaccines against diarrheal diseases. *Vaccine*, Vol. 23, No. 15, (March 2005), pp. 1866-1869, ISSN 0264-410X.

Thanavala, Y.; Huang, Z. & Mason, HS. (2006). Plant derived vaccines: a look back at the highlights and a view to the challenges on the road ahead. *Expert Review of Vaccines*, Vol. 5, No. 2, (February 2006), pp. 249-260, ISSN 1476-0584

Trabbic-Carlson, K.; Meyer, DE.; Liu, L.; Piervincenzi, R.; Nath, N.; LaBean, T. & Chilkoti, A. (2004) Effect of protein fusion on the transition temperature of an environmentally responsive elastin-like polypeptide: a role for surface hydrophobicity? *Protein Engineering Design and Selection*, Vol. 17, No. 1, (August 2004), pp. 57-66, ISSN 1741-0126

Valdez-Ortiz, A.; Rascón-Cruz, Q.; Medina-Godoy, S.; Sinagawa-García, S.R.; Valverde-González, M.E. & Paredes-López, O. (2005). One-step purification and structural characterization of a recombinant His-tag 11S globulin expressed in transgenic tobacco. *Journal of Biotechnology*, Vol. 115, No. 4, (February 2005), pp. 413-423, ISSN 0168-1656

Valiaev, A.; Lim, DW.; Schmidler, S.; Clark, RL.; Chilkoti, A. & Zauscher, S. (2008). Hydration and conformational mechanics of single, end-tethered elastin-like polypeptides. *Journal of the American Chemical Society*, Vol. 130, No. 33, (July 2008), pp. 10939-10946 , ISSN 0002-7863

Vaquero, C.; Sack, M.; Schuster, F.; Finnern, R.; Drossard, J.; Schumann, D.; Reimann, A.; & Fischer, R. (2002). A carcinoembryonic antigen-specific diabody produced in tobacco. *The FASEB Journal: Official Publication of the Federation of American Societies for Experimental Biology*, Vol. 16, No. 3, (January 2002), pp. 408-410, ISSN 0892-6638

Villani, M.E.; Morgun, B.; Brunetti, P.; Marusic, C.; Lombardi, R.; Pisoni, I.; Bacci, C.; Desiderio, A.; Benvenuto, E. & Donini, M. (2009). Plant pharming of a full-sized, tumour-targeting antibody using different expression strategies. *Plant Biotechnology Journal*, Vol. 7, No. 1, (January 2009), pp. 59-72, ISSN 1467-7644

Waugh, DS. (2005). Making the most of affinity tags. *Trends in Biotechnology*, Vol. 23, No. 6, (June 2005), pp. 316-320, ISSN 0167-7799

Webster, DE.; Wang, L.; Mulcair, M.; Ma, Ch.; Santi, L.; Mason, HS.; Wesselingh, S.L.; Ross L. & Coppel, R.L. (2009). Production and characterization of an orally immunogenic Plasmodium antigen in plants using a virus-based expression system. *Plant Biotechnology Journal* Vol. 7, No. 9, (December 2009), pp. 846–855, ISSN 14677644

Wiktorek-Smagur, A.; Hnatuszko-Konka, K.; Gerszberg, A.; Łuchniak, P.; Kowalczyk, T. & Kononowicz, AK. (2011). Expression of a staphylokinase, a thrombolytic agent in *Arabidopsis thaliana*. *World Journal Microbiology and Biotechnology*, Vol. 27, No. 6, (June 2011), pp. 1341–1347, ISSN 0959-3993

Witte, CP.; Noël, L.D.; Gielbert, J.; Parker, J.E. & Romeis, T. (2004). Rapid one-step protein purification from plant material using the eight-amino acid StrepII epitope. *Plant Molecular Biology*, Vol. 55, No. 1, (May 2004), pp. 135-147, ISSN 0167-4412

Wu, Cy.; Suzuki, A.; Washida, H. & Takaiwa, F. (1988). The GCN4 motif in a rice glutelin gene is essential for endosperm specific gene expression and is activated by Opaquae-2 in transgenic rice plants. *Plant Journal*, Vol. 14, No. 6, (June 1988), pp. 673-683, ISSN 1365-313X.

Xie, Y.; Liu, Y.; Meng, M.; Chen, L. & Zhu, Z. (2003). Isolation and identification of a super strong plant promoter from cotton leaf curl Multan virus. *Plant Molecular Biology* Vol. 53, No. 1-2, (July 2003), pp. 1-14, ISSN 0735-9640.

Xu, J.; Ge, X. & Dolan MC. (2011). Towards high-yield production of pharmaceuticals proteins with plant cell suspension culture. *Biotechnology Advances*, Vol. 29, No. 3, (January 2011), pp. 278-299, ISSN 0734-9750.

Yano, A.; Maeda, F. & Takekoshi M. (2004). Transgenic tobacco cells producing the human monoclonal antibody to hepatitis B virus surface antigen. *Journal of Medical Virology*, Vol. 73, No. 2, (June 2004), pp. 208-215 , ISSN 0146-6615

Yap, YK. & Smith DR. (2010). Strategies for the plant based expression of dengue subunit vaccines. *Biotechnology and Applied Biochemistry*, Vol. 57, No. 2, (December 2010), pp. 47-53, ISSN 0885-4513.

Yarmush, M.L.; Toner, M.; Plonsey, R. & Bronzino J.D. (2003). Monoclonal Antibodies and Their Engineered Fragments In: *Biotechnology for Biomedical Engineers*, 1-17, CRC Press; 1 edition (March 26, 2003), ISBN 0-8493-1811-4

Yu, J. & Langridge, W. (2003). Expression of rotavirus capsid protein VP6 in transgenic potato and its oral immunogenicity in mice. *Transgenic Research,* Vol. 12, No. 2, (April 2003), pp. 163–169, ISSN 1573-9368

Yusibov, V. & Rabindran, S. (2008). Recent progress in the development of plant derived vaccines. *Expert Review of Vaccines,* Vol. 7, No. 8, (October 2008), pp. 1173-1183, ISSN 1476-0584.

Zhang, D.; Nandi, S.; Bryan, P.; Pettit, S.; Nguyen, D.; Santos, MA. & Huang, N. (2010). Expression, purification, and characterization of recombinant human transferrin from rice (*Oryza sativa* L.). *Protein Expression and Purification,* Vol. 74, No. 1, (November 2010), pp. 69-70, ISSN 1046-5928

Zhou, A. & Li, J. (2005). Arabidopsis BRS1 is a secreted and active serine carboxypeptidase. *The Journal of Biological Chemistry,* Vol. 280, No. 42, (October 2005), pp. 35554-355561, ISSN 0021-9258

Zuo, J. & Chua, NH. (2000). Chemical-inducible systems for regulated expression of plant genes. *Current Opinion in Biotechnology,* Vol. 11, No. 2, (April 2000), pp. 146-151, ISSN 0958-1669.

3

Agrobacterium-Mediated Transformation of Wheat: General Overview and New Approaches to Model and Identify the Key Factors Involved

Pelayo Pérez-Piñeiro[1], Jorge Gago[1], Mariana Landín[2] and Pedro P. Gallego[1,*]

[1]*Applied Plant and Soil Biology, Dpt. Plant Biology and Soil Science,*
Faculty of Biology, University of Vigo, Vigo,
[2]*Dpt. Pharmacy and Pharmaceutical Technology, Faculty of Pharmacy,*
University of Santiago, Santiago de Compostela,
Spain

1. Introduction

Wheat is the world's second largest crop, supplying 19% of human calories; the largest volume crop traded internationally and grown on approximately 17% of the world's cultivatable land (over 200 million hectares) (Jones, 2005; Atchison et al., 2010). However, probably due to climate change, some adverse environmental conditions have caused a downward trend in world wheat production (FAO, 2003; 2011). In this context, developing new higher yielding wheat varieties more tolerant or resistant to abiotic and/or biotic stress, using all available plant biotechnology technologies available, should be considered as the major challenge.

The scientific community has made considerable efforts to understand and improve the goal of the integration of an exogenous T-DNA in the genome of a host plant cell and, subsequently, the regeneration into a whole plant. The most extended method for plant genetic transformation uses the *Agrobacterium* bacteria as the biological vector to transfer exogenous T-DNA into the plant cell. Although, *Agrobacterium*-mediated transformation became widely available for the routine transformation of most crops, cereals initially have been recalcitrant to this system, since these crops were not naturally susceptible to *Agrobacterium* sp (Potrykus, 1990, 1991). However, by the mid-1990s, improvements in technological development in *Agrobacterium*-mediated genetic transformation led to the desirable transformation of wheat (Cheng et al., 1997; Peters et al., 1999; Jones et al., 2007). These results "open the avenue" by avoiding the usage of gene direct transfer methods, such as biolistic, which is widely found more disadvantageous compared to *Agrobacterium*-mediated transformation (Jones, 2005; Jones et al., 2007; Khurana et al., 2008).

Developing an appropriate method for genetic *Agrobacterium*-mediated transformation is a highly complex task, because it is essential to understand the effect of all the factors

---

*Corresponding Author

influencing the T-DNA delivery into the tissue from which whole plant can be regenerated. After plant regeneration, further analyses were required to check the integration and stability of the T-DNA and to obtain the final transformation efficiency parameter. Artificial intelligence technologies are very successful in establishing relationships, in complex processes, between multiple processing conditions (variables or factors) and the results obtained, using networks approaches. Recently, several studies have demonstrated the effectiveness of artificial neural networks and neurofuzzy logic in modelling and optimizing different plant tissue culture processes. Neurofuzzy logic is a useful modeling tool that has been introduced to help the handling of complex models and to data mining. Data mining can be defined as the process of discovering previously unknown dependencies and relationships in datasets. It is a hybrid technology combining the strength and the adaptive learning capabilities from artificial neural networks (ANNs) and the ability to generalize rules of fuzzy logic. Neurofuzzy logic technology generates understandable and reusable knowledge in the way of IF (conditions) THEN (observed behavior) rules helping the researchers to understand the process or the phenomena they are studying (Gallego et al., 2011).

In this chapter we overview the recent advances in *Agrobacterium*-mediated transformation of the wheat, but we also proposed the utility of artificial intelligence technologies as a modeling tool used to understand the complex cause–effect relationships between the most common parameters used in *Agrobacterium*-mediated transformation of the cereals too. That information should help cereal researchers to gain in knowledge on the transformation process, which means determining the factors that favour the interaction between *Agrobacterium* and cereal plants in order to improve the transfer of T-DNA and afterwards to regenerate whole plants from transformed cells, improving final transformation efficiency. Moreover, in a near future, this technology could be easily adapted to the rest of cereals or even any crop.

2. *Agrobacterium*-mediated transformation: Main factors

From the early 1990s many efforts were carried out in order to achieve stable transformation of wheat via *Agrobacterium*-mediated transformation (Bhalla et al., 2006; Vasil, 2007). This methodology presents several advantages over other approaches including the ability to transfer large segments of DNA with minimal rearrangement of DNA, fewer copy gene insertion, higher efficiency and minimal cost.

Several factors were identified as influencing the efficiency of T-DNA delivery: primary source materials; *Agrobacterium* strains; plasmids vectors; *Agrobacterium* density; medium composition; transformation conditions such as temperature and time during pre-culture, inoculation and co-culture; surfactants or induction agents in the inoculation and co-culture; and antibiotics or selectable markers, among others (Jones et al., 2005; Bhalla et al., 2006; Opabode, 2006; Kumlehn & Hensel, 2009).

2.1 Plant material

A summary of the different plant sources reported as main factors for *Agrobacterium*-mediated transformation of wheat can be found in Table 1. Wheat recalcitrance to *in vitro* culture is one of the most important crucial steps for *Agrobacterium* mediated transformation protocols and directly correlated with the wheat source material. It was assessed that *in vitro* regeneration can be highly influenced by different factors such as plant growth regulators. In fact, auxins, polyamines and cytokinins were considered as essential to enhance the efficiencies on target explant and genotype (Khanna & Daggard, 2003; Przetakiewicz et al., 2003; Rashid et al., 2009).

2.1.1 Wheat genotype

Transformation and regeneration of the infected explants are highly genotype-dependent, the plant genotype has been revealed as a major factor influencing transformation efficiency. Indeed, the largest transformation efficiency compared to any other commercial wheat germplasm was reported when the highly regenerable wheat breeding line "Bobwhite" was used (Table 1).

The *Triticum aestivum* Spring "Bobwhite" is the most representative cultivar representing over 25% of the data reported of *Agrobacterium*-mediated transformation of wheat (Table 1), becoming "the genotype model" (Fellers et al., 1995; Sears & Deckard, 1982; He et al., 1988). It has a good response in tissue culture with a high rate of callus induction and regeneration (Janakiraman et al., 2002) making it a suitable cultivar for transformation, since a high ratio for both transformation and regeneration can be achieved. However, it would be highly desirable to transform genotypes other than the model ones (Kumlehn & Hensel, 2009) with much better agronomical and grain quality traits.

Other *T. aestivum* lines, cultivars or varieties such as "Turbo" (Hess et al., 1990); "Millewa" (Mooney et al., 1991); "Chinese" (Langridge et al., 1992); "Kedong 58", "Rascal" and "Scamp" (McCormac et al., 1998); "Lona" (Uze et al., 2000); "Baldus" (Amoah et al., 2001); "Fielder" (Weir et al., 2001); "Florida" and "Cadenza" (Wu et al., 2003); "Vesna" (Mitic et al., 2004); "Veery-5" (Khanna & Daggard, 2003; Hu el al., 2003) and so on (see the complete list in Table 1) were also tested.

Finally, some other commercial *Triticum sp* (different to *T.* aestivum) such as *Triticum dicoccum* (Chugh & Khurana, 2003), *Triticum durum* (Patnaik et al., 2006) or *Triticum turgidum* (Wu et al., 2008; Wu et al., 2009; He et al, 2010) were also being successfully used for Agrobacterium-mediated wheat transformation (see Table 1).

2.1.2 Target explants

The primary source of material is one of the main constraints for *Agrobacterium*-mediated wheat transformation. Regeneration is performed from highly regenerant tissues with active cell division. In these tissues embryogenic calli are induced and regeneration leads to the recovery new formed transgenic plants. Two types of explants are typically used for the recovery of fertile transgenic plants: immature inflorescences and the scutellum of immature zygotic embryos. Although other explants (Table 1) have been used for the same purpose such as reproductive-derived material (Hess et al., 1990; Liu et al., 2002), seeds (Zale et al., 2004); leaf (Wang & Wei, 2004) or shoot meristems (Ahmad et al., 2002), none of them were capable of reliably production of fertile adult transgenic wheat adult plants.

Wheat (Variety / Cultivar) Triticum aestivum	Explant Type	Strain / Plasmid	Promoter / Reporter Gene	Promoter / Selectable gene	Transformation efficiency (%)	Reference / Remarks
Turbo (Spring)	SPK	C58C1 / pGV3850:1103neo	-	nos / nptII	1	Hess et al., 1990 / No regeneration
Millewa (Spring)	IE	C58C / pGV3850:1103neo	-	nos / nptII	1 – 2 (based on kanamycin selection)	Mooney et al., 1991 / Gene integration was not demonstrated
Chinese (Spring)	SPK	LBA4404 / pPCV6NFHyg A281 / pPCV6NFHyg C58C1 / PCV6NFHyg GV3101 / pPCV6NFHyg	CaMV35S / gus	nos / nptII CaMV35S / hpt	0.8 – 4.7 (based on kanamycin selection)	Langridge et al., 1992 / No regeneration
Bobwhite (Spring)	IE PCIE IEdC	ABI / pMON18365	CaMV35S / gus	CaMV35S / nptII	1.12 (IE) 1.56 (PCIE) 1.55 (IEdC)	Cheng et al., 1997 / Salt strength test, surfactants & explants types
Bobwhite (Spring)	SDS	EHA105 / pIG121Hm	CaMV35S / gus	nos / nptII CaMV35S / hpt	28 foci/seed (GUS)	Trick et al., 1997 / Sonication test. Transient GUS expression
Rascal (Spring) Scamp (Spring) Kedong 58 (Winter)	IEdC	EHA101 / pBECKS.red	CaMV35S / gus CaMV35S / gfp CaMV35S / Lc/C1	nos / nptII	40 - 70 (based on reported genes)	McCormac et al., 1998 / gfp and Lc/C1 gene reporters
Chinese (Spring)	MSdC IEdC	GV3101 / pMVTBP GV3101 / pNFHK1 GV3101 / p35SGUSINT	CaMV35S / gus	CaMV35S / nptII	1.2 - 2.2 (GUS)	Peters et al., 1999 / Use of modular vector
Several (Chinese)	IEdC	AGL1 / pUNN2	-	ubi1 / nptII	3.7 – 5.9	Xia et al., 1999 / Stable transformation
Bobwhite (Spring) Lona (Spring)	PCIE	LBA4404 / pBin9UG EHA105 / pBin9UG C58C1 / pBin9UG LBA9402 / pBin9UG	ubi1 / gus	-	20 foci/callus (GUS)	Uze et al., 2000 / Several factors studied for transformation
Baldus (Spring)	INFdC	AGL1 / pAL154-pAL156 AGL1 / pAL155-pAL156 AGL1 / pSoup-pAL186	ubi1 / gus	ubi1 / bar	14 – 64 (GUS)	Amoah et al., 2001 / Inflorescence tissue. Sonication and vacuum infiltration
Fielder (Spring)	PCIE	AGL0 / pTO134	CaMV35S / gfp	CaMV35S / bar	1.8 PCIE	Weir et al., 2001 / Several factors studied for transformation
Nongda 146 (Spring)	IE PCIE	AGL1 / pAL155-pAL156	ubi1 / gus	ubi1 / bar	90 (GUS)	Ke et al., 2002/ Transient GUS expression
Sohag 2 (Durum)	SPK	LBA4404 / pBI-P5CS	CaMV35S / gus	nos / nptII	0.9	Sawahel and Hassan, 2002 / In planta transformation
Sourav (Spring) Gourav (Spring) Kanchan (Spring) Protiva (Spring)	IE ME IEdC MEdC	EHA105 / pCAMBIA1301	CaMV35S / gus	nos / nptII	75 – 85 (IE) 60 - 65 (ME) 80 – 87 (IEdC) 67 – 73 (MEdC)	Sarker & Biswas, 2002 / Transient GUS expression
Bobwhite (Spring)	PCIE IEdC	ABI / pMON18365	CaMV35S / gus	CaMV35S / nptII CaMV35S / aroA:CP4	3.1 (PCIE -glyphosate) 6.1 (PCIE-paromomy.) 10.5 (EC-paromomycin)	Cheng et al., 2003 / Large scale experiments
Bobwhite (Spring)	PCIE	C58C1 / pPTN115	CaMV35S / gus	CaMV35S / nptII	0.5 - 1.5	Haliloglu and Baenziger (2003) / Several factors were studied
Bobwhite (Spring)	PCIE	ABI / pMON30120 ABI / pMON30174 ABI / pMON30139	-	act1 / aroA:CP4 CaMV35S / aroA:CP4 ScBV / aroA:CP4	4.4	Hu et al., 2003 / Large-scale production. Roundup ready wheat
Veery5 (Spring)	IEdC	LBA4404 / pHK22 LBA4404 / pHK21	ubi1 / gus	ubi1 / bar	1.2 – 3.9	Khanna and Daggard, 2003 / Use of spermidine in regeneration
Bobwhite (Spring)	PCIE EC	n.d / PV-TXGT10	-	act1 / aroA:CP4 CaMV35S / aroA:CP4	-	Zhou et al., 2003 / Roundup ready wheat
Florida (Winter) Cadenza (Winter)	IE	AGL1 / pAL154-156	ubi1 / gus	ubi1 / bar	0.3 – 3.3	Wu et al., 2003 / Several factors studied for transformation
CPAN1676 (Bread) PBW343 (Bread)	MSdC	LBA4404 / pCambia3301	CaMV35S / gus	CaMV35S / bar	6.7 – 8.7	Chugh & Khurana, 2003 / Herbicide Resistance. Use of basal segment calli as target tissue
Vesna (Spring)	IE	LBA4404 / pTOK233 AGL1 / pDM805	CaMV35S / gus act1 / gus	ubi1 / bar nos / nptII CaMV35S / hpt	0.13 (LBA) 0.41 (AGL)	Mitic et al., 2004 / Use of super-binary vectors. Only PCR test
Kontesa (Winter) Torka (Winter) Eta (Winter)	IE	AGL1 / pDM805 EHA101 / pGAH LBA4404 / pTOK233	CaMV35S / gus act1 / gus	ubi1 / bar nos / nptII CaMV35S / hpt	1 (AGL1) 0.2 - 8.1 (EHA101) 0.2 - 2.3 (LBA4404)	Przetakiewicz et al., 2004 / Use of super-binary vectors and auxins
Hesheng3 (Winter) Yan103 (Winter) Yanyou361 (Winter)	IEdC	GV3101 / pROK2-AtNHX1	-	CaMV35S / nptII	1.3 – 2.9	Xue et al., 2004 / Survival tests in saline conditions. Field trial
Shannong 9956049 (Winter)	IEdC	LBA4404 / pROK2	-	nos / nptII	1.18	Bi et al., 2006 / Insect resistance
Yan361 (Winter) Yan2801 (Winter) H11 (Winter)	SDS	EH105 / pBLG	-	nos / nptII	8.62 – 11.2	Zhao et al., 2006 / Powdery Mildew resistance
HD2329 (Bread) CPAN1676 (Bread) PBW343 (Bread)	ME MEdC	LBA4404 / pBI101 LBA4404 / pCAMBIA3301	act1 / gus CaMV35S / gus	nos / nptII CaMV35S / bar	1.6	Patnaik et al., 2006 / Genotypic independence
Shiranekomugi	SDS	LBA4404 / pIG121Hm LBA4404 / pBI-res used LBA4404 / pBI-res2 used	CaMV35S / gus	nos / nptII CaMV35S / hpt	33 (PCR) 75 (Southern) 40 (plasmid rescue)	Supartana et al., 2006 / TE is referred to t1 progeny instead of inoculated explants
Yangmai158 (Winter)	PCIE	EHA105 / pCAMBIA3300	-	CaMV35S / bar	-	Yu and Wei, 2008 / Insect resistance
Een1 (Winter)	SDS	LBA4404 / n.d.	CaMV35S / gus	CaMV35S / nptII	3 - 31 (GUS)	Yang et al., 2008 / Use of seedling ages and inoculation time
Crocus (Spring)	SPK	C58C1 / pDs(Hyg)35S AGL1 / pBECKSred	CaMV35S / Lc/C1	nos / nptII nos / hpt	0.44	Zale et al., 2009 / In planta transformation
Certo (Winter)	IE	LBA4404 / pSB187	ubi1 / gfp	CaMV35S / hpt	2 – 10	Hensel et al., 2009 / Detailed protocols for transformation
EM12 (Chinese)	PCME	LBA4404 / pBI121	CaMV35S / gus	nos / nptII	0.27 - 2.5	Ding et al., 2009 / Optimization of transformation protocol

Agrobacterium-Mediated Transformation of Wheat: General Overview and New Approaches to Model and Identify the Key Factors Involved

71

Wheat (Variety / Cultivar) Triticum aestivum	Explant Type	Strain / Plasmid	Promoter / Reporter Gene	Promoter / Selectable gene	Transformation efficiency (%)	Reference / Remarks
Bobwhite (Sring) Yumai66 (Winter) Lunxuan208 (Winter)	ME, PCME	C58C1 / pUbiGN	ubi1 / gus	nos / nptII	0.06 – 0.89	Wang et al., 2009 / Use of mature embryos
Inqilab-91 (Bread)	ME	EHA101 / pIG121Hm	CaMV35S / gus	nos / nptII CaMV35S / hpt	6.25 – 15.62	Rashid et al., 2010 / Effect of AS & bacterial culture density
Triticum turgidum						
Durum (ofanto)	IE	AGL1 / pAL156-pAL154 AGL1 / pAL156-pAL155	ubi1 / gus	ubi1 / bar	0.6 – 9.7 (pAL154) 2.1 – 3.9 (pAL155)	Wu et al., 2008 / Super-binary vectors. First time durum transf
Durum (Stewart)	IE	AGL1 / pAL156- pAL154	ubi1 / gus	ubi1 / bar	2.8 - 6.3	He et al, 2010 / Effect super-binary vectors, AS & picloram
Triticum dicoccum						
DDK1001 (Emmer)	MSdC	LBA4404 / pCambia3301	CaMV35S / gus	CaMV35S / bar	6.9	Chugh & Khurana, 2003 / Herbicide Resistance. Use of basal segment calli as target tissue
Triticum durum						
PDW215 (Pasta)	ME	LBA4404 / pCAMBIA3301	CaMV35S / gus	CaMV35S / bar	3	Vishnudasan et al., 2005 / Nematode resistance
PDW215 (Pasta) PDW233 (Pasta) WH896 (Pasta)	ME MEdC	LBA4404 / pBI101	act1 / gus CaMV35S / gus	nos / nptII CaMV35S / bar	1.28	Patnaik et al., 2006 / Genotypic independence

Table 1. Summary of wheat materials, *Agrobacterium* strains and vectors, and marker genes used to investigate wheat transformation. Explant type: IE (immature embryo); PCIE (pre-cultured immature embryo); IEdC (immature embryo derived calli); ME (mature embryo); PCME (pre-cultured mature embryo); MEdC (mature embryo derived calli); INF (inflorescence); INFdC (inflorescence derived calli); SPK (spikelet); SDS (seedling); MSdC (mature seed derived calli). Promoters: *CaMV35S* (cauliflower mosaic virus); *ubi1* (maize ubiquitin); *act1* (rice actin), *nos* (nopaline synthase gene); *ScBV* (sugarcane bacilliform virus). Reporter genes: *gus* (β-glucuronidase); *gfp* (green fluorescent protein); *Lc/C1* (anthocyanin-biosynthesis regulatory). Selectable gene: *nptII* (neomycin phosphotransferase II) and *hpt* (hygromycin phosphotransferase) antibiotic resistance and *bar* (phosphinothricin acetyltransferase) and *aroA:CP4* (5-enolpyruvylshikimate-3-phosphate synthase (EPSPS)) herbicide resistance.

By far, the main target explant used to transform wheat was from immature embryos (IE). Concretely, the immature scutellum was used, a specialised tissue that forms part of the seed embryo, and it was recommended that embryo isolation was performed 11-16 days post-anthesis (Jones, 2005). Freshly isolated IE, pre-cultured IE or IE derived callus had been widely included in experiments to obtain transgenic wheat plants. Cheng et al. (1997) reported, for the first time, the success of *Agrobacterium*-mediated transformation in wheat using IE (freshly isolated and pre-cultured) and embryogenic calli producing fertile transgenic plants despite the experiments being limited to small-scale. Later, many attempts were carried out by several authors (McCormac et al., 1998; Xia et al., 1999; Uze et al., 2000; Ke et al., 2002, Sarker & Biswas, 2002) but no stable transgenic plants were reported until Weir et al. (2001), who confirmed results obtained previously by Cheng et al. (1997), transformed pre-cultured immature embryos, 9 day old. Large-scale experiments were carried out using immature embryos as the initiation tissue for both genetic transformation and plant regeneration (Cheng et al., 2003; Hu et al., 2003; Vasil, 2007; Jones et al., 2007; Rashid et al., 2009).

Immature inflorescences were also easier to isolate and can be collected earlier from younger plants in comparison to immature embryos. However, these explants present more specific-genotype requirements for its *in vitro* culture regeneration (Jones, 2005 and references therein). Seeds were also used as started explant for wheat in plant transformation (Trick & Finer, 1997; Supartana et al., 2006; Zhao et al., 2006; Yang et al., 2008; Razzaq et al., 2011) but only Supartana et al. (2006) and Zhao et al. (2006) demonstrated stable gene inheritance and integration in progeny by Southern blot analysis

(Table 1). Other initiation explants were also tested as tissue for wheat *Agrobacterium*-mediated transformation: mature embryo (ME) either freshly isolated, pre-cultured or derived calli (Sarker & Biswas, 2002; Vishnudasan et al., 2005; Patnaik et al., 2006; Ding et al., 2009; Wang et al., 2009; Rashid et al., 2010), inflorescence or inflorescence derived calli (Amoah et al., 2001) and mature seed derived calli (Peters et al., 1999; Chugh & Khurana, 2003). Mature embryos offer some advantage over the typically used immature embryos, as a low-cost procedure because immature embryos must be recollected from plants grown under a controlled environment, moreover the extraction of the embryos in a narrow developmental stage (i.e. 0.8–1.5 mm in diameter) is required (Wu et al., 2009; Wang et al., 2009).

In the early 1990s transgenic wheat materials were generated by inoculating florets with *Agrobacterium* at or near anthesis (Hess et al., 1990; Langridge et al. 1992) produced similar results since both failed to demonstrate gene integration in successive plant generations or successful plant regeneration (Table 1). Using the same protocol but changing the *Agrobacterium* strain and the plasmid construction, a floral dip efficient transformation of wheat was achieved by Sawahel & Hassan (2002). More recently (Zale et al., 2009) by performing transformation at an earlier stage of floral development than previously (i.e., Hess et al., 1990; Langridge et al. 1992; Sawahel & Hassan, 2002) successful transgene integration and expression were obtained when wheat ovules were used as target explants.

2.2 *Agrobacterium* and plasmids

It has been widely described in the literature that the combination of highly competent *Agrobacterium* strain with effective and suitable plasmid construction leading to improved successful wheat transformation efficiencies (Khanna & Daggard, 2003; Cheng et al., 2004). The most used *Agrobacterium* strains and plasmids are summarized in Table 1.

2.2.1 *Agrobacterium* strain

Cereals are not natural hosts for *Agrobacterium* and many studies have been carried out to match host strains with wheat genotypes (Jones et al., 2005). Mainly, only three strains of *Agrobacterium tumefaciens* are currently used in wheat transformation (Table 1) thus from the 41 reports reviewed: 44% used LBA4404, followed by C58C1 (24%) and AGL1 (24%). While other strains has been used with a less frequency (10%) including other *A. tumefaciens* strains such as: A281, GV3101, ABI, EHA101, EHA105, AGL0, M-21 and *A. rhizogenes* LBA9402 and Ar2626. Interestingly, most of those *Agrobacterium* strains share only two chromosomal backgrounds: the C58 type (C58C1, AGL1, GV3101, ABI, EHA101, EHA105 and AGL0) and TiAch5 (LBA4404) (Hellens et al., 2000; Jones et al., 2005).

The infection process of *Agrobacterium* include several chromosome-encoded genes involved in the attachment of bacteria to plant cells and Ti plasmid-encoded *vir* genes, that function in trans, helping the transfer and integration of T-DNA into the plant genome (Wu et al., 2008). Some of the above strains also contain a binary or helper plasmids, carrying further copies of virulence genes. Therefore, depending on agro construction, "standard or low virulent" strains as LBA4404 and C58C1 or "hyper-virulent strains" such as AGL have been designed to successful transformation of wheat.

Although rare, also some a-virulent *A. tumefaciens* mutant strain has also been used for wheat transformation studies as a reliable marker of transformation (Table 1). As an example, Supartana and co-workers (2006) employed the M-21 *Agrobacterium* mutant, in which the *iaaM* gene (tryptophan monooxygenase gene) - involved in IAA (indole acetic acid) biosynthesis in the T-DNA region - is destructed by transposon5 (Tn5) insertion. As a consequence, this mutant strain was capable of integrating its T-DNA into chromosomes of host plants, but no galls were produced. Wheat transformants obtained by the M-21 mutant strain were expected to synthesize a high cytokinin level (since all other genes including the *ipt* gene – involved in cytokinin biosynthesis in the T-DNA region – were intact and fully functional), resulting in a high altered phenotype due to hormone imbalance which can be easily detected (Supartana et al., 2006).

2.2.2 Plasmid and virulence

As stated previously, wheat is not a natural host for *Agrobacterium*, for this reason only a few genotypes (such as Bobwhite) can be transformed with standard strains, such as LBA4404 and binary vectors (Cheng et al., 1997; Hu et al., 2003). When other genotypes were tested, no successful transformation was obtained, only their virulence was increased by adding an extra binary plasmid (such as pHK21) with extra *vir* genes (Khanna & Daggard, 2003) that enhance the transformation.

Many other Ti vectors and helper plasmids, known as binary plasmids, which can include an extra copy of virulence genes in the namely "super-binary" vectors, have been incorporated in the selected *Agrobacterium* strain to enhance infection. Several combinations regarding virulence are possible: from a-virulent to hyper-virulent *Agrobacterium* strain.

The most common *Agrobacterium* strains used in wheat transformation below to hyper-virulent group and is the disarmed plasmid pTiBo542 from *A. tumefaciens* wild strain A281 harbouring additional virulence genes usually *vir* B, C and G, which confer the hyper-virulence character (Komari et al., 1990).

Two different constructs have been widely employed to carry extra *vir* region (Table 1): first, using the helper plasmid pAL155 which is a derivative of pSoup modified by the addition of *vir* G (Amoah et al., 2001; Ke et al., 2002; Wu et al., 2008); and second, using different plasmids as pAl154, pAL186 or pTOK233 carrying "15 kb Komari fragment" containing set of *vir* B, C and G (Amoah et al., 2001; Wu et al., 2003; Mitic et al., 2004; Przetakiewicz et al., 2004; Wu et al., 2008; Wu et al., 2009; He et al., 2010).

2.2.3 Promoters

Regarding the promoters (see Table 1), the most common were the constitutive "*CaMV35S*" (cauliflower mosaic virus) and "*ubi1*" (maize ubiquitin). Other promoters such as "*act1*" (rice actin promoter); "*nos*" (nopaline synthase gene) or "*ScBV*" (sugarcane bacilliform virus) (Hu et al., 2003) were also used with much less frequency.

A great challenge will be to identify specific promoters that would direct the expression of genes in a tissue-specific manner. This can be used not only with reporter genes in studies to optimize the *Agrobacterium*-meditated transformation protocols but also with agronomical importance genes, such as quality improvement, disease resistance or drought tolerance.

2.2.4 Reporter genes

Three reporter marker genes have been used to establish expression and/or integration of foreign DNA into wheat material (See Table 1).

The most usual one is *gusA* (*uidA*) gene encoding the enzyme β-glucuronidase (GUS); although *gfp* (green fluorescent protein) gene, (McCormac et al., 1998; Weir et al., 2001; Hensel et al., 2009) and *Lc/C1* (anthocyanin-biosynthesis regulatory) genes, that results in the accumulation of anthocyanin so creating the "red cell" phenotype (McCormac et al., 1998; Zale et al., 2009), were also used.

2.2.5 Selectable and interest genes

Antibiotic and herbicide resistance is by far the most widely used selection system in *Agrobacterium*-mediated transformation of wheat (See Table 1). As the selectable marker gene, the most common one is "*nptII*" (neomycin phosphotransferase II) gene (Table 2), which confers resistance to kanamycin antibiotic, although "hpt" (hygromycin phosphotransferase) gene conferring hygromycin B resistance has been recently employed (Zale et al., 2009; Rashid et al., 2010), which may be due to cereals being more sensitive to hygromycin B than to kanamycin (Janakiraman et al., 2002 and references therein).

Selectable marker gene	Encoded enzyme	Selective agent	Mode of action
nptII	neomycin phosphotransferase II	Aminoglycoside antibiotics: -kanamycin -neomycin -hygromycin - G418 (geneticin) - paromomycin	Binds 30S ribosomal subunit, inhibits translation
hpt	hygromycin phosphotransferease	Aminoglycoside antibiotics: -hygromycin	Binds 30S ribosomal subunit, inhibits translation
bar (pat)	phosphinothricin acetyl transferase	Herbicides: -phosphinothricin (PPT) -glufosinate ammonium -bialaphos (tripeptide antibiotic)	Inhibits glutamine synthase
aroA:CP4	5-Enolpyruvylshikimate-3-phosphate synthase	Herbicides: -glyphosate	Inhibits aromatic acid biosynthesis (EPSPS)

Table 2. Selectable marker genes most commonly used in wheat *Agrobacterium*-mediated transformation.

The other most popular selectable gene is "*bar*" (also called "pat", phosphinothricin acetyl transferase) gene that confers herbicide resistance to phosphinothricin (PPT) and glufosinate ammonium, the active ingredient being the herbicide Basta® by Hoechst AG and Liberty by AgroEvo®, respectively (Table 2; Rasco-Gaunt et al., 2001). Also, other resistance marker genes for wheat transgenic plants selection have been described (Table 2), such as" *aroA:CP4*" (5-enolpyruvylshikimate-3-phosphate synthase) gene that confers tolerance to glyphosate, the active ingredient of the RoundupReady® herbicide (Zhou et al., 2003; Hu et al., 2003).

2.3 Transformation conditions

Many variables have been pinpointed, and extensively reviewed (Janakiraman et al., 2002; Sahrawat et al., 2003; Bhalla et al., 2006; Jones, 2005), as the key factors in the *Agrobacterium*-mediated transformation process of wheat. Here, those variables are listed in Table 3 under heading that describe the factor, the type or stage studied, the range tested and the optimal value proposed for the highest transformation efficiency together with the main references related. Latter on those data are discussed step by step and we divided the *Agrobacterium*-mediated transformation protocol in four separates stages: preculture, inoculation, coculture and selection.

Factors	Type	Range tested / Higher efficiency	Some references
Time	Pre-culture	From 4 to 21 days. Optimal conditions varied among source explants	Haliloglu & Baenziger, 2003; Weir et al., 2001; Ding et al., 2009; Amoah et al., 2001
	Inoculation	From 30 min to 12 h. Optimal conditions at 30 min and 3 h.	Yang et al., 2008; Wu et al., 2003; Ding et al., 2009
	Coculture	From 1 to 5 days. Optimal conditions at 3 days.	Wu et al., 2003; Uze et al., 2000
Temperature	Inoculation	From 22 to 28 °C. Optimal condition at 24-25°C	Wu et al., 2003; Wu et al., 2008; Mitic et al., 2004
	Coculture	From 21 to 27°C. Optimal condition at 24-25°C.	Amoah et at., 2001; Weir et al., 2001; Khanna & Daggard, 2003; Xue et al., 2004; Wu et al., 2008
Auxins	Picloram	From 1 to 10 mg/L. Optimal conditions around 2- 2.2 mg/L	Weir et al., 2001; Ding et al., 2009; He et al., 2010; Jones et al., 2005
	2,4 D	From 0,5 to 10 mg/L. Optimal conditions at 0,5 and 2 mg/L.	Cheng et. al, 1997; Hu et al., 2003; Razzaq et al., 2011
Surfactans	Pluronic F68	From 0.01 to 0.05 %. Optimal conditions at 0.02%	Cheng et al., 1997; Cheng et al., 2003; Khanna & Daggard, 2003; Zhou et al., 2003
	Silwet L-77	From 0.001 to 0.5 %. Optimal conditions at 0.01-0.02%.	Cheng et al., 1997; Wu et al., 2003; Zale et al., 2009; Haliloglu & Baenziger, 2003
Sugars	Maltose	From 40 to 80 g/L. Optimal conditions at 40	He et al., 2010
	Glucose	From 10 to 36 g/L. Optimal conditions at 10-20 g/L.	Cheng et al., 1997; Khanna & Daggard, 2003
Optical Density		From 0.5 to 2 Optimal conditions at 0.6	Sarker & Biswas, 2002; Amoah et al., 2001; Ke et al., 2002; Haliloglu & Baenziger, 2003; Bi et al., 2006
Phenolic inducers	Acetosyringone	From 100 to 400 µM. Optimal conditions at 100-200 µM.	Cheng et al., 1997; McCormac et al., 1998; Amoah et al., 2001; Wu et al., 2003; Patnaik et al., 2006; He et al., 2010
Salt strength		From 0.1 to 2. Optimal conditions at 0.1 – 1 MS salts strength	Cheng et al., 1997; Ding et al., 2009

Table 3. Summary of current published data on main factors with positive effect on wheat *Agrobacterium*-mediated transformation efficiency.

2.3.1 Preculture

Most reports on *Agrobacterium*-mediated transformation include a first stage called "preculture" to increase the transformation efficiency. For example, survival rate was higher in explants precultured before inoculation than in freshly isolated explants (Cheng et al., 1997). Moreover, Uze et al. (2000) reported the highest T-DNA delivery ratio, based on transient GUS assay, of immature wheat embryos "Bobwhite" when precultured during 10 days; Amoah et al. (2001) found that inflorescence tissue precultured during 21d had the highest GUS activity and finally, Ding et al. (2009) obtained the best transformation rate when mature embryos were precultured for 14 days. However, other authors (Jones et al., 2005) described a successful protocol without pre-culture period or special inoculation treatments.

Some plant growth regulators, such as synthetic auxins picloram (4-amino-3, 5, 6-trichloropicolinic acid) and 2,4-D (2,4-dichlorophenoxyacetic acid), are commonly added to the preculture medium to increase regeneration and the recovery of transgenic explants. Przetakiewicz et al. (2004) demonstrated the promotion effect of 2,4-D for obtaining a higher number of transgenic plants than picloram, whereas, picloram promotes a higher regeneration frequency than 2, 4-D in other report (Ding et al., 2009). Taken into account those results, picloram and 2,4-D or both together have been widely employed in wheat transformation via *Agrobacterium* (Table 3).

2.3.2 Inoculation

The second step of any *Agrobacterium* mediated process is the inoculation of wheat explants in an *Agrobacterium* suspension during a quite variable period of time: 30 minutes to 12 hours (see references in Table 3) and several factors have been proposed as key for inoculation such as included as the most important inoculation stage such as: time, temperature, media strength or *Agrobacterium* optical density as well as some inducers of stable transformation, such as acetosyringone, sugars, auxins or surfactans.

Several authors (Amoah et al.; 2001; Yang et al., 2008) have described a direct relationship between increase of inoculation time and decrease in transformation efficiency after 2-3 h and there is a general consensus that the optimal time of inoculation for T-DNA delivery (Jones et al., 2005; Wu et al., 2008; Ding et al., 2009) should be around 3 h.

Although in the literature reviewed (Table 3), a wide range of inoculation temperatures have been tested: 22 – 28°C (Peters et al., 1999; Cheng et al., 2003; Mitic et al., 2004; Supartana et al., 2006) however, no clue on the optimal ones or significant differences has been clearly reported. Moreover, most reports do not indicate the inoculation temperature and it is assumed that room temperature has been applied (c.a. 25°C).

The use of surfactants and phenolic inducers in the media were widely assessed by different researchers (Table 3). Surfactants, like pluronic acid F68 and Silwet L-77, were first studied by Cheng et al. (1997) finding that either Silwet or pluronic enhance transient GUS expression, especially on the immature embryos because it is believed that the surface-tension-free cells favour the *A. tumefaciens* attachment. Several studies reported an optimal concentration for Silwet around 0.01% (Wu et al., 2003; Jones et al., 2005) and for pluronic around 0.02% (Cheng et., 1997). On the contrary, other authors (Haliloglu & Baenziger, 2003) have described that the presence of a surfactant in the inoculum medium makes no difference in terms of T-DNA delivery efficiency, even when concentrations as higher as 0.05% of Silwet have been used.

Acetosyringone was always pointed out to be the key factor in T-DNA delivery in a range of concentration from 100 to 400 µM (McCormac et al., 1998; Xue et al., 2004; He et al., 2010). Its presence, at 200 µM concentration, clearly increased transformation efficiency (Wu et al., 2003; Amoah et al., 2001).

The addition of some sugars, like maltose or glucose to the inoculation medium was essential to achieve efficient T-DNA delivery; in fact T-DNA delivery efficiency was significant reduced in the freshly isolated immature embryos when acetosyringone and glucose were absent in the inoculation media (Cheng et al., 1997, Wu et al., 2003).

Agrobacterium optical cell density at 600 nm around 0.5-0.6 (Cheng et al., 2003; Haliloglu & Baenziger, 2003; Bi et al., 2006); close to 1.0 (Khanna & Daggard, 2003; Jones et al., 2005) or even higher, such as 1.3 (Amoah et al., 2001) during inoculation were found to be crucial for transformation efficiency. However when *Agrobacterium* is inoculated at high density or when is cocultured with the explant at high temperatures or for long period conditions an overgrowth can occurs promoting the death of the explants. Several antibiotics can be used after coculture and the selection stage to control *Agrobacterium* overgrowth or to eliminate it completely, such as timentin (Hensel et al., 2009, Wu et al., 2009), carbenicillin (Cheng et al., 1997) and cefotaxime (Bi et al 2006, Chugh & Khurana, 2003).

2.3.3 Coculture

The third stage of any wheat *Agrobacterium*-tumefaciens transformation protocol starts, after the removal of excess of bacteria from the previous stage, when the explants are cocultivated for a period of 1-5 days (Table 3) in dark conditions at 23 -27°C. Again, during this period virulence inductors such as acetosyringone, osmoprotectors such as proline, carbon sources such as sugars, and plant growth regulators are added to the medium

Several studies have focused on time, temperature and media composition variables as important factors, during cocultivation stage, to transform wheat successfully. For example, Wu et al. (2003) found that a long cocultivation time (5d) promoted a reduction on the capacity of the transformed immature embryos to form embryogenic callus and regenerate when cocultivation was assessed for 1–5 days. Short periods (2-3 days) have been proposed as optimum for high transformation efficiency (Cheng et al., 1997; Amoah et al., 2001; Wu et al., 2003; Ding et al., 2009).

Also, the temperature during the cocultivation period could play an important role. Weir and coworkers (2001) obtained 83.9 and 81.4% of GFP expression at 21 and 24°C, respectively and concluded that transient GFP expression is not significantly affected by co-cultivation temperature. Although, an elegant assay demonstrated that coculture at two temperatures (1d at 27°C and 2d at 22°C) reduced the damage to the soft callus tissue due to the common overgrowth of *Agrobacterium* during coculture (Khanna & Daggard, 2003). More information about it can be found in 2.3.2 section.

As stated previously for inoculation condition, the addition of acetosyringone 200µM is also critical in the coculture media to increase the efficiency on T-DNA delivery (Cheng et al., 1998; Wu et al., 2003).

Finally, it has been described (Table 3) that the salt strength in both, the inoculation and co-culture media, had a significant influence on the T-DNA delivery. For example, transient GUS expression was higher on freshly isolated immature embryos when one tenth-strength MS salts were used than the full-strength MS salts (Cheng et al., 1997). Several medium strength 2x, 1x, 0.5x, and 0.1x media concentration were also assessed elsewhere (Khanna & Daggard, 2003) but no main conclusion has been drawn and MS media 1x has been generally employed in *Agrobacterium* mediated transformation of wheat (Weir et al., 2001; Ke et al., 2002; Sarker & Biswas, 2002; Wu et al., 2003; Patnaik et al., 2006; Ding et al., 2009)

2.3.4 Selection

Due to the most common selectable marker genes being *nptII, hpt* and *bar*, the most widely selected agents, to discriminate transformed explants , and not to transform explants, were kanamicyne, hygromycin and phosphinothricin (PPT) and their analogues G418 (geneticin) and paromomycin for *nptII* gen and Bialaphos when *bar* gene was used as selectable marker gene.

3. *Agrobacterium*-mediated genetic transformation: Time to model

As described in the previous section, plant genetic transformation is a really complex process to understand and, subsequently, to optimize. The reason behind this is the important number of variables (factors) involved in the whole process (plasmid or *Agrobacterium* strain, type of plant explant, preculture, inoculation, coculture and selection conditions, etc) together with the different scales of biological organization concerned (molecular, genetic, cellular, physiological and whole plant). Moreover, different kinds of data are generated in those studies: binary data (transformed- non transformed; alive-dead); discrete or categorical (number of GUS spots); continuous (length, weight, ...); image data (GUS or GFP) or even fuzzy data (callus colour: brown, brownish, yellowish and so on).

Traditionally, the effect of those variables on genetic transformation studies and particularly, wheat *Agrobacterium*-mediated transformation, is determined by analysis of variance (ANOVA). According to statistical theory (Mize et al., 1999), only continuous data normally or approximately normally distributed should be analysed with ANOVA. Discrete and binomial data should be analysed using Poisson and logistic regression, respectively. This type of methodology makes, the analysis of the results complicated and specialized, the biologist often being helped by statisticians. Finally, although statistics can be used for making predictions, normally this feature is not used in plant transformation studies.

Because of these limitations, plant genetic transformation studies include, usually, a small number of variables at the same time. Often, one variable at a time is studied; for example to study the effect of a variable (eg. effect of acetosyringone) on a selected response (eg. GUS transient expression), the experiments are performed at different concentrations (0, 100, 200 and 300 μM) keeping the rest of the variables constant. This "one-factor at a time" procedure is time consuming and has clear limitations when the best conditions for *Agrobacterium*-meditated transformation of wheat need to be achieved. The main limitation is that this procedure ignores the possible interactions between variables (the addition of acetosyringone can have a positive or negative interaction with any other variable kept constant during a particular experiment).

Finally, this kind of methodology enables the researcher to select the best combination of factors between the performed experiments and not to predict the best possible combination of factors or, in other words, to optimize the whole procedure.

The *Agrobacterium*-mediated transformation process is difficult to describe accurately by a simple stepwise algorithm or a precise formula and require a network (multivariable) approach using computational models. For developing a model several steps need to be followed: first, a clear identification of the process (including all kind of variables/factors) to be simulated, controlled and/or optimized; secondly, the selection of variables, and the definition of what the model is for; thirdly, the creation of the database with the most accurate and precise data of each variable and the selection of the type of model and finally, the model validation, to check if the distances between the observed and predicted data is low enough (Gallego et al., 2011).

To establish the key factors affecting the quality of an *Agrobacterium*-mediated transformation process an Ishikawa diagram can be developed (Fig. 1) using data from literature (Tables 1, 2 and 3). This cause-effect diagram helps in identifying the potential relationships among several factors, and provides an insight into the whole process. The main factors (causes) can be selected and grouped into major categories such as plant material, *Agrobacterium*, transformation conditions and selection conditions.

Initially both *Agrobacterium* characteristics (strain, plasmid, extra virulence gene, promoters, reporter and selectable marker gene) and plant material (genus and species, variety/cultivar/line and type of explant) should be defined. Within the transformation conditions (preculture, inoculation and coculture) several variables as process conditions (temperature and time); chemical properties as media composition (type, strength, vitamins, sugars, plant growth regulator (PGR) such as synthetic auxins) and/or transformation inductors (acetorysingone and surfactans) should be considered and interrelated. Finally, selection conditions (antibiotics and/or herbicides) need to be established.

From this diagram, it can be deduce that there are an enormous amount of variables involved in the transformation process. Moreover, variables of different types: numerical data (temperature, time, etc.) or nominal (strain, explant, etc.) should be considered. Once the key or main variables (inputs) are identified, their effects over the defined parameters (outputs) should be studied by the appropriate experimental design or model.

Different models and/or networks have been used to integrate all kind of biological components (Yuan et al., 2008). Both networks and model have become more and more accurate (and better at predicting outcomes of the complex biological process) by using new experimental and modelling tools (Giersch, 2000). Recent studies have pointed out the effectiveness of different artificial intelligence technologies, such as artificial neural networks (Gago et al., 2010a, 2010b, 2010c) combined with genetic algorithms and neurofuzzy logic (Gago et al., 2010d; 2011) in modelling and optimizing the complex plant biology process (Gallego et al., 2011).

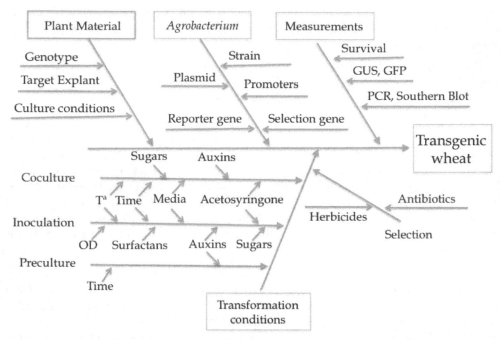

Fig. 1. Ishikawa diagram identifying the potential key variables of a wheat *Agrobacterium*-mediated transformation process.

4. Artificial Intelligence: A novel approach to model, understand and optimize cereals genetic transformation

Artificial intelligence approaches are based on the use of computational systems that simulate biological neural networks. They have been used not only for many industrial and commercial purposes since the 1950s (Russell & Norvig, 2003) but they have also been applied to fields more often related to biology, such as agricultural, ecological and environmental sciences (Jimenez et al., 2008; Huang, 2009). More detailed information about these technologies (Rowe & Roberts, 2005), and their applications to plant biology (Prasad & Dutta Gupta, 2008; Gallego et al., 2011) can be found elsewhere. Herein, we will briefly describe some relevant aspects of three of those technologies: Artificial Neural Networks (ANNs), genetic algorithms and neurofuzzy logic, which have been employed in plant science for modelling and optimizing different processes, in order to facilitate the understanding of its future applicability in cereal genetic transformation studies.

4.1 Artificial neural networks

Artificial Neural Networks (ANNs) are computational systems inspired in the biological neural systems. Information arrives to biological neurons through the dendrites. The neuron soma processes the information and passes it on via axon (Figure 2). In a similar way, ANNs use the processing elements called "artificial neurons", "single nodes" or

Agrobacterium-Mediated Transformation of Wheat: General Overview and New Approaches to Model and
Identify the Key Factors Involved

81

"perceptrons", that is, simple mathematical models (functions). Every perceptron receives information (inputs) from "neighbouring" nodes, then processes the information (either positive or negative) by multiplying each input by their associated weight (it is a measure of the strengths of the connection between perceptrons) giving a new result, which is adjusted by a previously assigned internal threshold (to simulate the output action), and produces an output to be transmitted to the next node. The perceptrons are organized into groups called layers. By connecting millions of perceptrons complex artificial neural networks can be achieved. The most used network architecture is called "multilayer perceptron" and consists in three simple layers: input, hidden and output layer (Rowe & Roberts, 2005).

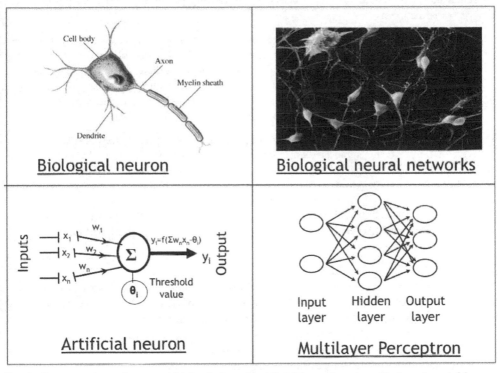

Fig. 2. Comparative schemes of biological and artificial neural system. X= input variable; W= weight of i_n input; θ= internal threshold value; f=transfer function.

Advantageously, while most conventional computer programs are explicitly programmed for each process, ANNs are able to learn, using algorithms designed to optimize the strength of the connections in the networks. For the network to learn it is necessary to use an example dataset (a collection of inputs and related outputs). Between 60 and 80% of the total data are chosen randomly, to perform the "training". In this process ANNs are able to search for a set of weight values that minimize the squared error between the data predicted by the model and the experimental data in the output layer. Furthermore, almost all the rest of the data set (10-20%) is used to "test" the model. Performance and predictability of the model can be demonstrated by statistical parameters like the correlation coefficient (R^2) and the f value of the ANOVA of the model. Values of both training and test sets over 75% and f values over the f critical value for the corresponding degrees of freedom are indicative of high predictability and good performance (Colbourn & Rowe, 2005; Shao et al., 2006). Validation of the model can be performed by using a set of unseen data (validation data set) After a validation of the model, the ANNs is able to quickly predict accurately the output for a specific never tested combination of inputs or, in other words to answer "what if" questions, saving costs and time. Predictions using ANNs technology have been demonstrated to be more accurate than ones derived from experimental design and traditional statistic methods (Landín et al., 2009; Gago et al., 2010a). In conclusion, the ANNs approach could be useful to data processing, modeling, predicting and optimizing wheat genetic transformation.

ANNs have also some limitations related to the difficulties of interpreting the results when large data sets are used (several inputs and outputs are fitted in the model) and a large number of 2D surface plots or even 3 D graphs are generated by the model. In this case, ANNs can be coupled with other artificial intelligence technologies, such as genetic algorithms or fuzzy logic, creating hybrid systems that help to handle complex models and/or to data mining (Colbourn, 2003).

Sometimes the objective of modelling a specific process is not to predict new results (outputs), such as, when wheat *Agrobacterium*-mediated transformation is used to estimate the transformation efficiency when more amount of acetosyringone is added in the coculture stage. Probably for most researchers the main question could be "how to get" the maximum transformation efficiency, and more generally in those cases the objective is to find the combination of inputs that will provide the "optimum/best/highest" output in other words: optimize the process. This can be achieved combining ANNs and genetic algorithms.

4.2 Generic algorithms

Genetic algorithms (GA) are also a bio-inspired artificial intelligence tool, specially design to select the best solution of a specific problem (optimization). They are based on the biological principles of genetic variation and natural selection (mutation, crossover, selection or inheritance), mimicking the basic ideas of evolution over generations. In a simple way: when combined with ANNs, the genetic algorithms randomly generate a set of inputs and their corresponding predicted outputs using the ANNs model, called "set of candidate solutions" to the problem. Candidate solutions are then selected according to their fitness to previous established criteria; the best ones are used for evolving new solution populations to the problem, using crossover and mutation. After few generations the optimum should be reached because the most suitable candidates have more chance of being reproduced. Using this approach, complex micropropagation processes have been modelled by ANNs and successfully optimized by genetic algorithms (Gago et al., 2010a, 2010b).

4.3 Neurofuzzy logic

Neurofuzzy logic is a hybrid system technology that combines the adaptive learning capabilities from ANNs with the generality of representation from fuzzy logic (Shao et al.,

2006). Fuzzy logic is also an artificial intelligence tool especially useful in problem solving and decisions making, helping with the understanding of the complex cause-effect relationships between variables. When coupled with ANN, it becomes a powerful technique in handling complex models by generating comprehensible and reusable knowledge through simple fuzzy rules: IF (condition) THEN (observed behaviour). This kind of rules facilitates the understanding of a specific process, in a semi-qualitative manner, in a similar way to how people usually analyse the real world (Babuska, 1998; Gallego et al., 2011 and references therein). Many times words are more important for making decisions, drawing conclusions or even solving problems than a collection of accurate data (Fig. 3). Human knowledge is normally built on linguistic tags, and not on quantitative mathematical data, even though sometimes words are imprecise or uncertain.

Fig. 3. Precision versus significance in the real world of researchers in the plant genetic transformation field.

The major capabilities of fuzzy logic are the flexibility, the tolerance with uncertainty and vagueness and the possibility of modelling non linear functions, searching for consistent patterns or systemic relationships between variables in a complex dataset, data mining and promoting deep understanding of the processes studied by generating comprehensible and reusable knowledge in an explicitly format (Setnes et al., 1998; Shao et al., 2006; Yuan et al., 2008). The neurofuzzy logic approach has been recently applied in modelling plant processes, such as *in vitro* direct rooting and acclimatization of grapevine (Gago et al., 2010d) or to gather knowledge of media formulation using data mining in apricot (Gago et al., 2011). In those cases, the authors found higher accuracy in identifying the interaction effects among variables of neurofuzzy logic than the traditional statistical analysis.

Moreover, neurofuzzy logic showed a considerable potential for data mining and retrieved knowledge from very large and highly complex databases.

5. Future perspectives

Agrobacterium-mediated transformation of wheat is a complex process although can be understood easily. It involves different scales of biological organization (genetic, biochemical, physiological, etc.) and many factors that influence the process. The storm of information generated by the analysis carried out during those processes would be useless if they could not be analysed together. Nowadays, artificial intelligence technologies give us the opportunity to handle a huge amount of biological data generated during the transformation process, with many advantages over traditional statistics. Artificial Intelligence technologies can solve common problems plant researchers associate to analysing, integrating variable information, extracting knowledge from data and predicting what will happen in a specific situation.

Different artificial intelligence approaches could be used for modeling, understanding and optimizing any *Agrobacterium*-mediated transformation procedure, either for wheat, cereals, fruit trees or any other biological process, giving results at least as good as, and less time consuming, those obtained by traditional statistics . More specifically, ANNs combined with genetic algorithms could predict the combination of variables (inputs) that would yield quality transformed wheat plants.

As a starting point a database can be obtained from historical results in the literature that can be modelled to find the more important variables affecting the *Agrobacterium*-mediated transformation procedure (data mining). On this knowledge, new experiments can be designed and performed and their results added to the database to fulfil the optimization processes (Gago et al., 2010a, 2011).

Great efforts have been made to improve the *Agrobacterium*-mediated transformation process, although the its full optimization is still far from being reached. In the future the application of modelling tools, such as those described here, could add a new insights into discovering the interactions between the variables tested and into understanding the regulatory process controlling molecular, cellular, biochemical, physiological and even developmental processes occurring during wheat *Agrobacterium*-mediated transformation.

6. Acknowledgments

We also want to thank Ms. J. Menis for her help in the correction of the English version of the work. This work was supported by Regional Government of Xunta de Galicia: exp.2007/097 and PGIDIT02BTF30102PR. PPG and ML thanks to Minister of Education of Spain for funding the sabbatical year at Faculty of Science, University of Utrecht, Netherlands.

7. References

Ahmad, A.; Zhong, H.; Wang, W.L. & Sticklen, M.B. (2002). Shoot apical meristem: *in vitro* regeneration and morphogenesis in wheat (*Triticum aestivum* L). *In vitro Cellular & Developmental Biology-Plant*, 38, 163–167.

Agrobacterium-Mediated Transformation of Wheat: General Overview and New Approaches to Model and
Identify the Key Factors Involved

85

Amoah, B.K.; Wu, H.; Sparks, C. & Jones, H.D. (2001). Factors influencing *Agrobacterium*-mediated transient expression of *uidA* in wheat inflorescence tissue. *Journal of Experimental Botany*, 52, 1135-1142.

Atchison, J.; Head, L.; Gates, A. (2010) Wheat as food, wheat as industrial substance; comparative geographies of transformation and mobility. *Geoforum*, 41, 236-246.

Babuska, R. (1998). Fuzzy modelling for control, In: *International Series in Intelligent Technologies*, Babuska R. (Ed.), 1-8, Kluwer Academic Publishers, Massachussets, USA.

Bhalla, P.L.; Ottenhof, H.H. & Singh, M.B. (2006). Wheat transformation–an update of recent progress. *Euphytica*, 149, 353-366.

Bi, R.M.; Jia, H.Y.; Feng, D.S. & Wang, H.G. (2006). Production and analysis of transgenic wheat (*Triticum aestivum* L.) with improved insect resistance by the introduction of cowpea trypsin inhibitor gene. *Euphytica*, 151, 351-360.

Cheng, M.; Fry, J.E.; Pang, S.; Zhou, H.; Hironaka, C.M.; Duncan, D.R.; Conner, T.W. & Wan, Y. (1997). Genetic transformation of wheat mediated by *Agrobacterium tumefaciens*. *Plant Physiology*, 115, 971-980.

Cheng, M.; Hu, T.; Layton, J.; Liu, C.N. & Fry, J.E. (2003). Desiccation of plant tissues post-*Agrobacterium* infection enhances T-DNA delivery and increases stable transformation efficiency in wheat. *In Vitro Cellular & Developmental Biology-Plant*, 39, 595-604.

Cheng, M.; Lowe, B.A.; Spencer, T.M.; Ye, X. & Armstrong, C.L. (2004). Factors influencing *Agrobacterium*-mediated transformation of monocotyledonous species. *In Vitro Cellular & Developmental Biology-Plant*, 40, 31-45.

Chugh, A. & Khurana, P. (2003). Herbicide-resistant transgenics of bread wheat (*T. aestivum*) and emmer wheat (*T. dicoccum*) by particle bombardment and *Agrobacterium* mediated approaches. *Current Science*, 84, 78-83.

Colbourn, E. (2003). Neural computing: enable intelligent formulations. *Pharmaceutical Technology Supplement*, 16-20.

Colbourn, E. & Rowe, R.C. (2005). Neural computing and pharmaceutical formulation. In: *Encyclopaedia of pharmaceutical technology*, Swarbrick, J. & Boylan, J.C. (Eds), Marcel Dekker, New York.

Ding, L.; Li, S.; Gao, J.; Wang, Y.; Yang, G. & He, G. (2009). Optimization of *Agrobacterium*-mediated transformation conditions in mature embryos of elite wheat. *Molecular Biology Reports*, 36, 29-36.

Fellers, J.P.; Guenzi, A.C. & Taliaferro, C.M. (1995) Factors affecting the establishment and maintenance of embryogenic callus and suspension cultures of wheat (*Triticum aestivum* L.). *Plant Cell Reports*, 15, 232-237.

Food and Agriculture Organization (2003). *World agriculture: towards 2015/2030: an FAO perspective*, Earthscan Publications Ltd, Longon, UK.

Food and Agriculture Organization (2011). Global cereal supply and demand brief. *Crop Prospects and Food Situation*, 1,4-5.

Gago, J.; Landín, M. & Gallego, P.P. (2010b). Artificial neural networks modeling the *in vitro* rhizogenesis and acclimatization of *Vitis vinifera* L. *Journal of Plant Physiology*, 167, 1226-1231.

Gago, J.; Landín, M. & Gallego, P.P. (2010c). Strengths of artificial neural networks in modelling complex plant processes. *Plant Signaling and Behavior* 5, 6, 1-3.

Gago, J.; Landín, M. & Gallego, P.P. (2010d). A neurofuzzy logic approach for modeling plant processes: a practical case of in vitro direct rooting and acclimatization of Vitis vinifera L. Plant Science, 179, 241-249.

Gago, J.; Martínez-Núñez, L.; Landín, M. & Gallego, P.P. (2010a). Artificial neural networks as an alternative to the traditional statistical methodology in plant research. Journal of Plant Physiology, 167, 23-27.

Gago, J.; Pérez-Tornero, O.; Landín, M.; Burgos, L. & Gallego, P.P. (2011) Improving knowledge on plant tissue culture and media formulation by neurofuzzy logic: a practical case of data mining using apricot databases. Journal Plant Physiology, 168, 1858-1865.

Gallego, P.P.; Landín, M. & Gago, J. (2011). Artiticial neural networks technology to model and predict plant biology process. In: Artificial Neural Networks- Methodological Advances and Biomedical Applications. Suzuki, K. (Ed), 197-216, Intech Open Access Publisher: Croatia.

Giersch, C. (2000) Mathematical modeling of metabolism. Current Opinion in Plant Biology, 3, 249-253.

Haliloglu, K. & Baenziger, P.S. (2003). Agrobacterium tumefaciens-mediated wheat transformation. Cereal Research Communications, 31, 9-16.

He, D. G.; Yang, Y. M.; Dahler, G.; Scott, K. J. (1988). A comparison of epiblast callus and scutellum callus induction in wheat. The effect of embryo age, genotype and medium. Plant Science, 57, 225–233.

He, Y.; Jones, H.D.; Chen, S.; Chen, X.M.; Wang, D.W.; Li, K.X.; Wang, D.S. & Xia, L.Q. (2010). Agrobacterium-mediated transformation of durum wheat (Triticum turgidum L. Var. Durum Cv Stewart) with improved efficiency. Journal of Experimental Botany, 61, 1567-1581.

Hellens, R.; Mullineaux, P. & Klee, H. (2000). A guide to Agrobacterium binary Ti vectors. Trends in Plant Science, 5, 446-451.

Hensel, G.; Kastner, C.; Oleszczuk, S.; Riechen, J. & Kumlehn, J. (2009). Agrobacterium-mediated gene transfer to cereal crop plants: current protocols for barley, wheat, triticale, and maize. International Journal of Plant Genomics, 835608, 1-9.

Hess, D.; Dressler, K. & Nimmrichter, R. (1990). Transformation experiments by pipetting Agrobacterium into the spikelets of wheat (Triticum aestivum L.). Plant Science, 72, 233-244.

Hu, T.; Metz, S.; Chay, C.; Zhou, H.; Biest, N.; Chen, G.; Cheng, M.; Feng, X.; Radionenko, M. & Lu, F. (2003). Agrobacterium-mediated large-scale transformation of wheat (Triticum Aestivum L.) using glyphosate selection. Plant Cell Reports, 21, 1010-1019.

Huang, Y. (2009). Advances in artificial neural networks - methodological development and application. Algorithms, 2, 973-1007.

Janakiraman, V.; Steinau, M.; McCoy, S.B. & Trick, H.N. (2002) Recent advances in wheat transformation. In Vitro Cellular & Developmental Biology-Plant, 38, 404-414.

Jiménez, D.; Pérez-Uribe, A.; Satizábal, H.; Barreto, M.; Van Damme, P. & Marco, T. (2008). A survey of artificial neural network-based modeling in agroecology. In: Soft Computing applications in industry, STUDFUZZ. B. Prasad (Ed), 247-269 Springer-Verlag, Berlin-Heidelberg Germany.

Jones, H.D. (2005). Wheat transformation: current technology and applications to grain development and composition. Journal of Cereal Science, 41, 137-147.

Agrobacterium-Mediated Transformation of Wheat: General Overview and New Approaches to Model and Identify the Key Factors Involved

87

Jones, H.D.; Doherty, A. & Wu, H. (2005). Review of methodologies and a protocol for the *Agrobacterium*-mediated transformation of wheat. *Plant methods*, 1, 5.

Jones, H.D.; Wilkinson, M.; Doherty, A. & Wu, H. (2007). High Throughput *Agrobacterium* transformation of wheat: a tool for functional genomics, In: *Wheat production in stressed environments. Proceedings of the 7th International Wheat Conference, 27 November - 2 December 2005, Mar Del Plata, Argentina*, H.T. Buck, J.E. Nisi & N. Salomón. (Ed.), 693-699, Springer, Netherlands (UE).

Ke, X.Y.; McCormac, A.C.; Harvey, A.; Lonsdale, D.; Chen, D.F. & Elliott, M.C. (2002). Manipulation of discriminatory T-DNA delivery by *Agrobacterium* into cells of immature embryos of barley and wheat. *Euphytica*, 126, 333-343.

Khanna, H. & Daggard, G. (2003). *Agrobacterium tumefaciens*-mediated transformation of wheat using a superbinary vector and a polyamine-supplemented regeneration medium. *Plant Cell Reports*, 21, 429-436.

Khurana, P.; Chauhan, H. & Desai, S.A. (2008). Wheat. In: *Compendium of transgenic crops: transgenic cereals and forage grasses*. Kole C. & Hall T.C. (Eds), pp. 83-100, Blachwell Publishing Ltd, New Delhi, India.

Komari, T. (1990) Transformation of cultured-cells of *Chenopodium quinoa* by binary vectors that carry a fragment of DNA from the virulence region of PTiBo542. *Plant Cell Reports*, 9, 303-306.

Kumlehn, J. & Hensel, G. (2009). Genetic transformation technology in the triticeae. *Breeding Science*, 59, 553-560.

Landín, M.; Rowe, R.C. & York, P. (2009). Advantages of neurofuzzy logic against conventional experimental design and statistical analysis in studying and developing direct compression formulations. *European Journal of Pharmaceutical Science*, 38, 325-331.

Langridge, P.; Brettschneider, R.; Lazzeri, P. & Lörz, H. (1992). Transformation of cereals *via Agrobacterium* and the pollen pathway: a critical assessment. *The Plant Journal*, 2, 631-638.

Liu, W.; Zheng, M.Y. & Konzak, C.F. (2002). Improving green plant production via isolated microspore culture in bread wheat (*Triticum aestivum L.*). *Plant Cell Reports*, 20, 821–824.

McCormac, A.C.; Wu, H.; Bao, M.; Wang, Y.; Xu, R.; Elliott, M.C. & Chen, D.F. (1998). The use of visual marker genes as cell-specific reporters of *Agrobacterium*-mediated T-DNA delivery to wheat (*Triticum aestivum L.*) and barley (*Hordeum vulgare L.*). *Euphytica*, 99, 17-25.

Mitić, N.; Nikolić, R.; Ninković, S.; Miljuš-Djukić, J. & Nešković, M. (2004). *Agrobacterium*-mediated transformation and plant regeneration of *Triticum Aestivum L. Biologia Plantarum*, 48, 179-184.

Mize, C.W.; Kenneth, J.K. & Compton, M.E. (1999). Statistical considerations for in vitro research: II – data to presentation. *In Vitro Cell & Developmental Biology – Plant*, 35, 122-6.

Mooney, P.A.; Goodwin, P.B.; Dennis, E.S. & Llewellyn, D.J. (1991). *Agrobacterium tumefaciens*-gene transfer into wheat tissues. *Plant Cell Tissue & Organ Culture*, 25, 209-218.

Opabode, J.T. (2006). *Agrobacterium*-mediated transformation of plants: emerging factors that influence efficiency. *Biotechnology and Molecular Biology Review*, 1, 12-20.

Patnaik, D.; Vishnudasan, D. & Khurana, P. (2006). Agrobacterium-mediated transformation of mature embryos of *Triticum aestivum* and *Triticum durum*. *Current science*, 91, 307-317.

Peters, N.R.; Ackerman, S. & Davis, E.A. (1999). A modular vector for *Agrobacterium*-mediated transformation of wheat. *Plant Molecular Biology Reporter*, 17, 323-331.

Potrykus, I. (1990). Gene transfer to cereals: as assessment. *Biotechnology*, 8, 535-542.

Potrykus, I. (1991). Gene transfer to plants: assessment of published approaches and results. *Annual Review of Plant Physiology and Plant Molecular Biology (Annual Review of Plant Biology since 2002)*, 42, 205-225.

Prasad, VSS. & Dutta Gupta, S. (2008). Applications and potentials of artificial neural networks in plant tissue culture. In: *Plant Tissue Culture Engineering*, Gupta, S.D. & Ibaraki, Y. (Eds). Springer-Verlag: Berlin, Germany pp 47-67.

Przetakiewicz, A., Orczyk, W. and Nadolska-Orczyk, A. (2003). The effect of auxin on plant regeneration of wheat, barley and triticale. *Plant Cell, Tissue and Organ Culture*, 73, 245-256.

Przetakiewicz, A.; Karas, A.; Orczyk, W. & Nadolska-Orczyk, A. (2004). Agrobacterium-mediated transformation of polyploid cereals. The efficiency of selection and transgene expression in wheat. *Cell & Molecular Biology Letters*, 9, 903-917.

Rasco-Gaunt, S.; Riley, A.; Cannell, M.; Barcelo, P. & Lazzeri, P.A. (2001). Procedures allowing the transformation of a range of European elite wheat (*Triticum aestivum* L.) varieties via particle bombardment. *Journal of Experimental Botany*, 52, 865–874.

Rashid, H.; Afzal, A.; Khan, M.H.; Chaudhry, Z. & Malik, S.A. (2010). Effect of bacterial culture density and acetosyringone concentration on *Agrobacterium* mediated transformation in wheat. *Pakistan Journal of Botany*, 42, 4183-4189.

Rashid, U.; Ali, S.; Ali, G.M.; Ayub, N. & Masood, M.S. (2009). Establishment of an efficient callus induction and plant regeneration system in pakistani wheat (*Triticum aestivum*) cultivars. *Electronic Journal of Biotechnology*, 12, 4-5.

Razzaq, A.; Hafiz, I.A.; Mahmood, I. & Hussain, A. (2011). Development of *in planta* transformation protocol for wheat. *African Journal of Biotechnology*, 10, 740-750.

Rowe, R.C. & Roberts, R.J. (2005). *Intelligent software for product formulation*, Taylor & Francis, London.

Russell, S.J. & Norvig, P. (2003). *Artificial Intelligence: A Modern Approach* (2nd ed.), , Prentice Hall, Upper Saddle River, NJ, USA.

Sahrawat, A.K.; Becker, D.; Lütticke, S. & Lörz, H. (2003). Genetic improvement of wheat *via* alien gene transfer, an assessment. *Plant Science*, 165, 1147-1168.

Sarker, R.H. & Biswas, A. (2002). *In vitro* plantlet regeneration and *Agrobacterium*-mediated genetic transformation of wheat (*Triticum aestivum* L.). *Plant Tissue Culture*, 12, 155-165.

Sawahel, W.A. & Hassan, A.H. (2002). Generation of transgenic wheat plants producing high levels of the osmoprotectant proline. *Biotechnology Letters*, 24, 721-725.

Sears, R. G. & Deckard, E. L. (1982). Tissue culture variability in wheat: callus induction and plant regeneration. *Crop Science*, 22, 546–550.

Setnes, M.; Babuska, R. & Verbruggen, H.B. (1998). Rule-based modelling: precision and transparency. *IEEE Transactions on Systems Man and Cybernet Part C-Applications and Reviews*, 28, 1, 165-169.

Shao, Q.; Rowe, R.C. & York, P. (2006). Comparison of neurofuzzy logic and neural networks in modelling experimental data of an immediate release tablet formulation. *European Journal of Pharmaceutical Science*, 28, 394-404.

Supartana, P.; Shimizu, T.; Nogawa, M.; Shioiri, H.; Nakajima, T.; Haramoto, N.; Nozue, M. & Kojima, M. (2006). Development of simple and efficient *in planta* transformation method for wheat (*Triticum aestivum* L.) using *Agrobacterium tumefaciens*. *Journal of bioscience and bioengineering*, 102, 162-170.

Trick, H.N. & Finer, J.J. (1997). SAAT: sonication-assisted *Agrobacterium*-mediated transformation. *Transgenic Research*, 6, 329-336.

Uze, M.; Potrykus, I. & Sautter, C. (2000). Factors influencing T-DNA transfer from *Agrobacterium* to precultured immature wheat embryos (*Triticum aestivum* L.). *Cereal Research Communications*, 28, 17-23.

Vasil, I.K. (2007). Molecular genetic improvement of cereals: transgenic wheat (*Triticum aestivum* L.). *Plant Cell Reports*, 26, 1133-1154.

Vishnudasan, D.; Tripathi, M.N.; Rao, U. & Khurana, P. (2005). Assessment of nematode resistance in wheat transgenic plants expressing potato proteinase inhibitor (*pin2*) gene. *Transgenic research*, 14, 665-675.

Wang, C.T. & Wei, Z.M. (2004). Embryogenesis and regeneration of green plantlets from wheat (*Triticum aestivum*) leaf base. *Plant Cell Tissue & Organ Culture*, 77, 149–156.

Wang, Y.L.; Xu, M.X.; Yin, G.X.; Tao, L.L.; Wang, D.W. & Ye, X.G. (2009). Transgenic wheat plants derived from *Agrobacterium*-mediated transformation of mature embryo tissues. *Cereal Research Communications*, 37, 1-12.

Weir, B.; Gu, X.; Wang, M.; Upadhyaya, N.; Elliott, A.R. & Brettell, R.I.S. (2001). *Agrobacterium tumefaciens*-mediated transformation of wheat using suspension cells as a model system and green fluorescent protein as a visual marker. *Functional Plant Biology*, 28, 807-818.

Wu, H.; Doherty, A. & Jones, H.D. (2008). Efficient and rapid *Agrobacterium*-mediated genetic transformation of durum wheat (*Triticum turgidum* L. Var. Durum) using additional virulence genes. *Transgenic research*, 17, 425-436.

Wu, H.; Doherty, A. & Jones, H.D. (2009). *Agrobacterium*-mediated transformation of bread and durum wheat using freshly isolated immature embryos, In: *Transgenic wheat, barley and oats*, H.D. Jones & P.R. Shewry. (Ed.), 93-103, Humana Press, New York (NY), USA.

Wu, H.; Sparks, C.; Amoah, B. & Jones, H. (2003). Factors influencing successful *Agrobacterium*-mediated genetic transformation of wheat. *Plant Cell Reports*, 21, 659-668.

Xia, G.M.; Li, Z.Y.; He, C.X.; Chen, H.M. & Brettell, R. (1999). Transgenic plant regeneration from wheat (*Triticum aestivum* L.) mediated by *Agrobacterium tumefaciens*. *Acta Phytophysiologica Sinica*, 25, 22-28.

Xue, Z.Y.; Zhi, D.Y.; Xue, G.P.; Zhang, H.; Zhao, Y.X. & Xia, G.M. (2004). Enhanced salt tolerance of transgenic wheat (*Triticum aestivum* L.) expressing a vacuolar Na /H antiporter gene with improved grain yields in saline soils in the field and a reduced level of leaf Na. *Plant Science*, 167, 849-859.

Yang, B.; Ding, L.; Yao, L.; He, G. & Wang, Y. (2008). Effect of seedling ages and inoculation durations with *Agrobacterium tumefaciens* on transformation frequency of the wheat wounded apical meristem. *Molecular Plant Breeding*, 6, 358-362.

Yu, Y. & Wei, Z. (2008). Increased oriental armyworm and aphid resistance in transgenic wheat stably expressing *Bacillus thuringiensis* (Bt) endotoxin and *Pinellia ternate* agglutinin (PTA). *Plant Cell, Tissue and Organ Culture,* 94, 33-44.

Yuan, J.S.; Galbraith, D.W.; Dai, S.Y.; Griffin, P. & Stewart, N. Jr. (2008). Plant systems biology comes of age. *Trends in Plant Science,* 13, 4, 165-171.

Zale, J.M.; Agarwal, S.; Loar, S. & Steber, C.M. (2009). Evidence for stable transformation of wheat by floral dip in *Agrobacterium tumefaciens*. *Plant Cell Reports,* 28, 903-913.

Zale, J.M.; Borchardt-Wier, H.; Kidwell, K.K. & Steber, C.M. (2004). Callus induction and plant regeneration from mature embryos of a diverse set of wheat genotypes. *Plant Cell Tissue and Organ Culture,* 76, 277–281.

Zhao, T.J.; Zhao, S.Y.; Chen, H.M.; Zhao, Q.Z.; Hu, Z.M.; Hou, B.K. & Xia, G.M. (2006). Transgenic wheat progeny resistant to powdery mildew generated by *Agrobacterium* inoculum to the basal portion of wheat seedling. *Plant Cell Reports,* 25, 1199-1204.

Zhou, H.; Berg, J.D.; Blank, S.E.; Chay, C.A.; Chen, G.; Eskelsen, S.R.; Fry, J.E.; Hoi, S.; Hu, T.; Isakson, P.J.; Lawton, M.B.; Metz, S.G.; Rempel, C.B.; Ryerson, D.K.; Sansone, A.P.; Shook, A.L.; Starke, R.J.; Tichota, J.M. & Valenti, S.A. (2003). Field efficacy assessment of transgenic roundup ready wheat. *Crop Science,* 43, 1072-1075

4

Bioactive Beads-Mediated Transformation of Rice with Large DNA Fragments Containing *Aegilops tauschii* Genes, with Special Reference to Bead-Production Methodology

Naruemon Khemkladngoen, Naoki Wada, Suguru Tsuchimoto,
Joyce A. Cartagena, Shin-ichiro Kajiyama and Kiichi Fukui
Osaka University
Japan

1. Introduction

Plant transformation is a technique that allows us to transfer genes from one species to another in order to introduce new characteristics into the recipient. The plant transformation technique has become widely adopted as a method both to understand plant physiology and to improve plant characteristics. There are now many established gene-transfer methods, both direct and indirect, for the stable introduction of novel genes into plant species. Examples include *Agrobacterium*-mediated transformation, particle bombardment (biolistic), and protoplast electroporation (Klein and Fitzpatrick-McElligott, 1993; Tzfira and Citovsky, 2006). Recently, a direct gene-transfer method called bioactive beads (BABs)-mediated transformation has been developed (Sone et al., 2002; Wada et al., 2011a, 2011b). This method involves immobilization of DNA molecules on alginate beads and their transfer to plant cells with the assistance of a polyethylene glycol (PEG) solution. Alginate, a hydrophilic polysaccharide that solidifies in the presence of Ca^{2+} ions, is utilized as a barrier membrane to produce calcium-alginate beads for immobilizing high-density DNA molecules. The original procedure for bead production was described in Sone et al. (2002) (Fig. 1). Firstly, a plasmid DNA solution was mixed with a $CaCl_2$ solution. Isoamyl alcohol was then added to a 1.5 ml micro-tube containing an aqueous phase 1% sodium alginate solution to form a water/oil mixture that was emulsified by sonication using an ultrasonic disrupter (UR-20P; Tomy Seiko, Tokyo, Japan) for 10 s. Bioactive beads (BABs) encapsulating DNA molecules were generated after immediately adding a $CaCl_2$ solution containing plasmid DNA into the emulsified solution. The DNA-immobilizing BABs were utilized in combination with PEG solution for protoplast transformation.

This method was successfully used for transformation of tobacco BY-2 protoplasts with a ten-fold higher transformation efficiency than the conventional PEG transformation method that is using naked plasmid DNA (Sone et al., 2002). The BABs transformation method has a wide applicability beyond tobacco BY-2 protoplasts to many organisms, both in plant cells, e.g., eggplant (*Solanum integrifolium*, Liu et al., 2004), tobacco (*Nicotiana tabaccum* SR-1, Liu et al., 2004) (Fig. 2), carrot (*Daucus carota*, Liu et al., 2004), rice (*Oryza sativa*, Wada et al., 2009) and in human cells, e.g., lymphocyte cell line K562 cells and human carcinoma HeLa cells (Higashi et al., 2004).

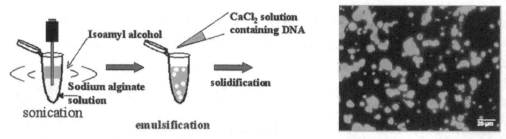

Fig. 1. Schematic diagram shows the preparation of bioactive beads (BABs) immobilizing DNA molecules

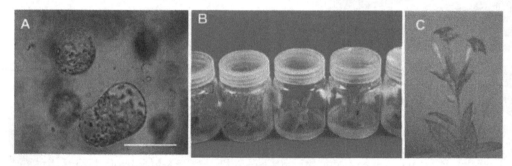

Fig. 2. Transgenic tobacco SR-1. A Cell division of protoplast 5 days after co-transformation. B Transgenic tobacco SR-1 regenerated plants 4 months after being transformed. C Flowering transgenic tobacco SR-1, 5.5 months after transformation. (With kind permission from Springer Science+Business Media: Journal of Plant Research, Obtaining transgenic plants using the bio-active beads method, 117, 2004b, 95-99, Liu, H.; Kawabe, A.; Matsunaga, S.; Murakawa, T.; Mizukami, A.; Yanagisawa, M.; Nagamori, E.; Harashima, S.; Kobayashi, A & Fukui, K., Figure 2)

Mizukami and associates (2003) have introduced yeast artificial chromosomes (YACs) into yeast spheroplasts using this method, and revealed the higher physical stability of 468 kb of YAC DNA embedded in and/or on the bioactive beads in solution compared to naked chromosome DNA molecules. The authors also checked whether YAC DNA molecules immobilized on bioactive beads would be intact even after vortex treatment. The result showed that naked YAC DNA molecules was degraded in the solution, resulting in no visible band after electrophoresis, but the band was clearly observed in the case of the YAC DNA molecules immobilized on the beads. This shows that BABs can stabilize large DNA fragments in solution. Moreover, yeast chromosomal DNA with YAC DNAs (128 kb, 256 kb and 468 kb) was embedded in and/or on the beads and transferred into recipient yeast cells lacking those YAC DNAs. Pulse field gel electrophoresis clearly showed that YAC DNAs up to 468 kb in size were successfully introduced into recipient

cells by PEG treatment with bioactive beads. This utility, coupled with the fact that the
method does not require any sophisticated equipment and is easy to practice, clearly
suggests the bioactive beads method as an alternative transformation method, especially
for large DNA molecules.

Despite the advantages of BAB transformation mentioned above, there is insufficient
information regarding the ideal production conditions, such as the shape and size of BABs,
or the concentration of DNA that is suitable for the most efficient BABs transformation.
Thus, the production conditions for BABs should be optimized. Here, improvement of the
BAB production system enabling uniform size and shape will be reported. Using this
system, various sizes of beads immobilizing the pUC18-sGFP construct could be produced.
Its applicability to plant transformation has already been examined by using tobacco BY-2
protoplasts.

As described above, the BAB method has the ability to transfer large DNA fragments into
yeast spheroplasts and into the protoplasts derived from tobacco BY-2 cells suspension.
Moreover, recently our group has successfully produced transgenic rice with large DNA
inserts containing *Aegilops tauschii* hardness genes by using this method. The detailed
transformation procedures and characterization of transgenic plants will be described in this
chapter.

2. Improvement of efficacy of a BAB transformation method

The capturing of DNA fragments by BABs can prevent their physical damage during
transformation processes. As a result, the BAB transformation method provides a feasible
large DNA fragment transfer method into many organisms (Mizukami et al., 2003; Wada
et al., 2009). However, the mechanism for DNA transfer is still obscure and experimental
conditions are not yet optimized, resulting in low DNA transfer efficiency (Liu et al.,
2004). During preparation of BABs by the original sonication method, $CaCl_2$-containing
DNA molecules were added into an alginate emulsion solution, and the beads were then
collected by centrifugation and used for transformation (Sone et al. 2002). It was,
however, found that large amounts of DNA remained in the $CaCl_2$ solution, resulting in a
poor efficiency, both in DNA immobilization and transformation. Therefore, a more
efficient BAB production system with higher DNA immobilization and DNA
transformation efficiencies is desired.

A new system for bead production was developed using an in-house device called a bead-
maker. The bead-maker has 2 major components; an automated micro-syringe, and a
vibrator (Fig. 3). The automated micro-syringe was assembled by placing a syringe (100 μl
gastight syringe, Hamilton, Nevada, USA) on a micro-syringe pump (MSP-RT, AS ONE) so
that the flow rate of the solution could be accurately controlled. The vibrator consists of a
loudspeaker (FR-8, 4Ω, Visaton, Germany), attached to a moveable rod fixed to a wooden
board and connected to a sine wave sound generator (AG-203D Kenwood, Tokyo, Japan). A
capillary tube (30 μm φ, GL Sciences, Tokyo, Japan) from the syringe was connected to the
moveable rod so it could be vibrated simultaneously when the rod was vibrated. The
frequency and amplitude of the sine wave sound generator are selectable. As a result, the

vibration of the moveable rod as well as the capillary tube linking the micro-syringe and the vibrator are controllable.

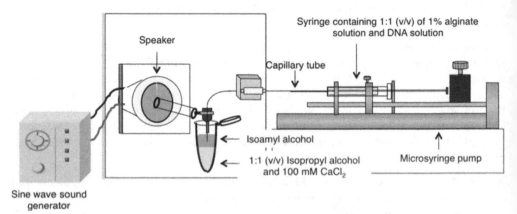

Fig. 3. Schematic diagram of a bead-maker

To test the effectiveness of this new system for plasmid DNA immobilization, bead production using 1% alginate containing 0.5 µg/µl of pUC18-sGFP was carried out. First, a DNA-containing alginate solution was prepared by mixing 100 µl of sodium alginate solution (1% w/v) with 100 µl of pUC18-sGFP (0.5 µg/µl), and then the freshly-prepared solution was slowly loaded into the micro-syringe. The syringe was connected to the capillary tube and placed on the micro-syringe pump. The solidifying solution was then prepared by firstly adding 750 µl of a mixture of 0.1 M $CaCl_2$ and isopropyl alcohol (1:1) in a 1.5 ml micro-centrifuge tube, followed by adding 750 µl of isoamyl alcohol and placing the tube in a plastic rack as shown in Fig. 3. As the bead-maker was working, the DNA-containing alginate solution was pumped out at a steady flow rate. Simultaneously, the sine wave sound generator produced sound waves at the speaker resulting in the vibration of the moveable rod connecting to the speaker. Consequently, vibration of the capillary tube linked to the moveable rod dropped alginate-DNA solution into the solidifying solution. The isoamyl alcohol kept the droplets spherical and the mixture of $CaCl_2$ and the isopropyl alcohol solidified the alginate-DNA droplets. The spherical beads were collected and washed at least twice in 0.1 M $CaCl_2$ solution by centrifugation (5000 g for 5 min) and re-suspended in a sufficient volume of 100 mM $CaCl_2$ solution. To verify the efficiency of DNA immobilization, the beads were stained by a DNA-staining dye, YOYO-1, and were observed under a fluorescence microscope (Zeiss). Images were captured using a CCD (charge-coupled device) camera (Fig. 4a-d). Observation of bead shape under the microscope revealed them to be spherical. Contrastingly, the shapes of beads made using the original system, with normal sonication, were irregular; many were not spherical. Qualitatively, the green fluorescence intensity emitted from DNA-immobilized beads, which correlates with the amount of DNA immobilized by the BABs, was measured using image analysis software, Image J (http://rsbweb.nih.gov/ij/), to further compare the efficiency of DNA immobilization of these two systems. The intensity of green fluorescence emitted from beads made using the new system was obviously

higher than that of the beads made by sonication. To quantify this result, the green
fluorescence intensities from more than 50 beads made using both the new system and the
sonication system were measured (Fig. 4e). The mean intensity from the improved beads
was much higher than that from beads made by sonication, indicating that the amount of
DNA immobilized by the improved beads was higher than that by beads made by the
original sonication system.

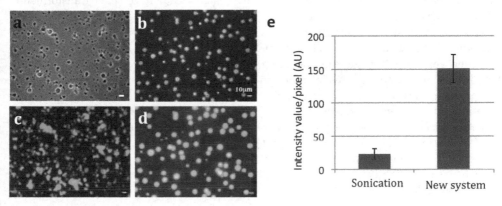

Fig. 4. Improvement of BAB production was achieved using the new BABs production
system. BABs made by the new system were spherical, size-controllable and more highly
efficient at DNA immobilization compared to BABs made by the original sonication system.
Phase contrast (a-b) and fluorescent images (c-d) of plasmid DNA-immobilized beads made
using the sonication system (a and c), and using the new system (b-d). Bars = 10 μm. Mean
intensity of green fluorescent intensity from plasmid-DNA immobilized beads stained by
YOYO-1 (e). Error bars: ±1 s.d.

We further investigated whether the bead size influenced transformation efficiency. As
described above, solution flow rate and the frequency and amplitude of the sine wave
were adjustable. Various combinations of these parameters were tested to obtain
uniformly sized BABs. Investigation of the effect of solution flow rate, vibration
frequency, and vibration amplitude on the size of beads indicated that a solution flow rate
at 0.4 μl/min was the most suitable for producing beads of a uniform size, compared to
other flow rates tested at 0.2, 0.8, 2 and 5 μl/min. However, the bead size was controllable
by changing the vibration frequency and amplitude. The vibration amplitude had a direct
effect on bead sizes: smaller beads are produced with a higher amplitude. The frequency
of the sine wave affected the bead size through the strength of the capillary tube
vibration; frequencies causing strong vibration produced smaller beads while weak
vibration produces larger beads.

Three different sizes of BABs immobilizing the same amounts of pUC18-sGFP were selected
(Fig. 5) and used for transformation into tobacco BY-2 protoplasts in combination with PEG
treatment (Fig. 6). Transient assays of GFP-expressing protoplasts were carried out 24 h and
48 h after transformation (Fig. 6c). The results showed a negative correlation between bead
size and transformation efficiency, that is, as the size of beads decreased, the transformation

efficiency increased. Moreover, with the new system, a *ca.* 10% transformation efficiency was achieved (Fig. 6 (⑥)). One to six μm diameter beads provided higher transformation efficiencies than beads made by sonication method. Thus it is likely that immobilization of DNA molecules on beads made by the new system is more efficient than that made by sonication system. Furthermore, it was also found that transformation efficiency obtained from one to six μm beads was higher than that obtained from seven to twelve μm beads. Two explanations are possible to reveal out why smaller beads have better transformation efficiency. One is that DNA is introduced into plant cells through physical uptake of bioactive beads. Smaller beads should be more easily incorporated into plant protoplasts. The other is because a higher number of smaller beads means a larger total surface area for the same volume of alginate solution being used. This means that a higher number or a larger total surface area of smaller beads might adhere to the protoplast membrane than when using larger beads, consequently enhancing the interaction between DNA-immobilized beads and protoplasts, and ultimately increasing DNA transfer between beads and protoplasts. From these results, it is concluded that, aside from DNA concentration, bead size optimization is also an important factor in achieving high transformation efficiency. This new system developed for BABs production with higher transformation efficiency should facilitate more applicability of BABs transformation and even enable multiple gene delivery into plant cells.

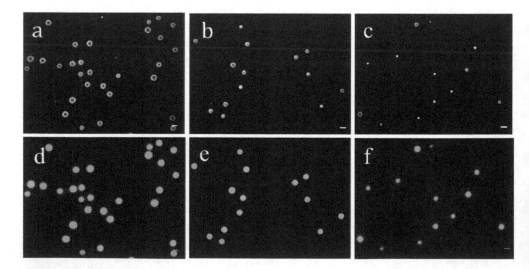

Fig. 5. Phase contrast (a-c) and fluorescence image (d-f) of BABs made under optimized conditions. BABs of 7-12 μm diameter made with 0.4 μl/min solution flow rate, frequency of 250 Hz, and amplitude set to 7 (a,d). BABs of 5-8 μm diameter made with 0.4 μl/min solution flow rate, frequency of 270 Hz, and amplitude set to 10 (b,e). BABs of 1-6 μm diameter made with 0.4 μl/min solution flow rate, frequency of 250 Hz, and amplitude set at 10 (c,f). Bars = 10 μm

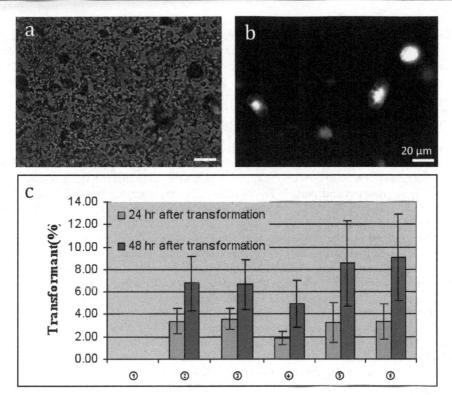

Fig. 6. Transformation of tobacco BY-2 protoplasts using BABs immobilizing pUC18-sGFP.
Phase contrast image of tobacco BY-2 protoplast transformed with the improved BABs (a),
tobacco BY-2 protoplasts expressing GFP protein (b). Bars = 20 μm. Transformation
efficiency after 24 h and 48 h of transformation: ①, negative control (distilled water); ②,
50μg naked pUC18-sGFP; ③, sonicated beads; ④, beads made using the new system, size: 7-
12 μm (68%); ⑤, beads made using the new system, size: 5-8 μm (62%), 2-3 μm (16%); ⑥,
beads made using new system, size: 1-6 μm (70%). Error bars: ±1 s.d.

3. Large DNA transfer into plants using the bioactive beads method

3.1 Transformation of rice with large DNA molecules using BAB method

Transformation of large DNA fragments is a promising approach to extend the reach of
plant genetic engineering. Until now, plant genetic engineering has been performed using
single or small numbers of genes, resulting in successful production of genetically
engineered plants such as herbicide- and insect-resistant plants (Gonsalves, 1998; Khan et
al., 2009; Song et al., 2003b; Tai et al., 1999; Wang et al., 2005). To produce transgenic plants
with more variety of phenotypes, however, multiple gene transfer will be required (Dafny-
Yellin et al., 2007; Daniell et al., 2002; Halpin, 2005; Naqvi et al., 2010). Even single traits are
often the result of expression of multiple genes. For example, a single metabolic pathway
may be related to several genes. Thus if we want to manipulate the metabolic pathway, it
will be best achieved by manipulating several genes simultaneously. Introducing these

genes at same time is preferable because: (1) introduction of multiple genes through crossing of different transgenic plants is time-consuming and laborious, and (2) co-transformation of different kinds of plasmid DNAs needs preparation of many constructs, which also takes time and needs different kinds of marker genes depending on the number of plasmids to be introduced. Therefore, transformation with large DNAs including multiple genes is promising to enhance the efficiency of transformation. In addition, large DNA transfer will enable the regulatory regions of transgenes to be transferred with the genes of interest. This will allow introduced genes to be expressed at the physiological level.

Although this approach is expected to have such advantages, it is still difficult to transfer large DNA fragments into plants. This difficulty is due to the lack of a reliable method of introducing large DNA fragments into plants. For general plant transformation, *Agrobacterium*-mediated transformation and particle bombardment are the methods normally used (Bhalla, 2006; Rakoczy-Trojanowska, 2002). However, these methods are difficult to apply to large DNA transfer into plants because of the instability of large DNAs in *Agrobacterium* cells and during the bombardment process. Song and associates (2003a) and Nakano and co-workers (2005) have recently reported that DNA fragments larger than 100 kb are unstable in *Agrobacterium* cells, resulting in deletion and/or rearrangements in *Agrobacterium* cells before their transformation into plants. Their data suggest that *Agrobacterium*-mediated transformation is not suitable for large DNA transfer. In particle bombardment, DNAs coated on metal particles generally suffer from physical damage during bombardment, resulting in fragmentation of large DNAs. However, some laboratories, have reported success in introducing DNA fragments over 100 kb in size by particle bombardment (Van Eck et al., 1995; Mullen et al., 1998; Phan et al., 2007). This is a promising improvement, but successful reports of large DNA transfers using this method are still limited. In addition, particle bombardment needs the equipment. Thus, an easy and inexpensive method needs to be developed for large DNA transfer. As described above, our group has recently developed such a new transformation method, namely the BAB method (Higashi et al., 2004; Liu et al., 2004a; Liu et al., 2004b; Mizukami et al., 2003; Sone et al., 2002; Wada et al., 2011a, 2011b). This method is easy and inexpensive. The method has been applied to yeast, mammalian cells (HeLa, K562 cells), and plant cells (tobacco BY-II, tobacco SR-I, carrot, egg plant, rice). Transformation efficiency is about 10 times higher than PEG treatment without bioactive beads. An important characteristic of the BAB method is its transformation capability with large DNA fragments. As mentioned above, 468 kb of YAC DNA was effectively transferred into yeast cells. This result clearly suggests that the BAB method is a suitable for large DNA transfer.

To examine the applicability of the BAB method to plant transformation with large DNA fragment, 124 kb of YAC DNA was introduced into cultured tobacco BY-2 cells (Liu et al., 2004b). The transient expression of introduced genes was detected in the transgenic suspension cells. To investigate this in more detail, the bioactive beads method was applied to the transformation of rice with *ca.* 100 kb BAC DNA (Wada et al., 2009). Rice (*Oryza sativa* L. ssp. *japonica* cv. Nipponbare) was used as the plant material and BAC DNA including hardness genes from *Agilops tauschii* was introduced using the bioactive beads method. The hardness genes consist of three genes (*puroindoline a, puroindoline b, GSP-1* gene) and are scattered in a *ca.* 100 kb region on the short arm of chromosome 5D in *Aegilops tauschi* (Turnbull et al., 2003). The BAC DNA encompasses the hardness locus and also contains the regulatory regions of each gene (Fig. 7). Because of this, it was expected that the expression of each gene would be shown at the regular physiological level.

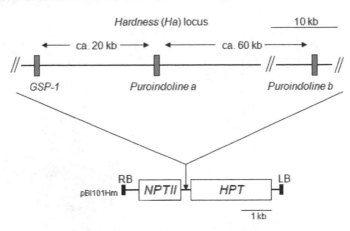

Fig. 7. Schematic diagram of the construct, used in this transformation, pBI BAC10-60.
(With kind permission from Springer Science+Business Media: Plant Cell Reports, Bioactive
beads-mediated transformation of rice with large DNA fragments containing *Aegilops
tauschii* genes, 28, 2009, 750-68, Wada, N.; Kajiyama, S.; Akiyama, Y.; Kawakami, S.; No, D.;
Uchiyama, S.; Otani, M.; Shimada, T.; Nose, N.; Suzuki, G.; Mukai, Y.; Fukui, K.; Figure 1)

As a result, nine transgenic plants were obtained and analyzed. The PCR analyses showed that
each gene was integrated into the rice genome (Table 1). Some transgenic plants contained
most transgenes, but transgenic plants with all the transgenes could not be obtained. This
indicates that rearrangement of introduced DNA molecules occurred during transformation.

	Lines	NPTII gene	Pinb gene	Pina gene	GSP-1 gene	HPT gene
Non-transgenic plant		–	–	–	–	–
Transgenic plants with pBI BAC10-60	9-1-1	–	+	–	+	+
	9-1-2	–	+	–	+	+
	9-1-3	+	+	–	+	+
	9-1-4	+	+	–	+	+
	9-1-6	–	+	–	+	+
	9-1-7	+	+	–	+	+
	9-1-8	+	+	–	+	+
	9-1-9	+	+	–	+	+
	9-1-10	–	–	–	+	–

Table 1. Profiles of T_0 transgenic plants as determined by PCR analysis. +: gene detected, – :
gene not detected. (With kind permission from Springer Science+Business Media: Plant Cell
Reports, Bioactive beads-mediated transformation of rice with large DNA fragments
containing *Aegilops tauschii* genes, 28, 2009, 750-68, Wada, N., Kajiyama, S., Akiyama, Y.,
Kawakami, S., No, D., Uchiyama, S., Otani, M., Shimada, T., Nose, N., Suzuki, G., Mukai, Y.,
Fukui, K.; Table 3)

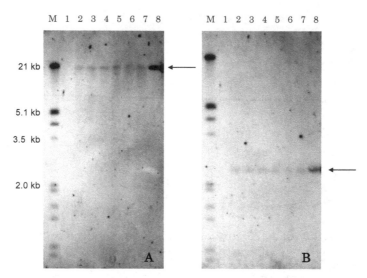

Fig. 8. Genomic Southern blot analysis of T_0 transgenic plants. Total DNA of rice plants was digested with HindIII and probed with (A) the HPT gene and (B) Pinb gene. Lane 1 : control plant, Lane 2 : transgenic plant 9-1-3, Lanes 3 to 7 : transgenic plants 9-1-6 to 9-1-10, Lane 8 : pBI BAC DNA digested with HindIII. The amount of BAC DNA corresponds to two copies insertion of the transgene in rice genome. Arrows indicate the locations of the observed bands (These figures, from Wada et al. (2011), are reproduced with permission from John Wiley and Sons, Inc.)

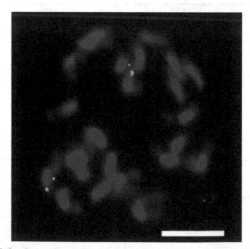

Fig. 9. FISH analysis of the kernels of the T_1 transgenic plant 9-1-6-3. The two paired green signals indicate the integration sites of pBI BAC 10-60. Bar = 10 μm. (With kind permission from Springer Science+Business Media: Plant Cell Reports, Bioactive beads-mediated transformation of rice with large DNA fragments containing Aegilops tauschii genes, 28, 2009, 750-68, Wada, N.; Kajiyama, S.; Akiyama, Y.; Kawakami, S.; No, D.; Uchiyama, S.; Otani, M.; Shimada, T.; Nose, N.; Suzuki, G.; Mukai, Y.; Fukui, K.; Figure 4)

Bioactive Beads-Mediated Transformation of Rice with Large DNA Fragments Containing
Aegilops tauschii Genes, with Special Reference to Bead Production Methodology
101

The copy numbers of *puroindoline b* (*Pinb*) and *HPT* genes were checked by genomic Southern blot analysis. The results showed that these two regions were integrated into the rice genome as a single copy of intact fragment (Fig. 8). T_0 plants showed sterility: 7 out of 9 transgenic plants did not produce any seeds. Two transgenic plants were partially fertile and produced some seeds. They recovered their fertility in successive generations. Segregation tests of the T_1 generation showed that transgenes were inherited into the next generation in Mendelian mode. Homozygous plants were also obtained. FISH analysis revealed that the transgenes were integrated into the telomeric region of a pair of rice chromosomes in the homozygous plants (Fig. 9). The expression of the introduced gene, *Pinb*, was also confirmed at the mRNA and protein levels. Thus, the promoter region of *Pinb* was functional even in rice cells. These results indicate that the BAB method can introduce multiple genes into plants and produce stable transgenic plants that can pass introduced transgenes on to successive generations.

To examine if the introduced gene is functional in transgenic rice, a phenotypic analysis was performed using scanning electron microscopy (SEM) and transmission electron microscopy (TEM) (Wada et al., 2010). The results indicated that the endosperm structure changed in the transgenic rice to a more loosely packed structure (Fig. 10). The hardness genes are known to affect the softness of wheat endosperms by giving the loosely packed endosperm structure. Thus, the EM observations clearly indicated that the introduced hardness gene functioned in similar manner as in wheat endosperm. Analysis of physico-chemical properties of the rice flour also indicated that the transgenic rice endosperm had the phenotype of soft textured seed. The results suggest that the PINB protein localized at the surface of the starch compounds, resulting in preventing the adhesion of each starch compound and changing some physico-chemical properties, such as flour particle size, and pasting properties. These results indicate that the hardness locus introduced was functional in transgenic rice. This suggests that introduction of a genomic locus that controls a trait could be a good strategy for adding desirable traits to plants.

The results obtained indicated that the BAB method is a promising method for plant transformation with large DNA fragments. Co-transformation experiments with two or three kinds of BAC DNAs simultaneously were also successful (data not shown). Co-transformation can increase the number of genes that can be introduced simultaneously. However, some aspects of the BAB method could be improved. First, the intactness of introduced DNA fragments should be examined. In our experiments, the deletion of some transgenes was observed. The rearrangement of introduced DNA fragments might also occur in other regions that were not checked. How often and to what extent such rearrangements occur with the BAB compared to conventional methods needs to be investigated to fully establish that the BAB method has advantages over other conventional methods for transformation with large DNA fragments. Second, further improvement in the transformation efficiency of the BAB method should be achieved. We have succeeded in immobilizing proteins on BABs (data not shown). The immobilization of large DNAs with proteins, such as VirE2, might target the introduced DNA into the host genome more efficiently because VirE2 is known to target T-DNA into nuclei and protect the T-DNA during *Agrobacterium*-mediated transformation (Gelvin, 2003; Gopalakrishna et al., 2003). Despite these points that could be improved, our results clearly indicate that the bioactive beads method could be an alternative way of producing stable transgenic plants with multiple transgenes.

Non-transgenic rice   Transgenic rice

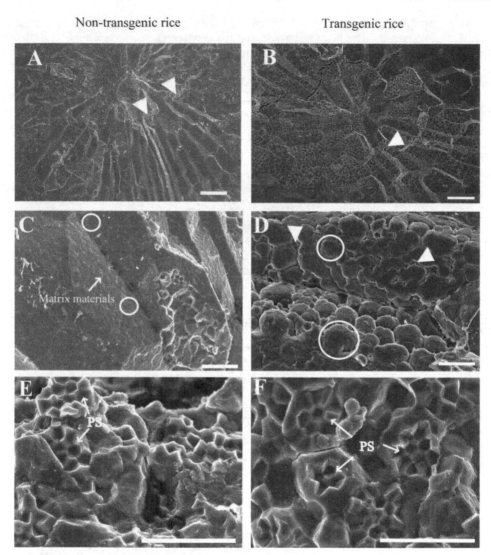

Fig. 10. SEM observation of rice endosperm cells. (A,B) Low magnification view of transversely fractured surface of milled rice of (A) non-transgenic rice and (B) transgenic rice. Arrowheads indicate intracellularly cleaved site. Bars: 100 μm. (C,D) Higher magnification view of intercellularly cleaved site of (C) non-transgenic rice and (D) transgenic rice. Bars: 20 μm. (C) Compound starch granules (circles) embedded within matrix material in non-transgenic rice. Intracellularly cleaved sites can also be seen. (D) Starch compound granules (circles) surrounded by airspaces (arrowheads). (E,F) Higher magnification view of intracellularly cleaved site of (E) non-transgenic rice and (F) transgenic rice. Partially split compound starch granules (PS) exposing individual starch granules with sharp angles and edges can be seen. Bars: 20 μm (Reproduced from Wada et al. (2010) with permission of Elsevier Science.)

In mammalian cells, chromosome engineering systems including artificial chromosomes, have been developed (Basu & Willard, 2005; Ikeno et al., 1998; Kazuki et al., 2011; Oshimura & Katoh, 2008). An artificial chromosome has a capacity to carry chromosomal fragments, with virtually no size limitation to the transgenes that can be transferred (Kuroiwa et al., 2000). In addition, microcell-mediated transfer (MMCT) has made it possible to introduce genes that cannot be transferred by conventional transfection. For example, a chromosomal region including the dystrophin gene (2.4 Mb) has been introduced into a mouse genome using human artificial chromosomes (HACs) and MMCT (Hoshiya et al., 2009). HACs can be engineered by recombination technology in cells that are the most suitable for each step (e.g. chicken DT40 cells for homologous recombination, hamster CHO cells for site-specific recombination, Kazuki et al. 2010). However, there is no such system for plants. Recently three reports have been published on the production of plant artificial chromosomes (Carlson et al., 2007; Yu et al., 2007; Ananiev et al., 2009). However, there is still no report of utilizing them for plant transformation with large DNA fragments. Thus, a plant transformation system using large DNA fragments has yet to be developed. The BAB method can be used to introduce large DNA fragments into a plant genome as a part of plant transformation system with large DNA fragments.

3.2 Further utility of bioactive beads

BABs can be not only applied to DNA transformation, but also to the immobilization of proteins (Zhou et al., 2009). BSA (Bovine serum albumin) protein was successfully entrapped by BABs and its interaction with FITC-labeled anti BSA was clearly observed. Moreover, the authors improved the efficacy of protein immobilization in BABs by treating the alginate solution with 1-ethyl-3-(3-dimethylaminopropyl) carbodiimide (EDC) and N-hydroxysulfosuccinimide (NHSS) to cross-link the desired protein (BSA) to the alginate carboxyl groups prior to solidification. It was found that cross-linking beads provided high protein-retention ability for up to 2 weeks after immobilization. Such improved protein-immobilizing beads with high retention capacity might have the potential to be an alternative choice for detecting antigen-antibody interactions.

BABs have also been successfully used in the immobilization of a single yeast cell displaying hydrolyzing enzymes to capture fluorescent molecules released after enzymatic reaction (Zhou et al., 2009). The retention of fluorescent products by yeast-encapsulating BABs enabled active and non-active cells to be differentiated by sorting in a flow cytometer. Using such a developed system, a library screening for novel enzymatic activities on the surface of yeast cells should be possible.

4. Conclusion

An alternative transformation method, BABs-mediated transformation, has been developed by applying a drug delivery system (DDS) in which highly concentrated DNA molecules are entrapped by small autonomously degradable alginate beads and transferred into plant cells in combination with polyethylene glycol (PEG) treatment. This transformation method is easy to perform, is applicable to a range of organisms, allows large-sized DNAs to be delivered, and facilitates the transportation of multiple genes of up to 468 kb size into yeast spheroplasts. Moreover, our latest results on transformation of BAC DNA containing *A. tauschii* hardness genes into rice protoplasts, along with an improvement in the efficiency of

DNA immobilization by BABs have verified that this method is capable of producing transgenic rice that carry large DNA fragments and can facilitate the production of useful transgenic plants by introduction of multiple genes simultaneously with high efficiency. Further development of the BAB method will contribute to the development of more flexible plant genetic engineering methodologies using large DNA fragments. This will open up new possibilities for plant genetic engineering and make it possible to produce a number of useful transgenic plants in the near future.

5. Acknowledgements

This work is supported by a fund from the Handai Frontier Research Center's Strategic Research Base supported by the Special Coordination Fund for Promoting Science and Technology of the Government of Japan to K. F. This research was partially supported by a Grant-in-Aid for Scientific Research (B) (No. 19380194) from the Ministry of Education, Culture, Sports, Science and Technology of Japan. We would also like to express special thanks to the Global COE (Center of Excellence) Program, "Global Education and Research Center for Bio-Environmental Chemistry", of Osaka University. N.K. would like to thank the Ministry of Education, Culture, Sports, Science and Technology, Japan (MEXT) and the International Graduate Program for Frontier Biotechnology, Graduate School of Engineering, Osaka University for fellowship support. N.W. was supported by a Research Fellowship from the Japanese Society for the Promotion of Science for Young Scientists.

6. References

Ananiev, E. V.; Wu, C.; Chamberlin, M. A.; Svitashev, S.; Schwartz, C.; Gordon-Kamm, W. & Tingey, S. (2009) Artificial chromosome formation in maize (Zea mays L.). *Chromosoma*, 118, 157–177

Basu, J. & Willard, H.F. (2005) Artificial and engineered chromosomes: non-integrating vectors for gene therapy. *Trends Mol Med*, 11, 251-258.

Bhalla, P. L. (2006) Genetic engineering of wheat--current challenges and opportunities. *Trends Biotechnol*, 24, 305-311.

Carlson, S. R.; Rudgers, G. W.; Zieler, H.; Mach, J. M.; Luo, S.; Grunden, E.; Krol, C.; Copenhaver, G. P. & Preuss, D. (2007) Meiotic transmission of an in vitro-assembled autonomous maize minichromosome. *PLoS Genet*, 3, 1965-1974.

Dafny-Yelin, M. & Tzfira, T. (2007) Delivery of multiple transgenes to plant cells. *Plant Physiology*, 145, 1118–1128.

Daniell, H. & Dhingra, A. (2002) Multigene engineering: dawn of an exciting new era in biotechnology. *Current Opinion in Biotechnology*, 13, 136–141.

Gelvin, B. (2003) Agrobacterium mediated plant transformation: The biology behind the "Gene-Jockeying" Tool. *Microbiology and Molecular Biology Reviews*, 67, 1, 16-37.

Gonsalves, D. (1998) Control of Papaya Ringspot Virus in Papaya: A Case Study. *Ann Rev Phytopath*, 36, 415-437.

Gopalakrishna, S.; Singh, P. & Singh, N. K. (2003) Enhanced transformation of plant cells following co-bombardment of VirE2 protein of agrobacterium tumefaciens with DNA substrate. *Current Science*, 85, 9, 1343-1347.

Halpin, C. (2005) Gene stacking in transgenic plants the challenge for 21st century plant biotechnology. *Plant Biotech. J.*, 3, 141-155.

Higashi, T.; Nagamori, E.; Sone, T.; Matsunaga, S. & Fukui, K. (2004) A novel transfection method for mammalian cells using calcium alginate microbeads. *J Biosci Bioeng*, 97, 191-195.

Hoshiya, H.; Kazuki, Y.; Abe, S.; Takiguchi, M.; Kajitani, N.; Watanabe, Y.; Yoshino, T.; Shirayoshi, Y.; Higaki, K.; Messina, G.; Cossu, G. & Oshimura, M. (2009) A highly stable and non-integrated human artificial chromosome (HAC) containing the 2.4Mb entire human dystrophin gene. *Mol Therapy*, 17, 309-17.

Ikeno, M.; Grimes, B.; Okazaki, T.; Nakano, M.; Saitoh, K.; Hoshino, H.; McGill, N.I.; Cooke, H. & Masumoto, H. (1998) Construction of YAC–based mammalian artificial chromosomes. *Nat Biotechnol*, 16, 431 – 439.

Kazuki, Y.; Hosoya, H.; Takiguchi, M.; Abe, S.; Iida, Y.; Osaki, M.; Katoh, M.; Hiratsuka, M.; Shirayoshi, Y.; Hiramatsu, H.; Ueno, E.; Kajitani, N.; Yoshino, T.; Kazuki, K.; Ishihara, C.; Takehara, S.; Tsuji, S.; Ejima, F.; Toyoda, A.; Sakaki, Y.; Larionov, V.; Kouprina, N. & Oshimura, M. (2010) Refined human artificial chromosome vectors for gene therapy and animal trasngenesis, *Gene Therapy*, 1-10.

Khan, E.U. & Liu, J.H.(2009) Plant biotechnological approaches for the production and commercialization of transgenic crops. *Biotechnol & Biotechnol Eq*, 23, 1281-1288.

Klein ,T. M. and Fitzpatrick-McElligott, S. (1993). Particle bombardment: A universal approach for gene transfer to cells and tissues. *Curr. Opin. Biotechnol.*, 4: 583-590.

Kuroiwa, Y.; Tomizuka, K.; Shinohara, T.; Kazuki, Y.; Yoshida, H.; Ohguma, A.; Yamamoto, T.; Tanaka, S.; Oshimura, M. & Ishida I.(2000) Manipulation of human minichromosomes to carry greater than megabase-sized chromosome inserts. *Nat Biotechnol*, 18, 1086–1090.

Liu, H.; Kawabe, A.; Matsunaga, S.; Kim, Y.; Higashi, T.; Uchiyama, S.; Harashima, S.; Kobayashi, A. & Fukui, K. (2004a) An Arabidopsis thaliana gene on the yeast artificial chromosome can be transcribed in tobacco cells. *Cytologia*, 69, 235-240.

Liu, H.; Kawabe, A.; Matsunaga, S.; Murakawa, T.; Mizukami, A.; Yanagisawa, M.; Nagamori, E.; Harashima, S.; Kobayashi, A & Fukui, K. (2004b) Obtaining transgenic plants using the bio-active beads method. *J Plant Res*, 117, 95-99.

Mizukami, A.; Nagamori, E.; Takakura, Y.; Matsunaga, S.; Kaneko, Y.; Kajiyama, S.; Harashima, S.; Kobayashi, A. & Fukui K. (2003) Transformation of yeast using calcium alginate microbeads with surface-immobilized chromosomal DNA. *Biotechniques*, 35, 734-740.

Mullen, J.; Adam, G.; Blowers, A. & Farle, E. (1998) Biolistic transfer of large DNA fragments to tobacco cells using YACs retrofitted for plant transformation. *Mol Breeding*, 4, 449-457.

Nakano, A.; Suzuki, G.; Yamamoto, M.; Turnbull, K.; Rahman, S.; Mukai, Y. (2005) Rearrangements of large-insert T-DNAs in transgenic rice. *Mol Genet Genomics*, 273,123-129.

Naqvi, S.; Farré, G; Sanahuja, G.; Capell, T.; Zhu, C. & Christou, P.(2010) When more is better: multigene engineering in plants. *Trends in Plant Science*, 15, 48–56.

Oshimura, M. & Katoh, M.(2008) Transfer of human artificial chromosome vectors into stem cells. *Reprod Biomed Online*, 16, 57–69.

Phan, B.H.; Jin, W.W.; Topp, C.N.; Zhong, C.X.; Jiang, J.M.; Dawe, R.K.& Parrott, W.A. (2007) Transformation of rice with long DNA-segments consisting of random genomic DNA or centromere-specific DNA. *Transgenic Res*, 16, 341-351.

Rakoczy-Trojanowska, M. (2002) Alternative methods of plant transformation - A short review. *Cell Mol Biol Lett*, 7, 849-858.

Sone, T.; Nagamori, E.; Ikeuchi, T.; Mizukami, A.; Takakura, Y.; Kajiyama, S.; Fukusaki, E.;
    Harashima, S.; Kobayashi, A. & Fukui, K. (2002) A novel gene delivery system in
    plants with calcium alginate micro-beads. *J Biosci Bioeng*, 94, 87-91.
Song, J; Bradeen, J.M.; Naess, S.K.; Helgeson, J.P. & Jiang, J. (2003a) BIBAC and TAC clones
    containing potato genomic DNA fragments larger than 100 kb are not stable in
    Agrobacterium. *Theor Appl Genet*, 107, 958-964.
Song, J.; Bradeen, J.M.; Naess, S.K.; Raasch, J.A.; Wielgus, S.M.; Haberlach, G.T.; Liu, J.;
    Kuang, H.; Austin-Phillips, S.; Buell, C.R.; Helgeson, J.P. & Jiang, J. (2003b) Gene RB
    cloned from Solanum bulbocastanum confers broad spectrum resistance to potato
    late blight. *Proc Natl Acad Sci USA*, 100, 9128–9133.
Tai, T.H.; Dahlbeck, D.; Clark, E.T.; Gajiwala, P.; Pasion, R.; Whalen, M.C.; Stall, R. E. &
    Staskawicz, B. J. (1999) Expression of the Bs2 pepper gene confers resistance to
    bacterial spot disease in tomato. *Proc Natl Acad Sci USA*, 96,14153–14158.
Turnbull, K.M.; Turner, M.; Mukai, Y.; Yamamoto, M.; Morell, M.K.; Appels, R. & Rahman,
    S. (2003) Theorganization of genes tightly linked to the Ha locus in *Aegilops tauschii*,
    the D-genome donor to wheat. *Genome*, 46, 330-338.
Tzfira, T. & Citovsky, V. (2006). Agrobacterium-mediated genetic transformation of plants:
    biology and biotechnology. *Curr. Opin. Biotechnol.*, 17, 147-154.
Van, E. J.M.; Blowers, A. & Earle, E. (1995) Stable transformation of tomato cell cultures after
    bombardment with plasmid and YAC DNA. *Plant Cell Rep*, 14, 299-304.
Wada, N.; Kajiyama, S.; Akiyama, Y.; Kawakami, S.; No, D.; Uchiyama, S.; Otani, M.;
    Shimada, T.; Nose, N.; Suzuki, G.; Mukai, Y. & Fukui, K. (2009) Bioactive beads-
    mediated transformation of rice with large DNA fragments containing *Aegilops
    tauschii* genes. *Plant Cell Rep*, 28, 759-768.
Wada, N.; Kajiyama, S.; Cartagena, J.A.; Lin, L.; Akiyama, Y.; Otani, M.; Suzuki, G.; Mukai,
    Y.; Aoki, N. & Fukui, K. (2010) The effects of puroindoline b on the ultrastructure of
    endosperm cells and physicochemical properties of transgenic rice plant. *J Cereal
    Sci*, 51,2, 182-188.
Wada, N.; Kajiyama, S.; Khemkladngoen, N. & Fukui, K. (2011a) A novel gene delivery
    system in plants with calcium alginate micro-beads, in *Plant Transformation
    Technologies*, Stewart Jr., C. N., Touraev, A., Citovsky, V., and Tzfira, T., pp. 73-82,
    Wiley-Blackwell Publishers.
Wada, N.; Cartagena, J.A.; Khemkladngoen, N. & Fukui K. (2011b) Bioactive beads-
    mediated transformation of plants with large DNA fragments, in *Transgenic Plants;
    Method and Protocols*, Dunwell, J. & Wetten, A., Springer, in press.
Wang, Y.; Xue, Y.& Li, J. (2005) Towards molecular breeding and improvement of rice in
    China. *Trends Plant Sci*, 10, 610–614.
Yu, W.; Han, F.; Gao, Z,.; Vega, J. M. & Birchler, J. (2007) Construction and behavior of
    engineered mini-chromosomes in maize. *Proc Natl Acad Sci USA*, 104, 8924–8929.
Zhou, Y.; Kajiyama, S.; Masuhara, H.; Hosokaw, Y.; Kaji T & Fukui K (2009). A new size and
    shape controlling method for producing calcium alginate beads with immobilized
    proteins. *JBiSE*, 2, 287-293.
Zhou, Y.; Kajiyama, S.; Itoh, K.; Tanino, T.; Fukuda, N.; Tanaka, T.; Kondo, A. & Fukui K
    (2009). Development of an enzyme activity screening system for beta-glucosidase-
    displaying yeasts using calcium alginate micro-beads and flow sorting. *Appl.
    Microbiol. Biotechnol.*, 84, 375-382.

5

Molecular Breeding of Grasses by Transgenic Approaches for Biofuel Production

Wataru Takahashi and Tadashi Takamizo

Forage Crop Research Division, NARO Institute of Livestock and Grassland Science

Japan

1. Introduction

Since the Industrial Revolution, fossil fuels have been consumed in massive amounts. However, little fossil fuels are estimated to remain, and the number of new oil fields in the world continues to decline. Thus, the future supply of fossil fuels will be tight and will not continue to support our levels of consumption for long. In addition, rapid consumption of fossil fuels for energy production continues to elevate the atmospheric carbon dioxide concentration, which was 280 ppm before the Industrial Revolution and is 380 ppm today. The emissions, including carbon dioxide, derived from the combustion of fossil fuels act as greenhouse gases and cause the serious problem of global warming.

In this context, society must develop beyond the consumption of our limited fossil fuels. Plant biomass, which is organic matter derived from the photosynthetic fixation of atmospheric carbon dioxide, holds much promise as a future energy source. The use of plant biomass for energy is called carbon neutral, because the combustion of plant biomass only releases carbon dioxide that was originally fixed from the atmosphere by plants. Because a sustainable society based on carbon-neutral energy must be constructed to preserve the global environment, we must develop related practical technology as soon as possible.

However, recent hikes in the global price of grain, caused by various political and economic events, has made us consider shifting toward using non-grain plant biomass to improve both energy and food-supply security. Thus, research into the production of biofuel from lignocellulose, which constitutes the bodies of woody and grass plant species, is a major field of study in modern plant science.

Needless to say, this research includes biotechnological approaches. In particular, genetic engineering is expected to be a key technology in herbaceous plant breeding, because most of the bioenergy crop species are not yet domesticated well. This chapter focuses on recent progress in the molecular breeding of herbaceous plants.

2. Herbaceous perennials for dedicated lignocellulosic biofuel crops

High biomass yields are indispensable to successful cultivate energy crops that can compete with and replace fossil fuels (Karpenstein-Machan, 2001; Lewandowski et al., 2003). Considering food-supply security, net energy yield, and environmental protection, large and fast-growing non-food grass species are among the most promising candidates for lignocellulose production. Perennial species are more attractive than annuals because they have minimal input requirements, commonly yield more aboveground biomass in the form of stems and leaves, and their well-developed root systems can serve as carbon sinks.

Name (Scientific name)[a]	Photosystem	Yield (t DM ha^{-1} yr^{-1})[b]	Location	Citation
Switchgrass (*Panicum virgatum* L.)	C4	15.9‡	U.S.A. (Oklahoma)	Aravindhakshan et al. (2010)
Sugarcane (*Saccharum* spp. hybrids)	C4	41.3c	Japan (South western islands)	Terajima et al. (2010)
Miscanthus species (*Miscanthus* × *giganteus*)	C4	28.7†	Central Italy (Pisa)	Angelini et al. (2009)
		12.4‡	U.S.A. (Oklahoma)	Aravindhakshan et al. (2010)
Erianthus species (*Erianthus arundinaceus* Retz.)	C4	58.4	Japan (Northern Kanto region)	Ando et al. (2011)
Giant reed (*Arundo donax* L.)	C3	37.7†	Central Italy (Pisa)	Angelini et al. (2009)

[a] Actual scientific names of experimental cultivars or lines used in the study cited are given for *Erianthus* species, *Miscanthus* species, and sugarcane.
[b] Values followed by the same obelisk were obtained by the same study.
[c] An average value calculated with the data from three different locations.

Table 1. Summary information for perennial grasses described in this review (modified and updated from Lewandowski et al., 2003)

C4 perennial grasses show promise as lignocellulosic biofuel crops because their highly efficient C4 photosynthesis often yields more biomass than C3 grasses (Jakob et al., 2009). There are many available candidate perennial grasses, which differ in their potential productivity, chemical and physical biomass properties, environmental demands and crop management requirements (Lewandowski et al., 2003). This review will focus on noteworthy C4 perennial grasses as dedicated lignocellulosic biofuel crops, including switchgrass (*Panicum virgatum* L.), *Miscanthus* species, *Erianthus* species, and sugarcane (*Saccharum* species). In addition, the high biomass C3 grass species giant reed (*Arundo donax* L.), which has a photosynthetic capacity comparable to or higher than that of C4 grasses (Angelini et al., 2009; Rossa et al., 1998), will be discussed as well. These perennial grasses are summarized in Table 1 and will be briefly described in sections 2.1 through 2.3.

2.1 Switchgrass

Switchgrass (*Panicum virgatum* L.) is an outcrossing perennial warm-season forage grass originally from the prairies of North America. In the 1990's, the United States Department of Energy sponsored a 10-year research project to evaluate and develop this grass as a dedicated herbaceous lignocellulosic energy crop because of its potential for high fuel yields, environmental enhancement, and ability to grow well on marginal cropland without heavy fertilizing or intensive management (Bouton, 2008; McLaughlin & Kszos, 2005; Wright & Turhollow, 2010).

Switchgrasses are divided into two ecotypes: lowland and upland (Bouton, 2008; Lewandowski et al., 2003). The lowland ecotypes are tall and robust plants adapted to wetter sites; they mature later and have longer growth periods and higher biomass yields than upland ecotypes. Although somatic chromosome numbers vary in switchgrass (2n from 2x = 18 to 12x = 108) (Hopkins et al., 1996), the lowland ecotypes are predominantly tetraploid (2n = 4x = 36) (Bouton, 2008; Lewandowski et al., 2003). The upland ecotypes have shorter and fine-stemmed morphology, and are adapted to drier habitats. They are commonly octoploid (2n = 8x = 72), or occasionally tetraploid (2n = 4x = 36) (Bouton, 2008) or hexaploid (2n = 6x = 54) (Lewandowski et al., 2003). Irrespective of ecotype, switchgrasses with the same ploidy level can be intercrossed (Lewandowski et al., 2003).

2.2 *Saccharum* complex

The term "Saccharum complex" was first coined by Mukherjee (1957) and originally encompassed four closely-related interfertile genera, *Saccharum*, *Erianthus*, *Sclerostachya*, and *Narenga*. Based on species richness and the geographic distributions of endemic species, India was considered the center of maximum variation of the *Saccharum* complex (Mukherjee, 1957). Eventually, the genus *Miscanthus* was added to the *Saccharum* complex, because it was thought to be involved in the origin of *Saccharum* (Daniels et al., 1975, as cited in Alwala et al., 2006; and Amalraj & Balasundaram, 2006).

Modern sugarcane varieties are mostly derived from interspecific hybridization within the genus *Saccharum* (Amalraj & Balasundaram, 2006), and intergeneric hybridization between *Saccharum* and other genera in the *Saccharum* complex is thought to be the primary gene pool for sugarcane breeding (Cheavegatti-Gianotto et al., 2011). Thus, the *Saccharum* complex can also be considered as the primary gene pool for breeding non-domesticated *Miscanthus* and *Erianthus* as well.

In sections 2.2.1 through 2.2.3, we will address individual genera within the *Saccharum* complex. However, a comprehensive molecular breeding system with intercrossing across the complex should ultimately be undertaken in the development of novel hybrid biofuel crops.

2.2.1 Sugarcane (*Saccharum* species)

Sugarcane is a tall perennial C4 grass that is cultivated in tropical and subtropical regions of the world. Notably, this grass stores high concentrations of sucrose in the stem. Approximately, 65 to 70% of global sugar production in the form of sucrose is derived from sugarcane (FAO, 2003). The potential of sugarcane as an important energy crop was argued because of the advent of large-scale sugarcane-based ethanol production in Brazil (Tew & Cobill, 2008).

Sugarcane belongs to the genus *Saccharum*. Although six polyploid species are recognized within *Saccharum* (Table 2), modern cultivars for sugar production are mostly derived from interspecific hybridization between *S. officinarum* (2n = 8x = 80) and *S. spontaneum* (2n from 5x = 40 to 16x = 128) and are thus complex polyploids with variable chromosome numbers

(2n approx. from 100 to 120) (Henry, 2010; Piperidis et al., 2010). Of the other four species, *S. robustum*, *S. barberi*, and *S. sinense* have also provided minor contributions to the breeding of some modern sugarcane cultivars (Cheavegatti-Gianotto et al., 2011).

Species	Chromosome number (2n)	Genomic contribution to modern interspecific hybrid cultivars and elite lines (%)[a]	Note
S. barberi	From 111 to 120	-	Semi-sweet Indian cane
S. edule	From 60 to 80	-	Cultivated for edible inflorescence
S. officinarum	80	From 70 to 80	Domesticated sweet cane
S. robustum	60 or 80	-	Putative ancestor of *S. officinarum*
S. sinense	From 81 to 124	-	Semi-sweet Chinese cane
S. spontaneum	From 40 to 128	From 10 to 23	Wild cane found throughout Asia

[a] Values are the proportions of total chromosome complement reported by Piperidis et al. (2010) where the proportion of chromosomes derived from interspecific exchanges between *S. officinarum* and *S. spontaneum* are shown to be 8-13%.

Table 2. Species of *Saccharum* and their characteristics (modified from Henry, 2010)

The breeding of sugarcane as a dedicated biomass crop called "energy cane" can be categorized into three strategies with different objectives: the "sugar model", the "sugar-and-fiber model", and the "fiber-only model". The fiber yield of energy cane is important because of its use for electricity generation, cellulosic ethanol production, and so forth; fiber is considered a by-product in the sugar and sugar-and-fiber models or the main product in the fiber-only model (Tew & Cobill, 2008). In the sugar model, improved sugar yield and sugar content are the main foci, so traditional sugarcane cultivars can be used. In the sugar-and-fiber and fiber-only models, Type I and Type II energy canes, respectively, would be used. Tew & Cobill (2008) defined Type I and Type II energy canes as follows:

- Type I energy cane is bred and cultivated to maximize both its sugar and fiber content.
- Type II energy cane is bred and cultivated primarily or solely for its fiber content.

Recently, Japanese breeders succeeded in developing a high-quality Type I energy cane cultivar 'KY01-2044' with 1.5 times the total biomass yield and 1.3 times the total sugar yield than the major Japanese sugar-producing cultivar (Asia Biomass Office, 2010; Terajima et al., 2010). The new cultivar allowed the establishment of an experimental system for simultaneous production of ethanol and sucrose from total sugar with residual fiber as a heat source. In addition to ethanol production, this system is designed to produce an amount of sucrose comparable to conventional sugar production systems.

2.2.2 *Miscanthus* species

Miscanthus is a genus of C4 perennial rhizomatous grasses widely distributed in Asia and the Pacific Islands. *Miscanthus* was thought to consist of 17 species divided into four sections (Deuter, 2000). However, recent taxonomic analyses using molecular markers revealed that the genus can be reduced to approximately 11-12 species. The species *M. sinensis* ssp. *condensatus* is sometimes recognized at specific rank as *M. condensatus* (Clifton-Brown et al., 2008) (Table 3).

Species	Chromosome number and ploidy level[a]
M. floridulus (Labill.) Warb.	2n = 2x = 38
M. intermedius (Honda) Honda	2n = 6x = 114
M. longiberbis Nakai	-
M. lutarioparius	-
M. oligostachyus Stapf.	2n = 2x = 38
M. paniculatus (B. S. Sun) Renvoize & S. L. Chen	-
M. sacchariflorus (Maxim.) Hack.	2n = 2x = 38 (in China)
	2n = 4x = 76 (in Japan)
M. sinensis Anderss.	2n = 2x = 38
M. sinensis ssp. condensatus (Hackel) T. Koyama	2n = 2x = 38
M. tinctorius (Steud.) Hack.	2n = 2x = 38
M. transmorrisonensis Hayata	2n = 2x = 38
M. × *giganteus* Greef & Deuter ex Hodkinson and Renvoize (Hybrid species between *M. sacchariflorus* and *M. sinensis*)	2n = 3x = 57

[a] Major cytotypes from Deuter (2000) are shown.

Table 3. Species in the genus *Miscanthus* and their chromosome numbers

The basic chromosome number of *Miscanthus* species is 19, and polyploids and aneuploids are observed (Deuter, 2000). Of the species in Table 3, *M. sacchariflorus*, *M. sinensis*, *M. sinensis* ssp. *condensatus*, *M. floridulus*, and *M.* × *giganteus* are of interest for biomass production (Deuter, 2000). In particular, the triploid hybrid species *M.* × *giganteus* shows superior characteristics, such as high biomass yield, and is thus thought to be the most practical *Miscanthus* species for bioenergy production, especially in Europe and North America (Lewandowski et al., 2000; Pyter et al., 2007). The existence of this promising hybrid will further encourage interspecific hybridization within *Miscanthus*, because the genus has a lot of genetic diversity within and between species. In addition, the frost tolerance, growth at low temperature, and robustness against pests and diseases of *Miscanthus* make it a potential gene pool for developing widely-adaptable stress-tolerant cultivars of sugarcane through intergeneric hybridization (Clifton-Brown et al., 2008).

2.2.3 *Erianthus* species

Erianthus is a tall C4 perennial rhizomatous grass. The genus *Erianthus* was erected by André Michaux in *Flora Boreali-Americana* in 1803 for the New World species *E. saccharoides* Michaux (Tagane et al., 2011). Old World species have distinct morphology from New World species (Grassl, 1972, as cited in Tagane et al., 2011) and are characterized by the presence of a distinctive luteolin, di-C-glycoside, that is not present in the New World species (Williams et al., 1974). The genus is widely distributed in the Americas (New World species) and in the Mediterranean, India, China, South East Asia, and New Guinea (Old World species) (Amalraj & Balasundaram, 2006).

The seven Old World species comprise the section *Ripidium* (Table 4). The basic chromosome number of these *Erianthus* species is x = 10, the same as *Saccharum officinarum*. Polyploids are also observed, as in other genera in the *Saccharum* complex.

Species	Chromosome number[a]
E. arundinaceus (Retz.) Jesw.	2n = 30, 40, 60
E. bengalense (Retz.)	2n = 20, 30, 40, 60
E. elephantinus Hook. f.	2n = 20
E. hostii Griseb.	2n = 20
E. kanashiroi Ohwi	2n = 60
E. procerus (Roxb.) Raizada	2n = 40
E. ravennae (L.)	2n = 20

[a] Chromosome numbers shown by Amalraj & Balasundaram (2006) are shown.

Table 4. Species in the genus *Erianthus* sect. *Ripidium* and their chromosome numbers

The Old World species are increasingly used in sugarcane breeding because of their high biomass yields, drought tolerance, and resistance to pests and diseases. Two species in particular, *E. arundinaceus* and *E. procerus,* are considered most useful due to their disease resistance, ratooning ability, vigor, and environmental stress tolerance (Tagane et al., 2011). Recently, Ando et al. (2011) reported the wintering ability and high biomass yields of the *E. arundinaceus* clone JW630 grown experimentally in Nasushiobara, Japan, where the monthly mean minimum air temperatures from December to March are below freezing, indicating its potential as breeding stock for cold-tolerant sugarcane and as a dedicated lignocellulosic biofuel crop (Table 1).

2.3 Giant reed

Giant reed (*Arundo donax* L.) is a rhizomatous C3 perennial grass that evolved in Asia but is also considered a native to the Mediterranean region (Lewandowski et al., 2003). The grass is used to make reeds for woodwind musical instruments, in construction, and as a source of cellulose for rayon manufacture and paper pulp production (Perdue, 1958).

Recently, giant reed has gained attention as a source of lignocellulose for bioenergy production because of its high yield and strong pest resistance (Lewandowski et al., 2003). Studies of the giant reed in Central Italy demonstrated biomass production equivalent to or higher than *Miscanthus* × *giganteus* (Angelini et al., 2009) (Table 1). Because giant reed exhibits heavy metal tolerance, it could be used simultaneously for both phytoremediation and bioenergy production (Papazoglou, 2007; Papazoglou et al., 2005).

Giant reed is highly polyploid, with a possible base chromosome number x = 12 [(2n = 10x = 120 - 8 or 2n = 9x = 108 + 4) (Gorenflot et al., 1972) or (2n = 110) (Lewandowski et al., 2003)]. Although isozyme and DNA analyses have revealed genetic diversity in giant reed (Khudamrongsawat et al., 2004; Lewandowski et al., 2003), the plant is thought to produce no viable seed, owing to aberrant division of the megaspore mother cell (Bhanwra et al., 1982). This behavior suggests that conventional breeding through sexual hybridization cannot be performed.

3. Plant regeneration systems

In most case, effective plant regeneration systems with *in vitro* cell cultures are required to establish reliable transformation systems in plants. In this section, we summarize the regeneration systems of switchgrass, sugarcane, *Miscanthus*, *Erianthus*, and giant reed.

3.1 Switchgrass regeneration systems

Biotechnology research on the molecular breeding of switchgrass began in 1992 at the University of Tennessee with the support of Oak Ridge National Laboratory (Vogel & Jung, 2001). This work yielded the first reported plant regeneration system through embryogenesis and organogenesis (Denchev & Conger, 1994). In the experiments, mature caryopses and young leaf segments were used as explants for callus induction. Then, the influences of the synthetic auxins 2,4-dichlorophenoxyacetic acid (2,4-D) and 4-amino-3,5,6-trichloropyridine-2-carboxylic acid (picloram) in combination with 6-benzyladenine (BA) were examined on the plant regeneration system. The combination of 2,4-D and BA led to best results in experiments with mature caryopses (Denchev & Conger, 1995). Alexandrova et al. (1996a) increased the efficiency of switchgrass regeneration by using immature inflorescences obtained from node cultures in aseptic conditions; one immature inflorescence can produce hundreds of spikelets of the same genotype in a single Petri dish that can be easily used as explants for callus induction. Most recently, a new medium, LP9, for the production, maintenance, and regeneration of switchgrass callus was reported (Burris et al., 2009). Callus produced on LP9 can be easily propagated, maintains its regenerability for at least six months, and is adaptable to *Agrobacterium*-mediated transformation (Burris et al., 2009).

Embryogenic suspension cultures initiated from embryogenic callus were also established in switchgrass (Gupta & Conger, 1999), and osmotic pre-treatment with 0.3 M each of sorbitol and mannitol was effective for plant regeneration from suspension cultures (Odjakova & Conger, 1999). Another type of culture system that is highly attractive for gene transfer experiments was also established: multiple shoot clumps were induced from intact seedlings with various combinations of 2,4-D and 1-phenyl-3-(1,2,3-thidiazol-5-yl) urea (TDZ) (Gupta & Conger, 1998).

Most of the research mentioned above used the lowland cultivar 'Alamo'. However, because switchgrass is outcrossing and self-infertile, genetic variation exists not only among cultivars but also within a cultivar, indicating that the establishment of a totipotent tissue culture is not highly dependent on the cultivar itself but rather on screening highly regenerable genotypes as in other outcrossing plants, like ryegrasses (Takahashi et al., 2004). Therefore, Odjakova & Conger (1999) and Burris et al. (2009) used specific single genotypes Alamo 2702 and Alamo 2, respectively, in their experiments, and Alexandrova et al. (1996b) developed a micropropagation procedure with node culture for the multiplication of such selected genotypes. Interestingly, the selected line HR8, having high somatic embryogenic capacity, was recently bred using recurrent tissue culture selection to allow easy induction of embryogenic callus from mature seeds (Xu et al., 2011).

3.2 Sugarcane regeneration systems

The first sugarcane regeneration system via callus culture was reported in 1964 (Nickell, 1964, as cited in Lakshmanan, 2006) and was followed by numerous reports resulting from a great deal of research activity. These early reports have been well-reviewed elsewhere (Lakshmanan, 2006; Suprasanna et al., 2008a). Although many papers on plant regeneration from sugarcane calli are still being published, they generally vary only in reporting specific culture conditions, which must be optimized for each cultivar and genotype because of the outcrossing nature of sugarcane. Most callus cultures are supplemented with the auxin 2,4-D, but sometimes other synthetic auxins, such as picloram and 3,6-dichloro-2-methoxybenzoic acid (dicamba) are used, with or without the cytokinins BA, TDZ, or kinetin. Research on sugarcane regeneration systems may be stalled. To spark innovation, here we update the previous reviews (Lakshmanan, 2006; Suprasanna et al., 2008a) by describing innovative reports published since 2007.

In addition to plant hormones, exogenous amino acids may positively or negatively influence somatic embryogenesis in plants. In sugarcane tissue culture, the addition of arginine to culture media was found to significantly induce somatic embryogenesis and to promote plant regeneration (Nieves et al., 2008). Similarly, glycine, arginine, and cysteine positively affected somatic embryogenesis and subsequent plant regeneration; 0.75 mM glycine was the most effective treatment studied (Asad et al., 2009).

The moisture status of *in vitro* cultures influences plant regeneration in sugarcane. Partial desiccation has been reported to enhance plant regeneration from calli (Garcia et al., 2007; Kaur & Gosal, 2009). A similar phenomenon was observed even in irradiated embryogenic cultures (Suprasanna et al., 2008b).

A fundamental but important aspect of sugarcane tissue culture is the selection of the initial explants. In most previous reports, explants were obtained from field-grown plants. However, Garcia et al. (2007) recommended using sugarcane plants grown *in vitro* because they provided a year-round source of physiologically uniform explants and could be prevented from releasing large amounts of phenolic compounds, which can hamper tissue culture (Garcia et al., 2007). Basnayake et al. (2011) recommended using explants obtained from donor plants with good water supplies, especially when working with genotypes that are recalcitrant to tissue culture due to phenolic compounds.

3.3 *Miscanthus* regeneration systems

To the best of our knowledge, the first preliminary regeneration experiments in *Miscanthus* were conducted with immature inflorescences, mature leaves, immature leaves, nodal segments, internodal segments, meristematic regions, and ovules as explants, which were cultured on media containing a wide range of auxin types (Gawel et al., 1987, as cited in Gawel et al., 1990). In these experiments, immature inflorescences were the only explants that produced calli, and only media containing 2,4-D or picloram with no cytokinin produced a regeneration response (Gawel et al., 1987, as cited in Gawel et al., 1990). The research group thus focused on immature inflorescences as explants and published that 9.0 µM of 2,4-D was effective for callus induction and subsequent plant regeneration in *M. sinensis* cultivars 'Gracillimus', 'Variegatus', and 'Zebrinus' (Gawel et al., 1990). In these experiments, callus induction and plant regeneration were not distinguished from one another; shoot formation occurred during successive culture on the same culture medium as callus induction (Gawel et al., 1990).

Many researchers have focused on the high-biomass species $M. \times giganteus$. The auxins 2,4-D or 2,4,5-trichlorophenoxyacetic acid in combination with the cytokinin 6-furfurylaminopurine were best for inducing regenerable embryogenic callus from immature inflorescences (Lewandowski & Kahnt, 1992). Although immature inflorescences emit fewer browning substances than shoot tips or leaf primordia (Lewandowski & Kahnt, 1993), browning substances from the explants were detrimental to tissue culture (Lewandowski & Kahnt, 1992). Ascorbic acid, cysteine, and watering the explants were not effective treatments, but liquid culture gave better results because the browning substances did not accumulate around the explants or calli (Lewandowski & Kahnt, 1992).

The effects of different explants, such as shoot apices, leaves and root sections of $in vitro$-propagated plants and leaf and immature inflorescence sections from greenhouse-grown plants, were examined in $M. \times giganteus$. Shoot apices had the highest percentage of embryogenic callus formation, while immature inflorescence-derived callus had the highest regenerability (Holme & Petersen, 1996). The growth conditions of donor plants influence tissue culture (Creemers-Molenaar et al., 1988), as has been observed in $M. \times giganteus$ (Holme & Petersen, 1996). Leaf explants from $in vitro$-propagated shoots and from greenhouse-grown plants showed differences in embryogenic callus formation; the best results were obtained from leaves of $in vitro$-propagated shoots and older leaves of greenhouse-grown plants (Holme & Petersen, 1996). However, supplying BA to the callus induction medium led to different results: a higher percentage of regenerable shoot-forming callus was formed on shoot apices compared with leaf sections of $in vitro$-grown shoots when 0.4 µM BA was supplied (Petersen, 1997). Also, small immature inflorescences of $M. \times giganteus$, between 2.5 and 8 mm, were more suitable for embryogenic callus formation than explants from shorter or longer inflorescences, shoot apices or leaf explants, indicating that the size and type of explant influence culture responses (Petersen et al., 1999). Furthermore, different carbon sources, and their sterilization methods, influenced $M. \times giganteus$ tissue culture. Significant differences were reported for carbon sources and their sterilization methods in tissue cultures of various explant-derived calli. Leaf explants were more affected by the carbon sources than were shoot apices or immature inflorescences, and both callus proliferation and plant regeneration were generally improved by the use of filter-sterilized carbon sources (Petersen et al., 1999).

As mentioned above, severe browning is a major problem in $Miscanthus$ tissue culture, but supplying proline to callus induction and suspension cultures effectively prevented browning in $M. \times giganteus$ (Holme et al., 1997). Proline is thought to inhibit polyphenol oxidase, which causes enzymatic browning of cultured tissues (Öztürk & Demir, 2002). The addition of proline to callus induction media increased embryogenic callus formation on shoot apices and leaf explants. However, results varied with proline concentration and the basal salts in the medium. Specifically, 12.5 to 50 mM proline in callus induction media with Murashige & Skoog (MS) salts (Murashige & Skoog, 1962) increased embryogenic callus formation more than media with N6 salts (Chu et al., 1975). The inhibitory effects of proline on tissue browning in $Miscanthus$ were confirmed by another group (Głowacka et al., 2010). Conversely, adding honey instead of sucrose to callus induction media inhibited browning of cultured immature inflorescences, and a combination of the honey and banana pulp was best for inducing regenerable callus in $M. \times giganteus$ (Płażek & Dubert, 2009, 2010).

Media component	Media		
(mg L⁻¹ final conc.)	HB[a]	MS	N6
Macroelements			
KNO_3	2000	1900	2830
NH_4NO_3	1500	1650	
$(NH_4)_2SO_4$			463
$CaCl_2 \cdot 2H_2O$	200	440	166
$MgSO_4 \cdot 7H_2O$	300	370	185
KH_2PO_4	300	170	400
Microelements			
Na_2-EDTA	37.25	37.3	37.25
$FeSO_4 \cdot 7H_2O$	27.85	27.8	27.85
$MnSO_4 \cdot 4H_2O$	11.25	22.3	4.4
$ZnSO_4 \cdot 7H_2O$	4.3	8.6	1.5
$CuSO_4 \cdot 5H_2O$	0.0125	0.025	
$CoCl_2 \cdot 6H_2O$	0.0125	0.025	
H_3BO_3	3.1	6.2	1.6
KI	0.415	0.83	0.8
Vitamins			
Inositol	100	100	
Nicotinic acid	0.5	0.5	0.5
Pyridoxine HCl	0.5	0.5	0.5
Thiamine HCl	0.4	0.1	1
Another organic			
Glycine	2	2	2

[a] Composition presented in Sun et al. (1999).

Table 5. Composition of HB, MS, and N6 basal media

Recently, studies involving tissue cultures of M. sinensis have been reported. Głowacka & Jeżowski (2009a, 2009b) reported culturing anthers of M. sinensis and subsequently demonstrated the effects of inflorescence developmental stage on callus induction and plant regeneration (Głowacka et al., 2010). Mature seeds can also serve as explants for tissue culture in M. sinensis. Mature seeds of 18 accessions from various sites in Japan were subjected to callus induction, and a combination of a relatively high 2,4-D concentration (5 mg L⁻¹) and a relatively low BA concentration (0.01 mg L⁻¹) efficiently induced regenerable calli that could be used to produce transgenic Miscanthus plants (Wang et al., 2011). Interestingly, there was a correlation between the average annual air temperature at accession collecting sites and the frequency of embryogenic callus induction; seeds collected from warmer regions formed a higher percentage of embryogenic calli (Wang et al., 2011). Most recently, Zhang et al. (2011) germinated seeds collected in China and compared explants from the epicotyls, young leaves, and radicles. The epicotyl was best for embryogenic callus formation and plant regeneration from the callus. In the experiments, Holley & Baker (HB) medium (Holley & Baker, 1963, as cited in Zhang et al., 2011) was better for callus induction than other common basal media, such as MS medium, N6 medium (Chu et al., 1975), or half strength of the MS medium. Zhang et al. surmised that HB medium would work well for callus induction in other grass species, because the medium had been effective for callus culture of wheat (Sun et al., 1999). The components of HB medium will be of interest to many researchers, so we compare HB, MS, and N6 media in Table 5. We were unable to obtain the original publication (Holley & Baker, 1963, as reported in Zhang et al., 2011), so we give the composition of HB basal medium presented in Sun et al. (1999).

3.4 *Erianthus* regeneration systems

To date, only one published report concerns the plant regeneration system of *Erianthus* species. Callus induction and plant regeneration from calli was achieved in *E. elephantinus* (Jalaja & Sreenivasan, 1999). Calli were induced from expanding leaves, leaf sheaths, and immature inflorescences as explants on MS basal medium containing 2 mg L⁻¹ 2,4-D. As in *Miscanthus* species, tissue browning occurred, especially in explants from leaf sheaths (Jalaja & Sreenivasan, 1999). Unfortunately, detailed data on the frequencies of callus induction and plan regeneration from the callus were not reported because the study focused mainly on somaclonal variation, such as morphology, pollen fertility, and chromosome aberrations, of the regenerants (Jalaja & Sreenivasan, 1999).

3.4.1 Protocol for plant regeneration system of *Erianthus arundinaceus*

We have recently succeeded in establishing a plant regeneration system in *E. arundinaceus*. Here, we present the simplified protocol. All culture media used were based on MS medium containing 3% (w v⁻¹) sucrose, adjusted to pH 5.8, and solidified with 0.25% (w v⁻¹) Gelrite (Wako, Osaka, Japan). Components of the culture media are shown in Table 6.

Media	Components[a]
Plant maintenance medium	0.3% (w v⁻¹) activated charcoal 3%(w v⁻¹) sucrose
Callus induction medium	5 mg L⁻¹ 2,4-D 25 mM L-proline 750 mg⁻¹ $MgCl_2 \cdot 6H_2O$ 3%(w v⁻¹) sucrose
Multiple shoot formation medium	1 mg L⁻¹ BA 3 mg L⁻¹ NAA 3%(w v⁻¹) sucrose
Shoot elongation medium	0.3 mg L⁻¹ GA 3%(w v⁻¹) sucrose

[a] Additives added to MS basal medium are shown for each medium.

Table 6. Culture media used in our plant regeneration system for *Erianthus arundinaceus*

- Establishment of *in vitro*-grown plants

We obtained *in vitro*-grown plants of the *E. arundinaceus* clone JW630 (Ando et al., 2011). Shoot tips, each containing an apical meristem, of greenhouse-grown plants (approx. 3 cm long) were used as explant sources. After removing the mature outer leaves, the shoot tips were submerged in 70% ethanol for 1 min, surface-sterilized in 10% (v v⁻¹) sodium hypochlorite solution (1% available chlorine) for 20 min, rinsed twice in sterile distilled water under aseptic conditions, aseptically stripped of more outer leaves, and cultured on plant maintenance medium under long-day conditions [16 h light (80 µmol m^{-2} s^{-1})/8 h dark; 28°C]. The explants were subcultured every two weeks until rooting.

- Callus induction from axillary buds of *in vitro*-grown plants

Axillary buds at the proximal end of each leaf were isolated from *in vitro*-grown plants under a stereomicroscope. The isolated buds were placed on callus induction medium, and cultured under continuous fluorescent light (40 µmol m^{-2} s^{-1}) at 25°C. They were subcultured every two weeks.

- Plant regeneration from calli

For plant regeneration, calli were transferred to the plant maintenance medium and cultured under long-day conditions [16 h light (80 µmol m^{-2} s^{-1})/8 h dark; 28°C].

- Results and tips

As with other C4 plants, such as switchgrass, sugarcane, and *Miscanthus*, explant browning was severe at first. Rooting of the explants took about two months, but once rooted, the *in vitro* plants grew vigorously on the plant maintenance medium (Figure 1A). Calli were easily induced from axillary buds of *in vitro*-grown plants (Figure 1B), and plant regeneration was observed from the calli (Figure 1C). However, calli induced from explants from field-grown plants were rare due to severe bacterial contamination and tissue browning, even when we used apical meristems, immature inflorescences, and young leaf rolls (data not shown). The *in vitro*-grown plants produced multiple shoot clumps when cultured on multiple shoot formation medium (Figure 2A), and the newly formed short shoots maintained their morphological states and could elongate to normal morphology after being transferred to shoot elongation medium containing gibberellic acid (GA) (Figure 2B).

Fig. 1. Plant regeneration system for *Erianthus arundinaceus*. An *in vitro*-grown plant cultured for three months (A), a callus induced from an axially bud of the *in vitro*-grown plants (B), and plant regeneration from calli (C).

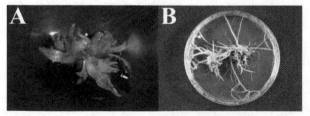

Fig. 2. Multiple shoot formation in *Erianthus arundinaceus*. A multiple shoot clump derived from an *in vitro*-grown plant (A), and elongated shoots from the multiple shoot clumps (B).

3.5 Giant reed regeneration system

To the best of our knowledge, only one scientific article on the regeneration of giant reed has been published so far (Takahashi et al., 2010). In the system, *in vitro* propagation was first optimized for the year-round production of explants (Figure 3).

Fig. 3. An *in vitro*-grown giant reed plant. The photo was taken two months after subculture (adapted from Takahashi et al., 2010).

Calli were induced from axillary buds isolated from the *in vitro*-grown plants. Several combinations of 2,4-D and BA were used for callus induction. They influenced both callus induction frequency and later plant regeneration on hormone-free media because of possible carryover effects of plant hormones in the original callus induction media. Media supplemented with 3 mg L^{-1} 2,4-D appeared to stabilize callus formation and subsequent shoot formation on hormone-free medium (Figure 4).

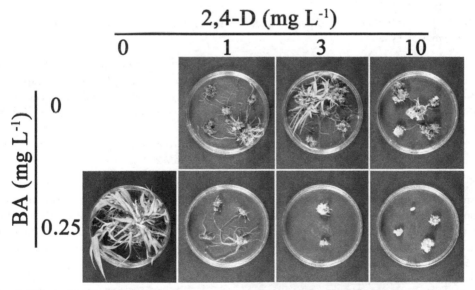

Fig. 4. Effect of original callus induction media on subsequent shoot formation on hormone-free media in giant reed. The concentrations of 2,4-D and BA are those in original callus induction media (modified and adapted from Takahashi et al., 2010).

In total, 11 genotypes were treated under optimized conditions, and data suggest genotypic effects in tissue culture response, although the giant reed is thought to propagate asexually because its seeds are non-viable. Ex-callus induction frequencies and shoot formation frequencies ranged from 82.9 to 100% and from 0 to 37.5%, respectively.

4. Transformation systems

We summarize here several earlier reports on transformation systems in switchgrass, sugarcane, *Miscanthus*, and giant reed. We know of no such reports for *Erianthus*.

4.1 Switchgrass transformation systems

There are some reports on the establishment of transformation systems in switchgrass. Gene transfer methods and names of transgenes used so far are shown in Table 7.

Transgenes[a]	Gene transfer methods	Reference
sgfp, [*bar*]	Gene gun	Richards et al. (2001)
uidA, [*bar*]	*Agrobacterium*	Somleva et al. (2002)
pporRFP, [*hph*]	*Agrobacterium*	Burris et al. (2009)
uidA, *GUSPlus*, [*hph*]	*Agrobacterium*	Xi et al. (2009)
sgfp, [*hph*]	*Agrobacterium*	Li & Qu (2011)
sgfp, *GUSPlus*, [*bar*][b], [*hph*]	*Agrobacterium*	Xu et al. (2011)
uidA, [*bar*], [*hph*], [*nph*]	*Agrobacterium*	Song et al. (2011)

[a] Selectable marker genes in brackets.
[b] Not used as a selectable marker.

Table 7. A list of transgenes and gene transfer methods in used in reported switchgrass transformation systems

The first transgenic switchgrass plants were produced by the gene gun method (Richards et al., 2001). Tungsten particles coated with a plasmid harboring the *sgfp* and *bar* genes were introduced into immature inflorescence-derived embryogenic calli. In total, 2,430 calli were subjected to the gene transfer, and 97 plants eventually showed tolerance, conferred by the *bar* gene, to 0.1% Basta herbicide (Richards et al., 2001). Basta tolerance was also observed in T_1 transgenic progeny resulting from crosses between transgenic and nontransgenic control plants, indicating inheritance of the *bar* gene (Richards et al., 2001).

All subsequent reports on switchgrass transformation have used *Agrobacterium*-mediated transformation (Table 7). The first such report involved sophisticated experiments that examined the effects of various explants of different genotypes for callus induction and of various concentrations of acetosyringone during inoculation and cocultivation on gene transfer efficiency (Somleva et al., 2002). The gene transformation efficiencies ranged from 0 to 97.3%, but the efficiencies were clearly depended on genotype and kind of explant. Improvements in basal media should lead to efficient transformation in switchgrass. Recently, switchgrass Type II calli were induced from newly-developed LP9 medium (see section 3.1) and were subjected to the *Agrobacterium*-mediated transformation with a final transformation efficiency of 4.4% (Burris et al., 2009).

Because *Agrobacterium* is the causal agent of crown gall, interactions between *Agrobacterium* and plant cells during gene transfer often trigger an undesired plant defense mechanism, called hypersensitive cell death, that leads to lower gene transfer efficiency. Thus, highly compatible *Agrobacterium* strains must be selected for plant transformation. Indeed, different strains of *Agrobacterium*-strain varied in their transient transgene expression efficiencies (Chen et al., 2010) and in stable transformation rates in switchgrass (Song et al., 2011; Xi et al., 2009). For stable transformation of switchgrass, the most noteworthy *Agrobacterium* strain so far has been EHA105; two different research groups found it to be more effective than other strains, such as AGL1, GV3101, and LBA4404 (Song et al., 2011; Xi et al., 2009).

Li & Qu (2011) recently developed the most high-through-put system to date, with a stable transformation efficiency of over 90%. Although most previous transformation systems employed the cultivar 'Alamo', the authors a found higher transformation efficiency with the cultivar 'Performer'. Their modifications included infection under vacuum, co-cultivation in desiccation conditions, resting between co-cultivation and selection, and L-proline supplementation in the callus culture and selection media (Li & Qu, 2011). Interestingly, this system also employed *Agrobacterium* strain EHA105, suggesting that it may be highly compatible with 'Performer'. Switchgrass transformation using a EHA105 with the switchgrass selected line HR8, which has high somatic embryogenic capacity (Xu et al., 2011) (see section 3.1), may be a promising avenue for future research.

Transgene inheritance and phenotypic expression in progeny derived from *Agrobacterium*-mediated transformation were observed in some of these reports (Li & Qu, 2011; Somleva et al., 2002; Xi et al., 2009).

4.2 Sugarcane transformation systems

Numerous reports on transformation systems in sugarcane have been previously reviewed (Hotta et al., 2010; Lakshmanan et al., 2005; Suprasanna et al., 2008a). Here, we summarize two key studies published in 2011.

Sophisticated adjustment of culture conditions for several cultivars have resulted in transformation systems widely-adaptable to different genotypes. Basnayake et al. (2011) investigated the effects of 2,4-D levels during callus proliferation, geneticin concentrations during selection, and/or light intensity during regeneration in 16 Australian sugarcane cultivars destined for microprojectile-mediated transformation. This study will be a useful guide for the rapid optimization of key tissue culture variables for efficient genetic transformation of other sugarcane cultivars.

Microprojectile-mediated gene transfer is the most common gene transfer method in sugarcane (Hotta et al., 2010), but the method often causes complex transgene integration that may result in gene silencing. However, minimal expression cassettes lacking vector backbone sequences have been reported to support simple transgene integration in plants. Most recently, a linear minimal expression cassette for the *npt* gene was introduced into embryogenic callus by microprojectile-mediated gene transfer of different amounts of the expression cassette DNA (Kim et al., 2011). Genomic DNA from transgenic sugarcane plants displayed two to 13 *npt* hybridization signals on Southern blots, and the authors observed a trend toward reduced transgene integration complexity and reduced transgene expression levels when lower amounts of the minimal expression cassette were used per shot. This suggests that backbone free minimal expression cassettes might be efficiently integrated and expressed in sugarcane and other plant species.

4.3 *Miscanthus* transformation systems

The first transgenic *Miscanthus* plants produced via particle bombardment were reported in *Miscanthus sacchariflorus* (Zili et al., 2004, as cited in Jakob et al., 2009), although we were unable to obtain the original reference and cannot report the details. Later, transgenic *M. sinensis* plants were produced by particle bombardment methods (Wang et al., 2011); four transgenic plants containing a foreign *hph* gene with or without the *gfp* were recovered from 120 bombarded calli, and foreign gene expression was confirmed by reverse transcription-PCR analyses. So far, there are no other reports on *Miscanthus* transformation.

4.4 Giant reed transformation systems

An optimized particle bombardment protocol for gene transfer with embryogenic calli was recently reported in giant reed (Dhir et al., 2010). In the study, embryogenic calli were induced from inflorescence segments collected from field-grown mature plants, and several physical parameters, such as helium pressure, bombardment distance to target tissue, and vacuum pressure, together with other factors such as gold microparticle size, DNA concentration, and the number of bombardments, were examined with transient expression of beta-glucuronidase (GUS) and green fluorescent protein (GFP) genes (Dhir et al., 2010). Unfortunately, however, no transgenic plants were obtained due to the lack of regeneration potential of the tissue culture system (Dhir et al., 2010).

We also tried to produce transgenic giant reed plants by using particle bombardment-mediated gene transfer in combination with the plant regeneration system described above (see section 3.5). We obtained more than 100 hygromycin-resistant calli from ca. 5,000 bombarded calli (Figure 5A). Transformation of these calli with foreign *hph* was confirmed by PCR analyses (data not shown). However, we could not obtain transgenic plants from the resistant calli, because the calli had lost their shoot formation potential and produced only adventitious roots on regeneration medium (Figure 5B). Thus, improvements in these tissue culture systems are needed to produce transgenic giant reed plants.

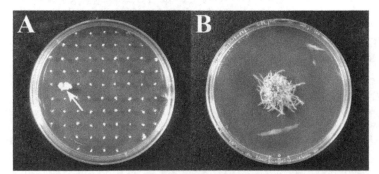

Fig. 5. Transgenic calli of giant reed. A hygromycin-resistant callus propagated during selective culture with the antibiotic hygromycin (arrow) (A), and adventitious root formation from the calli on plant regeneration medium (B).

5. Transgenic plants with improved traits

Here, we summarize reports of transgenic switchgrass and sugarcane plants having distinctive improved traits that will be of significance for biofuel production. In switchgrass, we know of only four such reports, as shown in Table 8. However, in sugarcane, several reports exist on the metabolic engineering of value-added sugarcane via carbohydrate biosynthesis and increased sucrose accumulation; the target genes and the resulting transgenic plants have been discussed in previous reviews (Hotta et al., 2010; Lakshmanan et al., 2005; Suprasanna et al., 2008a; Waclawovsky et al., 2010). Thus, for transgenic sugarcane, we focus here on reports published from 2007 to 2011 (Table 8). No studies have yet been reported for improved traits in *Miscanthus*, *Erianthus*, or giant reed.

Transgenes[a]	Gene transfer methods	Reference
Switchgrass		
phaA, phaB, phaC, [*bar*]	*Agrobacterium*	Somleva et al. (2008)
PvCAD[b] [*hph*]	*Agrobacterium*	Fu et al. (2011b)
PviCAD2[b] [*hph*]	*Agrobacterium*	Saathoff et al. (2011)
COMT[b] [*hph*]	*Agrobacterium*	Fu et al. (2011a)
Sugarcane		
phaA, phaB, phaC, [*npt*]	Gene gun	Petrasovits et al. (2007)
mds6pdh, zmglk, [*npt*]	Gene gun	Chong et al. (2007)
*HIS*CaneCPI-1, [*npt*]	Gene gun	Ribeiro et al. (2008)
avidin[c], [*adhA*]	Gene gun	Jackson et al. (2010)
mulB, [*npt*]	Gene gun	Hamerli & Birch (2011)

[a] Selectable marker genes in brackets.
[b] Partial inverted repeat sequences for RNAi technology.
[c] Several signal peptides were fused for subcellular localization.

Table 8. A list of transgenic plants with improved traits

5.1 Transgenic switchgrass plants

A pioneering study of transgenic switchgrass involved the successful production of value-added transgenic plants that could synthesize polyhydroxybutyrate (PHB), a biodegradable alternative to standard consumer plastic (Somleva et al., 2008). Primary transgenic plants containing up to 3.72% dry weight of PHB in leaf tissues and 1.23% dry weight of PHB in whole tillers were obtained. Polymer accumulation was also analyzed in the T1 generation. This achievement resulted from the successful expression of a functional multigene pathway involving complex metabolic engineering (Somleva et al., 2008).

In the first half of 2011, three reports were published on the downregulation of lignin biosynthesis genes using RNAi technology to improve ethanol production from lignocellulosic biomass (Fu et al., 2011a; Fu et al., 2011b; Saathoff et al., 2011). The first two reports were published at almost the same time and both targeted the gene encoding cinnamyl alcohol dehydrogenase (CAD), a key enzyme for catalyzing the last step of lignin monomer biosynthesis. The resulting transgenic plants showed significantly fewer transcripts of the target gene, reduced CAD activity, lower lignin content (Fu et al., 2011b; Saathoff et al., 2011), and altered lignin composition (Fu et al., 2011b). Furthermore, these modified lignin biosynthesizers had improved sugar release from cell walls with or without chemical pre-treatment (Fu et al., 2011b; Saathoff et al., 2011). In another study, the expression of the lignin biosynthesis-related O-methyltransferase gene was similarly downregulated, and the resultant transgenic plants had reduced lignin content, digestibility by cellulase, and up to 38% more efficient ethanol yield using conventional biomass fermentation processes (Fu et al., 2011a).

5.2 Transgenic sugarcane plants

PHB production is also targeted in sugarcane molecular breeding. The same gene set and the same strategy used with transgenic switchgrass (Somleva et al., 2008) were also employed for producing transgenic sugarcane (Petrasovits et al., 2007). In the study, PHB accumulated in leaves of transgenic plants to a maximum of 1.88% of dry weight without obvious deleterious effects. The PHB concentration in culm internodes of the transgenic plants was much lower (0.0033%) than in leaves (Petrasovits et al., 2007).

The same research group also examined sugar manipulation in sugarcane by engineering a new carbon sink for the six-carbon sugar alcohol sorbitol; sorbitol has intrinsic value as a non-caloric sweetener and is also used to manufacture ascorbic acid (Chong et al., 2007). Transgenic sugarcane plants expressing the *mds6pdh* gene accumulated the sorbitol. The average amounts of sorbitol detected in the most productive line were 120 mg g^{-1} dry weight (equivalent to 61% of the soluble sugars) in the leaf lamina and 10 mg g^{-1} dry weight in the stalk pith, but the accumulation caused evident necrosis in expanding leaves and reduced growth (Chong et al., 2007). More recently, another group focused on production of the sucrose isomer trehalulose (Hamerli & Birch, 2011). The transgenic sugarcane plants accumulated the trehalulose up to 600 mM in juice from mature nodes. Contrary to the case of the sorbitol-accumulating sugarcane, the trehalulose-accumulating transgenic plants were vigorous, and trehalulose production in selected lines was retained over multiple vegetative generations under glasshouse and field conditions (Hamerli & Birch, 2011).

A high value-added dedicated lignocellulosic biofuel crop could also be developed by accumulating heterologous functional proteins that could be used for therapeutic, industrial, or other purposes. This concept is adaptable to the production of cystatin, a natural inhibitor of cysteine proteinases, that can be used to protect against insect attacks (Ribeiro et al., 2008). A transformed sugarcane plant expressing high levels of the His-tagged cystatin gene $_{HIS}CaneCPI-1$ was reported to be useful for production and purification of functional $_{HIS}$CaneCPI-1 protein (Ribeiro et al., 2008). The $_{HIS}$CaneCPI-1 protein purified through affinity chromatography in a nickel column was able to inhibit the catalytic activity of midgut cysteine proteinase purified from the sugarcane weevil *Sphenophorus levis* and human cathepsin L in nanomolar amounts, indicating that this system can be used for the production of functional recombinant proteins (Ribeiro et al., 2008). The accumulated recombinant proteins may disturb cell metabolism. If so, subcellular targeting of the recombinant proteins would be necessary to optimize protein yield and avoid detrimental effects of the accumulated protein. Subcellular targeting peptides are thus considered a useful tool for compartmentalization of recombinant proteins. To find ideal subcellular targeting for recombinant protein accumulation in sugar cane, Jackson et al. (2010) used the glycoprotein avidin, which is potentially toxic to cells, as a test protein fused with several subcellular targeting peptides. Accumulation of avidin was directed to the apoplast, endoplasmic reticulum, and lytic and delta type vacuoles. The study identified the delta type vacuole as a promising target, but the efficiency may depend on tissue type. If the protein is resistant or can be protected by proteolytic attack, the lytic vacuole would be a preferable target (Jackson et al., 2010).

6. Conclusions

We described recent progress in the identification of candidate herbaceous perennials for dedicated lignocellulosic biofuel crops. The biotechnological approaches for molecular breeding of these plants as dedicated biofuel crops are still in immature stages. Biotic and abiotic stress tolerance and herbicide resistance are the minimum required traits. Other traits and related technologies will be important and will make large impacts in the molecular breeding of biofuel crop; these include low-fertilizer needs, sophisticated manipulation technology of secondary cell wall biosynthesis, high-efficiency photosynthesis for high productivity, promoted and synchronized flowering for hybridization, accelerated generation and, especially, organellar transformation for the effective accumulation of high-value chemicals and proteins. These traits and candidate target genes are reviewed in other works (Jakob et al., 2009; Vega-Sánchez & Ronald, 2010; Vogel & Jung, 2001).

7. Acknowledgments

This work was supported by the Japan Society for the Promotion of Science (JSPS) Grant-in-Aid for Scientific Research (C) (23580027).

8. References

Alexandrova, K. S., Denchev, P. D. & Conger, B. V. (1996a). In vitro development of inflorescences from switchgrass nodal segments. *Crop Science*, Vol. 36, No. 1, pp. 175-178, ISSN 0011-183X

Alexandrova, K. S., Denchev, P. D. & Conger, B. V. (1996b). Micropropagation of switchgrass by node culture. *Crop Science*, Vol. 36, No. 6, pp. 1709-1711, ISSN 0011-183X

Alwala, S., Suman, A., Arro, J. A., Veremis, J. C. & Kimbeng, C. A. (2006). Target region amplification polymorphism (TRAP) for assessing genetic diversity in sugarcane germplasm collections. *Crop Science*, Vol. 46, pp. 448-455, ISSN 0011-183X

Amalraj, V. A. & Balasundaram, N. (2006). On the taxonomy of the members of 'Saccharum complex'. *Genetic Resources and Crop Evolution*, Vol. 53, No. 1, pp. 35-41, ISSN 0925-9864

Ando, S., Sugiura, M., Yamada, T., Katsuta, M., Ishikawa, S., Terashima, Y., Sugimoto, A. & Matsuoka, M. (2011). Overwintering ability and dry matter production of sugarcane hybrids and relatives in the Kanto region of Japan. *Japan Agricultural Research Quarterly*, Vol. 45, No. 3, pp. 259-267, ISSN 0021- 3551

Angelini, L. G., Ceccarini, L., Nassi o Di Nasso, N. & Bonari, E. (2009). Comparison of *Arundo donax* L. and *Miscanthus* x *giganteus* in a long-term field experiment in

Central Italy: Analysis of productive characteristics and energy balance. *Biomass and Bioenergy*, Vol. 33, No. 4, pp. 635-643, ISSN 0961-9534

Aravindhakshan, S. C., Epplin, F. M. & Taliaferro, C. M. (2010). Economics of switchgrass and miscanthus relative to coal as feedstock for generating electricity. *Biomass and Bioenergy*, Vol. 34, No. 9, pp. 1375-1383, ISSN 0961-9534

Asad, S., Arshad, M., Mansoor, S. & Zafar, Y. (2009). Effect of various amino acids on shoot regeneration of sugarcane (*Sacchrum officinarum* L.). *African Journal of Biotechnology*, Vol. 8, No. 7, pp. 1214-1218, ISSN 1684-5315

Asia Biomass Office (2010). To expand domestic biofuel production in Japan. In: *Biomass Topics*, 17.09.2011, Available from:
<http://www.asiabiomass.jp/english/topics/1005_02.html>

Basnayake, S. W. V., Moyle, R. & Birch, R. G. (2011). Embryogenic callus proliferation and regeneration conditions for genetic transformation of diverse sugarcane cultivars. *Plant Cell Reports*, Vol. 30, No. 3, pp. 439-448, ISSN 0721-7714

Bhanwra, R. K., Choda, S. P. & Kumar, S. (1982). Comparative embryology of some grasses. *Proceedings of the Indian National Science Academy*, Vol. 48, pp. 152-162

Bouton, J. (2008). Improvement of switchgrass as a bioenergy crop, In: *Genetic Improvement of Bioenergy Crops*, W. Vermerris (Ed.), pp. 295-308, Springer, ISBN 978-0-387-70804-1, New York, USA

Burris, J. N., Mann, D. G. J., Joyce, B. L. & Stewart, C. N., Jr. (2009). An improved tissue culture system for embryogenic callus production and plant regeneration in switchgrass (*Panicum virgatum* L.). *Bioenergy Research*, Vol. 2, No. 4, pp. 267-274, ISSN 1939-1234

Cheavegatti-Gianotto, A., de Abreu, H. M. C., Arruda, P., Bespalhok Filho, J. C., Burnquist, W. L., Creste, S., di Ciero, L., Ferro, J. A., de Oliveira Figueira, A. V., de Sousa Filgueiras, T., Grossi-de-Sá, M. d. F., Guzzo, E. C., Hoffmann, H. P., de Andrade Landell, M. G., Macedo, N., Matsuoka, S., de Castro Reinach, F., Romano, E., da Silva, W. J., de Castro Silva Filho, M. & César Ulian, E. (2011). Sugarcane (*Saccharum* X *officinarum*): A reference study for the regulation of genetically modified cultivars in Brazil *Tropical Plant Biology*, Vol. 4, No. 1, pp. 62-89, ISSN 1935-9756

Chen, X., Equi, R., Baxter, H., Berk, K., Han, J., Agarwal, S. & Zale, J. (2010). A high-throughput transient gene expression system for switchgrass (*Panicum virgatum* L.) seedlings. *Biotechnology for Biofuels*, Vol. 3, No. 9, Retrieved from:
<http://www.biotechnologyforbiofuels.com/content/3/1/9>

Chong, B. F., Bonnett, G. D., Glassop, D., O'Shea, M. G. & Brumbley, S. M. (2007). Growth and metabolism in sugarcane are altered by the creation of a new hexose-phosphate sink. *Plant Biotechnology Journal*, Vol. 5, No. 2, pp. 240-253, ISSN 1467-7644

Chu, C. C., Wang, C. C., Sun, C. S., Hsu, C., Yin, K. C., Chu, C. Y. & Bi, F. Y. (1975). Establishment of an efficient medium for anther culture of rice through comparative experiments on nitrogen-sources. *Scientia Sinica*, Vol. 18, No. 5, pp. 659-668

Clifton-Brown, J., Chiang, Y.-C. & Hodkinson, T. R. (2008). *Miscanthus*: Genetic resources and breeding potential to enhance bioenergy production, In: *Genetic Improvement of Bioenergy Crops*, W. Vermerris (Ed.), pp. 273-294, Springer, ISBN 978-0-387-70804-1, New York, USA

Creemers-Molenaar, J., Loeffen, J. P. M. & Van der Valk, P. (1988). The effect of 2,4-dichlorophenoxyacetic acid and donor plant environment on plant regeneration from immature inflorescence-derived callus of *Lolium perenne* L. and *Lolium multiflorum* L. *Plant Science*, Vol. 57, No. 2, pp. 165-172, ISSN 0168-9452

Denchev, P. D. & Conger, B. V. (1994). Plant-regeneration from callus-cultures of switchgrass. *Crop Science*, Vol. 34, No. 6, pp. 1623-1627, ISSN 0011-183X

Denchev, P. D. & Conger, B. V. (1995). In-vitro culture of switchgrass - influence of 2,4-D and picloram in combination with benzyladenine on callus initiation and regeneration. *Plant Cell, Tissue and Organ Culture*, Vol. 40, No. 1, pp. 43-48, ISSN 0167-6857

Deuter, M. (2000). Breeding approaches to improvement of yield and quality in *Miscanthus* grown in Europe, In: *European miscanthus improvement – final Report September 2000*, I. Lewandowski & J. C. Clifton-Brown (Eds.), pp. 28-52, Institute of Crop Production and Grassland Research, University of Hohenheim, Stuttgart, Germany

Dhir, S., Knowles, K., Pagan, C. L., Mann, J. & Dhir, S. (2010). Optimization and transformation of *Arundo donax* L. using particle bombardment. *African Journal of Biotechnology*, Vol. 9, No. 39, pp. 6460-6469, ISSN 1684-5315

FAO (2003). Important commodities in agricultural trade: sugar. In: *FAO fact sheets: Input for the WTO ministerial meeting in Cancún*, 16.09.2011, Available from: <http://www.fao.org/docrep/005/y4852e/y4852e11.htm>

Fu, C. X., Mielenz, J. R., Xiao, X. R., Ge, Y. X., Hamilton, C. Y., Rodriguez, M., Chen, F., Foston, M., Ragauskas, A., Bouton, J., Dixon, R. A. & Wang, Z. Y. (2011a). Genetic manipulation of lignin reduces recalcitrance and improves ethanol production from switchgrass. *Proceedings of the National Academy of Sciences of the United States of America*, Vol. 108, No. 9, pp. 3803-3808, ISSN 0027-8424

Fu, C. X., Xiao, X. R., Xi, Y. J., Ge, Y. X., Chen, F., Bouton, J., Dixon, R. A. & Wang, Z. Y. (2011b). Downregulation of cinnamyl alcohol dehydrogenase (CAD) leads to improved saccharification efficiency in switchgrass. *Bioenergy Research*, Vol. 4, No. 3, pp. 153-164, ISSN 1939-1234

Garcia, R., Cidade, D., Castellar, A., Lips, A., Magioli, C., Callado, C. & Mansur, E. (2007). In vitro morphogenesis patterns from shoot apices of sugar cane are determined by light and type of growth regulator. *Plant Cell, Tissue and Organ Culture*, Vol. 90, No. 2, pp. 181-190, ISSN 0167-6857

Gawel, N. J., Robacker, C. D. & Corley, W. L. (1990). In vitro propagation of *Miscanthus sinensis*. *Hortscience*, Vol. 25, No. 10, pp. 1291-1293, ISSN 0018-5345

Głowacka, K. & Jeżowski, S. (2009a). Genetic and nongenetic factors influencing callus induction in *Miscanthus sinensis* (Anderss.) anther cultures. *Journal of Applied Genetics*, Vol. 50, No. 4, pp. 341-345, ISSN 1234-1983

Głowacka, K. & Jeżowski, S. (2009b). *In vitro* culture of *Miscanthus sinensis* anthers. *Acta Biologica Cracoviensia Series Botanica*, Vol. 51, Suppl. 1, p. 17, ISSN 0001-5296

Głowacka, K., Jeżowski, S. & Kaczmarek, Z. (2010). The effects of genotype, inflorescence developmental stage and induction medium on callus induction and plant regeneration in two *Miscanthus* species. *Plant Cell, Tissue and Organ Culture*, Vol. 102, No. 1, pp. 79-86, ISSN 0167-6857

Gorenflot, R., Raicu, P., Cartier, D. & Plantefol, L. (1972). Caryologie de la canne de Provence (*Arundo donax* L.). *Comptes rendus de l'Académie des Sciences*, Vol. 274, No. 3, pp. 391-393, (in French)

Gupta, S. D. & Conger, B. V. (1998). *In vitro* differentiation of multiple shoot clumps from intact seedlings of switchgrass. *In Vitro Cellular & Developmental Biology-Plant*, Vol. 34, No. 3, pp. 196-202, ISSN 1054-5476

Gupta, S. D. & Conger, B. V. (1999). Somatic embryogenesis and plant regeneration from suspension cultures of switchgrass. *Crop Science*, Vol. 39, No. 1, pp. 243-247, ISSN 0011-183X

Hamerli, D. & Birch, R. G. (2011). Transgenic expression of trehalulose synthase results in high concentrations of the sucrose isomer trehalulose in mature stems of field-grown sugarcane. *Plant Biotechnology Journal*, Vol. 9, No. 1, pp. 32-37, ISSN 1467-7644

Henry, R. J. (2010). Basic information on the sugarcane plant, In: *Genetics, Genomics and Breeding of Sugarcane*, R. Henry & C. Kole (Eds.), pp. 1-7, CRC Press, ISBN 978-1-57808-684-9, Boca Raton, Florida, USA

Holme, I. B., Krogstrup, P. & Hansen, J. (1997). Embryogenic callus formation, growth and regeneration in callus and suspension cultures of *Miscanthus* x *ogiformis* Honda Giganteus' as affected by proline. *Plant Cell, Tissue and Organ Culture*, Vol. 50, No. 3, pp. 203-210, ISSN 0167-6857

Holme, I. B. & Petersen, K. K. (1996). Callus induction and plant regeneration from different explant types of *Miscanthus* x *ogiformis* Honda 'Giganteus'. *Plant Cell, Tissue and Organ Culture*, Vol. 45, No. 1, pp. 43-52, ISSN 0167-6857

Hopkins, A. A., Taliaferro, C. M., Murphy, C. D. & Christian, D. (1996). Chromosome number and nuclear DNA content of several switchgrass populations. *Crop Science*, Vol. 36, No. 5, pp. 1192-1195, ISSN 0011-183X

Hotta, C. T., Lembke, C. G., Domingues, D. S., Ochoa, E. A., Cruz, G. M. Q., Melotto-Passarin, D. M., Marconi, T. G., Santos, M. O., Mollinari, M., Margarido, G. R. A., Crivellari, A. C., dos Santos, W. D., de Souza, A. P., Hoshino, A. A., Carrer, H., Souza, A. P., Garcia, A. A. F., Buckeridge, M. S., Menossi, M., Van Sluys, M.-A. & Souza, G. M. (2010). The biotechnology roadmap for sugarcane improvement. *Tropical Plant Biology*, Vol. 3, No. 2, pp. 75-87, ISSN 1935-9756

Jackson, M. A., Nutt, K. A., Hassall, R. & Rae, A. L. (2010). Comparative efficiency of subcellular targeting signals for expression of a toxic protein in sugarcane. *Functional Plant Biology*, Vol. 37, No. 8, pp. 785-793, ISSN 1445-4408

Jakob, K., Zhou, F. S. & Paterson, A. (2009). Genetic improvement of C4 grasses as cellulosic biofuel feedstocks. *In Vitro Cellular & Developmental Biology-Plant*, Vol. 45, No. 3, pp. 291-305, ISSN 1054-5476

Jalaja, N. & Sreenivasan, T. (1999). *In vitro* regeneration and cytological studies on somaclones of *Erianthus elephantinus* Hook. F. *Sugar Tech*, Vol. 1, No. 4, pp. 132-138, ISSN 0972-1525

Karpenstein-Machan, M. (2001). Sustainable cultivation concepts for domestic energy production from biomass. *Critical Reviews in Plant Sciences*, Vol. 20, No. 1, pp. 1-14, ISSN 0735-2689

Kaur, A. & Gosal, S. S. (2009). Desiccation of callus enhances somatic embryogenesis and subsequent shoot regeneration in sugarcane. *Indian Journal of Biotechnology*, Vol. 8, No. 3, pp. 332-334, ISSN 0972-5849

Khudamrongsawat, J., Tayyar, R. & Holt, J. S. (2004). Genetic diversity of giant reed (*Arundo donax*) in the Santa Ana River, California. *Weed Science*, Vol. 52, No. 3, pp. 395-405, ISSN 0043-1745

Kim, J. Y., Gallo, M. & Altpeter, F. (2011). Analysis of transgene integration and expression following biolistic transfer of different quantities of minimal expression cassette into sugarcane (*Saccharum* spp. hybrids). *Plant Cell, Tissue and Organ Culture*, ISSN 0167-6857, (Published online: September 21, 2011)

Lakshmanan, P. (2006). Somatic embryogenesis in sugarcane - An addendum to the invited review 'Sugarcane biotechnology: The challenges and opportunities,' In Vitro Cell. Dev. Biol. Plant 41(4): 345-363; 2005. *In Vitro Cellular & Developmental Biology-Plant*, Vol. 42, No. 3, pp. 201-205, ISSN 1054-5476

Lakshmanan, P., Geijskes, R. J., Aitken, K. S., Grof, C. L. P., Bonnett, G. D. & Smith, G. R. (2005). Sugarcane biotechnology: The challenges and opportunities. *In Vitro Cellular & Developmental Biology-Plant*, Vol. 41, No. 4, pp. 345-363, ISSN 1054-5476

Lewandowski, I., Clifton-Brown, J. C., Scurlock, J. M. O. & Huisman, W. (2000). Miscanthus: European experience with a novel energy crop. *Biomass and Bioenergy*, Vol. 19, No. 4, pp. 209-227, ISSN 0961-9534

Lewandowski, I. & Kahnt, G. (1992). Development of a tissue culture system with unemerged inflorescences of *Miscanthus* 'Giganteus' for the induction and regeneration of somatic embryoids. *Beiträge zur Biologie der Pflanzen*, Vol. 67, No. 3, pp. 439-451, ISSN 0005-8041

Lewandowski, I. & Kahnt, G. (1993). Possibilities for the establishment of an *In-vitro-*propagationsystem for *Miscanthus* "Giganteus" by using different parts of the plant as donor-tissue. *Bodenkultur*, Vol. 44, No. 3, pp. 243-252, ISSN 0006-5471, (in German)

Lewandowski, I., Scurlock, J. M. O., Lindvall, E. & Christou, M. (2003). The development and current status of perennial rhizomatous grasses as energy crops in the US and Europe. *Biomass and Bioenergy*, Vol. 25, No. 4, pp. 335-361, ISSN 0961-9534

Li, R. & Qu, R. (2011). High throughput *Agrobacterium*-mediated switchgrass transformation. *Biomass and Bioenergy*, Vol. 35, No. 3, pp. 1046-1054, ISSN 0961-9534

McLaughlin, S. B. & Kszos, L. A. (2005). Development of switchgrass (*Panicum virgatum*) as a bioenergy feedstock in the United States. *Biomass and Bioenergy*, Vol. 28, No. 6, pp. 515-535, ISSN 0961-9534

Mukherjee, S. K. (1957). Origin and distribution of *Saccharum*. *Botanical Gazette*, Vol. 119, pp. 55-61, ISSN 0006-8071

Murashige, T. & Skoog, F. (1962). A revised medium for rapid growth and bioassays with tobacco tissue cultures. *Physiologia plantarum*, Vol. 15, No. 3, pp. 473-497, ISSN 0031-9317

Nieves, N., Sagarra, F., González, R., Lezcano, Y., Cid, M., Blanco, M. A. & Castillo, R. (2008). Effect of exogenous arginine on sugarcane (*Saccharum* sp.) somatic embryogenesis, free polyamines and the contents of the soluble proteins and proline. *Plant Cell, Tissue and Organ Culture*, Vol. 95, No. 3, pp. 313-320, ISSN 0167-6857

Odjakova, M. K. & Conger, B. V. (1999). The influence of osmotic pretreatment and inoculum age on the initiation and regenerability of switchgrass suspension cultures. *In Vitro Cellular & Developmental Biology-Plant,* Vol. 35, No. 6, pp. 442-444, ISSN 1054-5476

Öztürk, L. & Demir, Y. (2002). *In vivo* and *in vitro* protective role of proline. *Plant Growth Regulation,* Vol. 38, No. 3, pp. 259-264, ISSN 0167-6903

Papazoglou, E. G. (2007). *Arundo donax* L. stress tolerance under irrigation with heavy metal aqueous solutions. *Desalination,* Vol. 211, No. 1-3, pp. 304-313, ISSN 0011-9164

Papazoglou, E. G., Karantounias, G. A., Vemmos, S. N. & Bouranis, D. L. (2005). Photosynthesis and growth responses of giant reed (*Arundo donax* L.) to the heavy metals Cd and Ni. *Environment International,* Vol. 31, No. 2, pp. 243-249, ISSN 0160-4120

Perdue, R. E. (1958). *Arundo donax*—Source of musical reeds and industrial cellulose. *Economic Botany,* Vol. 12, No. 4, pp. 368-404, ISSN 0013-0001

Petersen, K. K. (1997). Callus induction and plant regeneration in *Miscanthus* x *ogiformis* Honda 'Giganteus' as influenced by benzyladenine. *Plant Cell, Tissue and Organ Culture,* Vol. 49, No. 2, pp. 137-140, ISSN 0167-6857

Petersen, K. K., Hansen, J. & Krogstrup, P. (1999). Significance of different carbon sources and sterilization methods on callus induction and plant regeneration of *Miscanthus* x *ogiformis* Honda 'Giganteus'. *Plant Cell, Tissue and Organ Culture,* Vol. 58, No. 3, pp. 189-197, ISSN 0167-6857

Petrasovits, L. A., Purnell, M. P., Nielsen, L. K. & Brumbley, S. M. (2007). Production of polyhydroxybutyrate in sugarcane. *Plant Biotechnology Journal,* Vol. 5, No. 1, pp. 162-172, ISSN 1467-7652

Piperidis, G., Piperidis, N. & D'Hont, A. (2010). Molecular cytogenetic investigation of chromosome composition and transmission in sugarcane. *Molecular Genetics and Genomics,* Vol. 284, No. 1, pp. 65-73, ISSN 1617-4615

Płażek, A. & Dubert, F. (2009). Optimization of medium for callus induction and plant regeneration of *Miscanthus* x *giganteus*. *Acta Biologica Cracoviensia Series Botanica,* Vol. 51, Suppl. 1, pp. 56-56, ISSN 0001-5296

Płażek, A. & Dubert, F. (2010). Improvement of medium for *Miscanthus* x *giganteus* callus induction and plant regeneration. *Acta Biologica Cracoviensia Series Botanica,* Vol. 52, No. 1, pp. 105-110, ISSN 0001-5296

Pyter, R., Voigt, T., Heaton, E., Dohleman, F. & Long, S. (2007). Giant miscanthus: biomass crop for Illinois. *Proceeding of the 6th National New Crops Symposium,* pp. 39-42, San Diego, California, Octorber 14-18, 2007

Ribeiro, C. W., Soares-Costa, A., Falco, M. C., Chabregas, S. M., Ulian, E. C., Cotrin, S. S., Carmona, A. K., Santana, L. A., Oliva, M. L. V. & Henrique-Silva, F. (2008). Production of a his-tagged canecystatin in transgenic sugarcane and subsequent purification. *Biotechnology Progress,* Vol. 24, No. 5, pp. 1060-1066, ISSN 8756-7938

Richards, H. A., Rudas, V. A., Sun, H., McDaniel, J. K., Tomaszewski, Z. & Conger, B. V. (2001). Construction of a GFP-BAR plasmid and its use for switchgrass transformation. *Plant Cell Reports,* Vol. 20, No. 1, pp. 48-54, ISSN 0721-7714

Rossa, B., Tüffers, A. V., Naidoo, G. & von Willert, D. J. (1998). *Arundo donax* L. (Poaceae) - a C-3 species with unusually high photosynthetic capacity. *Botanica Acta,* Vol. 111, No. 3, pp. 216-221, ISSN 0932-8629

Saathoff, A. J., Sarath, G., Chow, E. K., Dien, B. S. & Tobias, C. M. (2011). Downregulation of cinnamyl-alcohol dehydrogenase in switchgrass by RNA silencing results in

enhanced glucose release after cellulase treatment. *PLoS ONE*, Vol. 6, No. e16416, Retrieved from:
http://www.plosone.org/article/info%3Adoi%2F10.1371%2Fjournal.pone.0016416

Somleva, M. N., Snell, K. D., Beaulieu, J. J., Peoples, O. P., Garrison, B. R. & Patterson, N. A. (2008). Production of polyhydroxybutyrate in switchgrass, a value-added co-product in an important lignocellulosic biomass crop. *Plant Biotechnology Journal*, Vol. 6, No. 7, pp. 663-678, ISSN 1467-7652

Somleva, M. N., Tomaszewski, Z. & Conger, B. V. (2002). *Agrobacterium*-mediated genetic transformation of switchgrass. *Crop Science*, Vol. 42, No. 6, pp. 2080-2087, ISSN 0011-183X

Song, G., Walworth, A. & Hancock, J. (2011). Factors influencing *Agrobacterium*-mediated transformation of switchgrass cultivars. *Plant Cell, Tissue and Organ Culture*, ISSN 0167-6857, (Published online: September 24, 2011)

Sun, G. Z., Ma, M. Q., Zhang, Y. Q., Xian, X. L., Cai, X. L. & Li, X. P. (1999). A medium adapted to embryogenic callus induction and subcultures of wheat. *Journal of Hebei Agricultural Sciences*, Vol. 3, No. 2, pp. 24-26, ISSN 1008-1631, (in Chinese)

Suprasanna, P., Patade, V. Y. & Bapat, V. A. (2008a). Sugarcane biotechnology - A perspective on recent developments and emerging opportunities, In: *Advances in plant biotechnology*, G. P. Rao, Y. Zhao, V. V. Radchuk & S. K. Bhatnagar (Eds.), pp. 313-342, Studium Press LLC, ISBN 19-3369-938-8, Houston, Texas, USA

Suprasanna, P., Rupali, C., Desai, N. S. & Bapat, V. A. (2008b). Partial desiccation augments plant regeneration from irradiated embryogenic cultures of sugarcane. *Plant Cell, Tissue and Organ Culture*, Vol. 92, No. 1, pp. 101-105, ISSN 0167-6857

Tagane, S., Ponragdee, W., Sansayawichai, T., Sugimoto, A. & Terajima, Y. (2011). Characterization and taxonomical note about Thai *Erianthus* germplasm collection: the morphology, flowering phenology and biogeography among *E. procerus* and three types of *E. arundinaceus*. *Genetic Resources and Crop Evolution*, ISSN 0925-9864

Takahashi, W., Komatsu, T., Fujimori, M. & Takamizo, T. (2004). Screening of regenerable genotypes of Italian ryegrass (*Lolium multiflorum* Lam.). *Plant Production Science*, Vol. 7, No. 1, pp. 55-61, ISSN 1343-943X

Takahashi, W., Takamizo, T., Kobayashi, M. & Ebina, M. (2010). Plant regeneration from calli in giant reed (*Arundo donax* L.). *Grassland Science*, Vol. 56, No. 4, pp. 224-229, ISSN 1744-6961

Terajima, Y., Terauchi, T., Sakaigaichi, T., Hattori, T., Fujisaki, N., Teruya, H., Naitou, T., Daiku, M., Matsuoka, M., Ohara, S., Irei, S., Ujihara, K. & Sugimoto, A. (2010). Productivities of a large biomass sugarcane line KY01-2044 in Nansei islands. *Proceeding of the 229th Meeting of the Crop Science Society of Japan*, pp. 124-125, Utsunomiya, Japan, March 30-31, 2010, (in Japanese)

Tew, T. L. & Cobill, R. M. (2008). Genetic improvement of suagarcane (*Saccharum* spp.) as an energy crop, In: *Genetic Improvement of Bioenergy Crops*, W. Vermerris (Ed.), pp. 249-272, Springer, ISBN 978-0-387-70804-1, New York, USA

Vega-Sánchez, M. E. & Ronald, P. C. (2010). Genetic and biotechnological approaches for biofuel crop improvement. *Current Opinion in Biotechnology*, Vol. 21, No. 2, pp. 218-224, ISSN 0958-1669

Vogel, K. P. & Jung, H.-J. G. (2001). Genetic modification of herbaceous plants for feed and fuel. *Critical Reviews in Plant Sciences*, Vol. 20, No. 1, pp. 15-49, ISSN 0735-2689

Waclawovsky, A. J., Sato, P. M., Lembke, C. G., Moore, P. H. & Souza, G. M. (2010). Sugarcane for bioenergy production: an assessment of yield and regulation of sucrose content. *Plant Biotechnology Journal*, Vol. 8, No. 3, pp. 263-276, ISSN 1467-7644

Wang, X., Yamada, T., Kong, F. J., Abe, Y., Hoshino, Y., Sato, H., Takamizo, T., Kanazawa, A. & Yamada, T. (2011). Establishment of an efficient *in vitro* culture and particle bombardment-mediated transformation systems in *Miscanthus sinensis* Anderss., a potential bioenergy crop. *Global Change Biology Bioenergy*, Vol. 3, No. 4, pp. 322-332, ISSN 1757-1693

Williams, C. A., Harborne, J. B. & Smith, P. (1974). Taxonomic significance of leaf flavonoids in *Saccharum* and related genera. *Phytochemistry*, Vol. 13, No. 7, pp. 1141-1149, ISSN 0031-9422

Wright, L. & Turhollow, A. (2010). Switchgrass selection as a "model" bioenergy crop: A history of the process. *Biomass and Bioenergy*, Vol. 34, No. 6, pp. 851-868, ISSN 0961-9534

Xi, Y., Fu, C., Ge, Y., Nandakumar, R., Hisano, H., Bouton, J. & Wang, Z.-Y. (2009). *Agrobacterium*-mediated transformation of switchgrass and inheritance of the transgenes. *Bioenergy Research*, Vol. 2, No. 4, pp. 275-283, ISSN 1939-1234

Xu, B., Huang, L., Shen, Z., Welbaum, G. E., Zhang, X. & Zhao, B. (2011). Selection and characterization of a new switchgrass (*Panicum virgatum* L.) line with high somatic embryogenic capacity for genetic transformation. *Scientia Horticulturae*, Vol. 129, No. 4, pp. 854-861, ISSN 0304-4238

Zhang, Q. X., Sun, Y., Hu, H. K., Chen, B., Hong, C. T., Guo, H. P., Pan, Y. H. & Zheng, B. S. (2011). Micropropagation and plant regeneration from embryogenic callus of *Miscanthus sinensis*. *In Vitro Cellular & Developmental Biology - Plant*, ISSN 1054-5476, (Published online: Augst 20, 2011)

6

Genetic Transformation of Immature Sorghum Inflorescence via Microprojectile Bombardment

Rosangela L. Brandão, Newton Portilho Carneiro, Antônio C. de Oliveira,
Gracielle T. C. P. Coelho and Andréa Almeida Carneiro*
Embrapa Maize and Sorghum
Brazil

1. Introduction

Sorghum bicolor is one of the most important cereals in the world after rice, maize, wheat and barley. In 2010, more than 60 million tons were produced from approximately 50 million ha around the world. It is an important crop in the arid and semi-arid regions of the world and it is primarily used in Brazil as a supply for an increasing livestock market. However, the Brazilian sorghum productivity is low (1,500 to 2,500 kg/ha) and extremely variable along the years, typical of a culture sowed in marginal climate conditions and mainly without the use of high input technologies. Conventional breeding programs have already done a great deal of research to increase the sorghum productivity, though in some fields, the gains obtained by these programs are reaching stationary levels due to the lack of genetic variability (Nwanze *et al.*, 1995). Alternatively, recombinant DNA technology and the generation of transgenic plants can increment conventional breeding programs through the amplification of the gene pool that can be used to improve sorghum environmental fitness and nutritional qualities.

However, unlike others Poaceae, sorghum transformation has been a challenge mainly due to recalcitrance in tissue culture and long periods of selection required for the recovery and regeneration of putative transgenic plants (Casas *et al.*, 1993, Zhao *et al.*, 2000; Jeoung *et al.*, 2002; Howe *et al.*, 2005). Since the earliest 90's, laboratories around the world have generated improvements in sorghum regeneration and transformation that are ensuing in more consistent protocols. Transgenic sorghum plants have been generated via biolistic (Casas *et al.*, 1993; Casas *et al.*, 1997; Zhu *et al.*, 1998; Able *et al.*, 2001; Emani *et al.*, 2002; Jeoung *et al.*, 2002; Devi and Sticklen, 2003; Tadesse *et al.*, 2003; Girijashankar *et al.*, 2005) or *Agrobacterium* mediated transformation (Zhao *et al.*, 2000; Carvalho *et al.*, 2004; Gao *et al.*, 2005; Howe *et al.*, 2006; Van Nguyen *et al.*, 2007; Nguyen et al., 2007; Gurel et al., 2009; Lu et al. 2009; Mall et al., 2011).

Even though the efficiency of sorghum transformation using the microprojectile bombardment had been improved, by all this studies, since the initial experiments (from 0,3 to 1,3%), it is still low if compared with the efficiency of sorghum transformation mediated by *Agrobacterium tumefaciens* from 2.1% – 4.5% (Zhao et al., 2000; Gao et al., 2005; Howe et

*Corresponding Author

al., 2006). So, there is still need for more improvements in the microprojectile bombardment, once this technique can be used for genotypes and explants not susceptible to the transformation mediated by *Agrobacterium tumefaciens*.

2. Genetic transformation of *Sorghum bicolor*

2.1 *In vitro* regeneration of transgenic sorghum cells

Sorghum tissue culture is reported to be highly recalcitrant mainly because the release of toxic phenolics compounds in culture media, lack of regeneration in long term *in vitro* cultures, and high degree of genotype dependence. Consequently, cell transformation followed by plant regeneration remains extremely complicated in sorghum transgenic technology. So, the establishment of sorghum regeneration systems from somatic cells constitutes a prerequisite of extreme importance within the process of transgenic sorghum plants production.

The regeneration of sorghum *in vitro* has been achieved from a variety of tissues, such as immature embryos (Bhat and Kuruvinashetti 1995; Bai et al. 1995; Elkonin and Pakhomova 1996), immature inflorescences (Kaeppler and Pedersen 1997), young leaves (Han et al. 1997) and shoot tips (Nahadi and de-Wet 1995; Patil and Kuruvinashetti 1998; Shyamala and Devi, 2003). Immature zygotic embryos and inflorescences are the explants with higher embryogenic competence and frequently used to regenerate various cultivars of sorghum. Gupta and co-workers (2006) compared the regeneration in tissue culture of immature embryos and immature inflorescences from five genotypes of *S. sudanense* (SDSL 98984, SDSL 98988, SDSL 981125, SDSL 981142 and SDSL 981144) and three genotypes of *S. bicolor* (2219B, GD 68727 and 981013). They indicated that the regeneration potential of immature inflorescences was much superior to that of immature embryos and their performance was almost equivalent across the genotypes tested. The superiority of immature inflorescences can be due to its higher proportion of meristematic tissues (floral meristems, rachis, rachillae, and primordial of various floral organs) in comparison to immature embryos (mainly scutellum) according to authors.

However, the main explants used in the transformation of sorghum are immature zygotic embryos between 1.5 and 2.0 mm in length (Casas et al, 1993; Emani et al, 2002; Jeoung et al, 2002; Gao et al, 2005, Howe et al, 2006; Gurel et al, 2009). One of the constrains in working with immature embryos is the intensive labor to generate large quantities of explants to be used in the transformation procedures. In this sense, immature inflorescences are easier to isolate, show very good regeneration rates in tissue culture and, morphogenetic competence over a wider size range (1–5 cm) than immature embryos. Besides, it is faster to grow donor plants for the production of immature inflorescence than for immature embryo (Cai and Butler, 1990; Kaepler and Pedersen, 1997; Jogeswar et al 2007; Brandão et al. 2007).

Even though outstanding studies aiming to identify sorghum genotypes able to produce high quality callus from immature inflorescence have been conducted, the efficiency to produce transgenic sorghum plants using this type of explant is still very low.

2.2 Genetic transformation of *Sorghum bicolor* via microprojectile bombardment

The bombardment of plant cells with the DNA of interest is a direct method of transformation designed (Taylor and Fauquet, 2002) in the late 80's to manipulate the genome of plants recalcitrant to transformation via *Agrobacterium*, including the cereals (Klein et al 1987; Taylor and Fauquet, 2002). In the transformation via particle bombardment, microprojectiles of metal physically covered with the gene of interest are launched toward the target cells, using equipment known as "gene gun" (Sandford et al. 1987). The velocity of these particles is fast enough (1500 km/h) to penetrate the cell wall of a target tissue and does not cause cell death. The precipitated DNA on the microprojectiles is released progressively into the cell after the blast, and integrated into the genome (Taylor and Fauquet, 2002). The acceleration of microprojectiles is obtained by a high voltage electrical discharge or compressed gas (helium). The metal particles used must be non-toxic, non-reactive, and lower than the diameter of the target cell. The most commonly used are gold or tungsten. Several physical parameters correlated with the biolistic equipment such as pressure, distance of the macro and micro-carrier flight, and vacuum, must be optimized for successful transformation. In addition to these parameters, the biology of the plant material and the gene of interest (GOI) should also be studied in preliminary experiments (Sandford et al., 1993). Since the 90's, the biolistic was used to transform a wide variety of plants, including sorghum.

Some advantages of the microprojectile bombardment are related to its efficiency in the transformation of monocots, the use of simple vectors, easier to handle, as well as the possibility of inserting more than one GOI into cells efficiently (Chen et al. 1998; Wu et al. 2002). Although considered a very efficient method in cereals, one drawbacks of this technique is the occurrence of multiple copies of the GOI in the transgenic plant and complex integration patterns (Wang and Frame, 2004).

The biolistic has proved to be an efficient method for introducing new features in sorghum and a few transformation protocols are already available (Casas et al. 1993; Casas et al. 1997; Zhu et al. 1998; Able et al. 2001; Devi and Sticklen, 2002 ; Emani et al. 2002; Tadesse et al., 2003; Girijashankar et al., 2005). The optimization of physical and biological parameters was the subject of most of the work published about sorghum transformation via bombardment. The pioneer work was done by Casas and collaborators between 1993 and 1997. Initially, using anthocyanin (*R* and *C1* genes) and β-glucuronidase (*uidA* gene) as reporter genes, they established an ideal pressure and microprojectile flying distance to transform immature sorghum embryos via bombardment. In this work it was also shown that the bialaphos was suitable to select transgenic cells of sorghum. Major problems in the protocol were the low efficiency (0,3%) and long time to select transgenic callus (7 months). In the next work (1997), the group tried to improve the transformation efficiency by using immature sorghum inflorescence, an explant with higher morphogenetic potential compared to immature embryos. At this time a greater number of transgenic plants were obtained but the overall efficiency of the process was low and the selection protocol generated many escapes.

Optimization of physical and biological parameters to produce transgenic sorghum plants was also the purpose of the work by Able et al. (2001) and Tadesse et al. (2003). Able and co-workers (2001) analyzed the transient expression of the reporter genes GUS and GFP over different physical bombardment parameters to identify the best conditions to generate transgenic sorghum plants using the particle inflow gun (PIG). Three transgenics events were confirmed by molecular analyses. Tadesse and associates (2003) also used reporter genes to test different acceleration pressures, target distances, gap widths and macroprojectile travel distances to bombard immature and mature embryos, shoot tips and embryogenic calli. The strength of four different promoters (*ubi1*, *act1D*, *adh1* and *CaMV35S*) was also analyzed in transient assays. The optimization of the transformation conditions generated a protocol with an efficiency of 0.5 to 1.3% of transgenic sorghum production from shoots tips and immature embryos, respectively.

Currently few sorghum events expressing genes with agronomical interest were developed by the laboratories working with transformation of this specie via microprojectile bombardment. The gene *chi II*, encoding rice chitinase under the control of the constitutive CaMV35S promoter, has been transferred to sorghum for resistance to stalk rot (*Fusarium thapsinum*) by Zhu et al. (1998). Chitinases are proteins produced by plants as defense against pathogen attack; they function by degrading the fungal cell wall. Their work was done with calli developed from immature zygotic embryos as target tissue for microprojectile bombardment and, six independent events that were bialaphos-resistant and containing the chitinase gene were reported.

Another chitinase, the gene *ECH2*, and the *bar* gene were used to produce disease and herbicide resistant transgenic plants, respectively. Devi and Sticklen (2003) transformed shoots clumps, originated from mature sorghum seeds cultivated in presence of N^6-benzyladenine (BA), via microjectile bombardment with these genes. Shoot clumps were used with the purpose to develop a transformation protocol using an explant easier to be obtained throughout the year. Only five different events were generated, but the proficiency of sorghum shoot meristems for regeneration and transformation was demonstrated.

The *cry1Ac* gene from *Bacillus thuringiensis* under the control of the wound-inducible promoter from the maize protease inhibitor gene (*mpiC1*) was inserted in the genome of three independent transgenic sorghum events. Shoot apices were bombarded and subcultured in a MS medium supplemented with benzylaminopurine (BAP) and naphthalene acetic acid (NAA). Leaf damage by the spotted stem borer (*Chilo partellus*) was reduced up to 60% in the sorghum transgenic plants generated (Girijashankar et al., 2005).

To withstand toxic aluminum concentrations present in acidic soils, sorghum was genetically modified to express the *ALMT1* (Sasaki et al. 2004) gene from wheat (Brandão, 2007). The *ALMT1* gene codes a malate transporter Al^{+3} activated protein that is highly expressed in wheat root apices of aluminum tolerant cultivars. Transgenic sorghum plants grown in hydroponic culture under stress of Al^{+3}, showed a higher level of aluminum tolerance when compared with isogenic non-transgenic control plants.

3. Genetic transformation of immature sorghum inflorescence

Here, we report improvements made in the transformation process via microprojectile bombardment that enable us to obtain a protocol where putative transgenic plants can be produced with an efficiency ranging from 1.01 to 3.33% using immature inflorescences of sorghum.

3.1 Material and methods

3.1.1 Plant material and explants preparation

Seeds from nine *Sorghum bicolor* (Moench L.) accessions were obtained from the Embrapa Maize and Sorghum National Research Center – Brazil. Shoots were collected at different developmental stages (3 to 5 cm in length), from field plants prior to the appearance of the flag leaf (Figure 1). The outermost leaf blades were removed and shoots rinsed with 70% ethanol and sterile distilled water. After that, immature inflorescences were dissected, chopped in approximately 5 mm long segments and cultivated on callus induction medium (CIM) as described by Tadesse et al. (2003), with minor modifications [MS salts (Murashige and Skoog, 1962), 1 mg.L⁻¹ thiamine HCl, 7.5 mg.L⁻¹ glycine, 100 mg.L⁻¹ DL-asparagine, 100 mg.L⁻¹ myo-inositol, 0.2 mg.L⁻¹ kinetin, 2.5 mg.L⁻¹ 2,4-D and 30 g.L⁻¹ sucrose]. Medium pH was adjusted to 5.8 with 1 N potassium hydroxide prior to autoclaving.Cultures were incubated at 25°C in dark for three to four weeks.

For the biolistic experiments 30 calli pieces of approximately 3 mm diameter (Figure 2B) were uniformly distributed within a 35 mm diameter circle of 60 x 15 mm Petri dishes containing CIM media in which a higher osmotic value was achieved by the addition of 12% sucrose.

Fig. 1. Harvest and cultivation of immature sorghum inflorescence. (A and B) Sorghum plant used to isolate immature inflorescence; (C) Rinsing of sorghum shoots, without the outermost leaves, with 70% ethanol; (D) Rinsing of shoots with sterile distilled water; (E and F) Dissection of immature inflorescence; (G) Cutting the immature inflorescence in 5mm pieces; (H) Culturing inflorescence pieces.

3.1.2 Plasmid constructs

The genetic cassettes p35S::*C1* and p35S::*Bperu* (Goff *et al.* 1990), used in the transient transformation experiments, were kindly provided by Dr. Vicki Chandler from the Department of Plant Science, University of Arizona, Tucson, Arizona. These plasmids contain the CaMV35S promoter, directing the expression of *B-peru* (1.9Kb) and *C1* (1.1 Kb) cDNAs, maize *Adh1* intron and nopaline synthase terminator. For the stable transformation experiments, the plasmid pCAMBIA3301 (Cambia, Canberra, Australia) which contains β-glucuronidase (GUS) reporter gene (Jefferson et al., 1987) and the *bar* gene (De Block et al., 1987) that encodes phosphinothricin acetyltransferase (PAT), both driven by the CaMV35S promoter was used.

3.1.3 Particle bombardment

Embryogenic calli were bombarded with tungsten microprojectiles using a biolistic particle helium acceleration device (Biomics – Brasília / Brazil). For the transient experiments 3 μL of each plasmid (stock 1 μg/μL), p35S::*C1* and p35S::*Bperu*, were co-precipitated with tungsten particles, while for the stable transformation 10 μL pCAMBIA3301 plasmid (stock 1 μg/μL) were used. To precipitate DNA onto the microparticles, plasmid DNA were mixed with 50μL (60 mg.mL^{-1}) tungsten particle M10 (Sylvania, GTE Chemicals/ Towanda – USA) under low agitation. Next, 50μL of 2.5M CaCl$_2$ and 20μL of 0.1M spermidine were sequentially added and homogenized. The mixture was kept for three minutes under low agitation and for an additional three minutes without agitation. Particles coated with DNA were centrifuged for five seconds and the supernatant was removed. DNA-coated particles were washed carefully once with 70% ethanol, twice with 100% ethanol, and suspended in 100% ethanol. Eight microliters of DNA-coated particle were deposited at the center of sterile macro-carries membrane (Ficael, São Paulo, SP).

Treatments	Osmotic Media (hours)	Helium Pressure (psi)	Microcarrier Flying Distance (cm)
1	0	1 000	3
2	0	1 000	6
3	0	1 000	9
4	4	1 000	3
5	4	1 000	6
6	4	1 000	9
7	24	1 000	3
8	24	1 000	6
9	24	1 000	9
10	0	1 200	3
11	0	1 200	6
12	0	1 200	9
13	4	1 200	3
14	4	1 200	6
15	4	1 200	9
16	24	1 200	3
17	24	1 200	6
18	24	1 200	9

Table 1. Conditions tested in transient transformation experiments.

Eighteen treatments (Table 1) were designed to test the permanence of explants on osmoticum prior to bombardment (0, 4 and 24 hs), pressure of the accelerating helium pulse (1000 and 1200 psi), and microprojectile flying distance (3, 6 and 9 cm), in transient sorghum transformation. The distance between the high pressure chamber and the macro-carrier membrane (8 mm), the distance between the macro-carrier membrane and the retention screen (17 mm) and the vacuum pressure (27 mmHg) were maintained constant. For each treatment three plates containing 30 calli pieces were bombarded once.

3.1.4 Expression analysis

Anthocyanin: For the anthocyanin expression studies, bombarded calli were incubated at 27°C for 2 d in the darkness. The number of anthocyanin spots was scored under a stereoscope (Zeiss Stemi SV11, Germany).

GUS: GUS expression was detected after explants were incubated at 37°C for 20 h in a solution containing 1 mM 5-bromo-4-chloro-3-indolyl β-D glucuronide (X-Gluc, Sigma Chem. Co., São Paulo, Brazil), 50 mM phosphate buffer pH 6.8, 20% methanol, 1% Triton X-100 (Rueb and Hensgens, 1998). Chlorophyll was extracted from leaf tissue in 70% ethanol for 30-60 min.

PAT: Seeds from four independent transgenic T_0 events tested positive by PCR and PAT activity were propagated, and the T_1 plants screened for PAT and GUS enzyme activity inheritance. From each T_0 line 50 seeds were propagated in greenhouse. Seedlings with four leaves were sprayed with 0.6% aqueous solution of the commercial herbicide Finale® (Bayer, São Paulo, Brazil) to confirm expression of the *bar* gene. Control non-transgenic plants or segregating seedlings showed symptoms in 2–5 d and died within 2 weeks. Leaf material from PPT-resistant plants was GUS stained as described above. Chi-square goodness of fit was used to test the significance of observed versus expected ratios.

3.1.5 Selection procedures

Explants were cultured on solid CIM media at 25°C in the dark for one week and transferred to selective SE media (modified CIM supplemented with 0.5 mg.L^{-1} kinetin and without DL-asparagine) containing 15 µL.L^{-1} of the herbicide Finale® (3 mg.L^{-1} 4-hidroxi(methyl) phosphynol-DL-homoalanine ammonium salt) for one week. After that, the explants were transferred to a media containing 30 and 45 µL.L^{-1} herbicide every week. Growing calli were cultured for one more week in a SE media supplemented with 45 µL.L^{-1} herbicide, and subsequently transferred to a callus maturation media RM [MS salts and vitamins (Murashige and Skoog, 1962), 60 g.L^{-1} sucrose, 100 mg. L^{-1} myo-inositol, 0.2 mg.L^{-1} NAA, 3 g/L phytagel, pH 5.8) supplemented with 30 µL.L^{-1} herbicide and cultured in the dark at 25°C for somatic embryo maturation. Approximately 2 to 4 weeks later, mature somatic embryos showing a white and opaque coloration were transferred to Magenta boxes (Sigma, São Paulo, Brazil) containing germination media composed by MS media without plant growth regulators, supplemented with 15 µL.L^{-1} herbicide and placed in a lighted (16 h / 60 µmol m^{-2} s^{-1}) growth room. Germinated plantlets (4-6 cm) were cultured in soil, for the first week under a plastic lid, in a greenhouse.

3.1.6 Plant DNA extraction, polymerase chain reaction (PCR) and Southern blot hybridization analysis

Total genomic DNA was isolated from leaf tissue of primary transformants using a CTAB protocol described by Saghai-Maroof *et al.* (1984). The presence of *bar* and *uid*A genes were detected, initially, by the polymerase chain reaction (PCR). The 407 bp coding region of *bar* gene was amplified using primers (AGA AAC CAC GTC ATG CC and TGC ACC ATC GTC AAC CAC). The 406 bp coding region of *uid*A gene was amplified using the primers (TCG TGC TGC GTT TCG ATG and GCA TCA CGC AGT TCA ACG). Each 25 uL amplification reactions containing 50 ng of template DNA, 5 µM each primer, 500 µM dNTP mixture, 2.5 µl TAQ DNA polymerase reaction buffer and 1 unit Taq DNA polymerase (Invitrogen, São Paulo, Brazil) were carried out using a thermal cycler (Eppendorff Mastercycler, Hamburg, Germany) under the following conditions: 94°C for 5 min; 30 cycles 94°C for 60 s; 55°C for 30s, and a final extension at 72°C for 10 min. The amplified products were separated by electrophoresis on a 1.2% agarose gel and visualized with ethidium bromide.

For Southern blot analysis 10 µg of total genomic DNA from each T_0 plant were completely digested with *Xho*I at 37°C overnight. Digested DNA fragments were separated by electrophoresis in 0.8% agarose gel, and then transferred to a Hybond-N$^+$ nylon membrane (Amersham, São Paulo, Brazil) according to Sambrook *et al.* (1989). The blot was hybridized with a P^{32} labeled *bar* gene coding region. Negative control samples consisted of non-transgenic genomic DNA. After overnight hybridization at 65°C, the membranes were washed in 2X SSC, 0.1% SDS at room temperature and in 1X SSC, 0.1% SDS at 65°C and exposed to Kodak™ XAR-5 film at 75°C for 3 d.

3.1.7 Statistical data analysis

The data obtained from the transient experiment was collected in Microsoft Excel (Version 5).The experimental design was based on randomized blocks, factorial 2x3x3 (2 pressure of the accelerating helium pulse, 3 time of explants on osmoticum and 3 microprojectile flying distance), with triplicates, totalizing 54 experimental units. The data were subjected to ANOVA and means compared by Tukey test (p< 0.05), using the statistical program SISVAR 4.0 (Ferreira, 2000).

4. Results

4.1 Explant preparation and selection of transformed calli *in vitro*

Nine *Sorghum bicolor* accessions were selected and screened for the quality of callus produced four weeks after cultivation of immature inflorescence on callus induction media (CIM). Embryogenic calli were produced at different efficiency levels by all of the accessions (data not shown). For this study, the genotype CMSXS102B, which gave the highest percentage of immature inflorescence sections producing embryogenic callus (85%), was used.

Embryogenic calli from immature inflorescence of sorghum were bombarded three to four weeks after cultivation in CIM media (Fig. 2B), with the plasmids containing the *B-peru* and *C1* or *bar* and GUS genes for transient and stable transformation experiments, respectively. Within this cultivation period, explants started to expand and enlarge at the cut edges.

Initially, the bombarded embryogenic calli had a compact appearance and then they developed into more friable structures in the next weeks of cultivation. After bombardment, calli were transferred back to CIM media without selection pressure for one week, and then moved to selection medium.

Fig. 2. Genetic transformation of calli from immature sorghum inflorescence. (A) Immature sorghum inflorescence; (B) Calli used for the particle bombardment; (C) Embryogenic callus induced from immature inflorescence; (D) Herbicide resistant calli after 6 weeks particle bombardment; (E) Maturation of calli after selection; (F) Regeneration of herbicide resistant callus, (G) Transient expression of anthocyanin; (H) Putative transgenic *Sorghum* plants in greenhouse; (I) Transformed (left) and wild type (right) sorghum plants weeks after application of herbicide Finale®; (J) GUS expression of wild type (left) and transgenic (right) germinated seeds. *Bar* = 1mm.

Growth of bombarded callus was slightly inhibited and some of them turned brown on selection medium supplemented with 3.0 mg.L-1 of PPT, compared to non-bombarded ones. After one week of cultivation the concentration of PPT on selection medium was increased to 6 mg.L-1 and most of the bombarded calli turned black. In order to reduce escapes within chimerical clusters, surviving bombarded clumps were carefully divided and cultured, at one week intervals, onto selection medium supplemented with 9 mg.L-1 PPT, during four weeks. At this herbicide concentration complete inhibition of non-transgenic calli growth was observed, most calli turned dark, necrotic and died (Fig. 2D). After six weeks of selection, surviving calli were transferred onto maturation media supplemented with 6 mg.L^{-1} PPT. On this medium, as soon as the yellowish calli become white and opaque, between 2 to 4 weeks of cultivation, they were transferred onto germination medium (Figure 2E). A concentration of 3 mg.L^{-1} PPT was used for the differentiation and germination of mature calli that occurs around 20 d of cultivation. Control non-bombarded explants did not survive on selection medium containing 6 or 9 mg.L^{-1} PPT. This selection procedure has been used successfully in different transformation experiments with calli derived from immature inflorescence of sorghum. The overall time for selection and regeneration of putative transgenic plants using this protocol is around 16 weeks.

4.2 Transient and stable transformation

The optimization of DNA delivery parameters was initially performed by using the transient expression of maize *R* and *C1* transcriptional activators. All bombardments were carried out with plasmidial DNA from the same stock, and the number of anthocyanin spots ranged from 3 to 348 depending on the particle bombardment conditions used (Fig. 2G).

Statistical analyses of the transient anthocyanin expression (Table 2) identified interactions among the different factors studied. Embryogenic calli submitted to 1000 psi of helium accelerating pressure, cultivated during 4 h in a higher osmotic medium and positioned at 3 cm from the micro-carrier launch platform (Treatment 4) presented a larger number of cells expressing anthocyanin than when the explants were positioned at 6 or 9 cm (Treatment 5 and 6). Without the pre-cultivation of explants in an osmotic medium, there were no differences among the positions (3, 6 or 9 cm) of explants (Treatments 1 to 3) and the overall amount of anthocyanin spots was lower. At 1200 psi acceleration pressure, treatments 10 to 18, the best results obtained were when calli were cultivated in osmotic medium for 24 h and positioned at 3 or 6 cm from the micro-carrier launch platform (Treatments 16 and 17).

Time in osmotic medium	TARGET DISTANCE Pressure 1000 psi		
	3 cm	6 cm	9 cm
Without pre-treatment	(1) 30,283 aA	(2) 47,246 aA	(3) 37,013 aA
4 hours	(4) 91,356 bB	(5) 52,786 aAB	(6) 31,170 aA
24 hours	(7) 74,930 aA	(8) 79,240 aA	(9) 42,813 aA
Time in osmotic medium	Pressure 1200 psi		
Without pre-treatment	(10) 43,183 aA	(11) 41,686 aA	(12) 44,626 aA
4 hours	(13) 44,736 aA	(14) 74, 263 aA	(15) 44,370 aA
24 hours	(16) 88,576 bB	(17) 59,086 aB	(18) 18,343 aA

Note. Means followed by the same small letter in the vertical and capital letter in the horizontal are not significantly different at 5% level according to Tukey's multiple range test. Numbers in parentheses represent the different treatments.

Table 2. Mean number of anthocyanin spots induced transiently by p35S::*C1* and p35S::*Bperu* constructs in embryogenic callus of the sorghum.

The highest number of cells expressing the anthocyanin genes was obtained with embryogenic calli cultivated in osmotic medium during 4 h before the bombardment, positioned at 3 cm distant from the microcarrier release platform and using 1000 psi accelerating pressure. Therefore, these conditions were used in six independent experiments to test stable transformation of embryogenic calli obtained from immature sorghum inflorescences, with a cassette containing the *bar* and *iud*A genes. The transformation efficiency for these sets of experiments ranged from 1.01% to 3.33 % (Table 3).

Experiment number	Number of transgenic events	Number of calli bombarded	% Efficiency
RLB3962008	1	93	1.07
RLB5420703	3	120	2.5
RLB5430703	2	60	3.33
RLB5222202	2	180	1.11
RLB020106	3	296	1.01
RLB120705	2	150	1.33

Table 3. Biolistic transformation of calli from immature sorghum inflorescence

4.3 Evaluation of transgenic material

To estimate the transgene copy number and the inheritance of the *bar* gene, T_1 progenies were tested for their tolerance to the herbicide Finale ®. Germinated T_1 transgenic and control seedlings at the stage of five leaves were sprayed with herbicide and scored for damage 7 and 14 d after the application. Transgenic seedlings segregated for the presence of the *bar* gene, there were plants with and without tissue damage, while all control non-transgenic plants presented necrosis and died (Fig. 2I).

Transgenic Lines Number	Herbicide-resistant plants	Herbicide-sensitive plants	Segregation ratio	Chi-square
1 (RLB3962008)	28	16	3:1	$\chi^2 = 3,03; P>0,05$
2 (RLB5222202)	07	01	3:1	$\chi^2 = 0,67; P>0,05$
3 (RLB5430703)	34	12	3:1	$\chi^2 = 0,029; P>0,05$
4 (RLB5420703)	37	8	3:1	$\chi^2 = 1,25; P>0,05$
5 (RLB2109056A)	23	25	1:1	$\chi^2 = 0,83; P>0,05$
6 (RLB2109053B)	28	20	1:1	$\chi^2 = 1,33; P>0,05$
Wild-type plant (CMSXS102B)	0	50	ND[a]	ND[a]

[a] – ND, Not determined

Table 4. Independent transgenic T_1 generation plants analyzed for the inheritance and segregation of herbicide resistance

Segregation data obtained from six T_1 progenies sprayed with herbicide is presented in Table 4. Among the progenies of self -pollinated T_0 transgenic plant lines, Chi-square tests showed a Mendelian segregation ratio of 3:1 in four lines. This ratio indicated that the *bar* gene was inserted in a single locus, efficiently inherited and transcribed in T_1 progeny plants. Two lines (Lines 5 and 6) showed a 1:1 ratio, suggesting semi-dominance.

Leaves of all T_0 transgenic events tested negative for the β-glucuronidase expression. However, GUS expression could be detected in germinated seeds (T_1) of event RLB5420703 (Fig. 2J).

The presence of *uid*A and *bar* genes in genomic DNA of all independent T_0 lines was confirmed by PCR analysis of genomic DNA. The results revealed the presence of 406 bp band of *uid*A and 407bp band of *bar* genes in all of the plants tested (Fig. 3A and B).

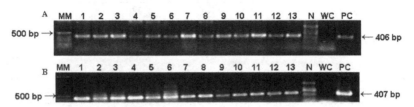

Fig. 3. PCR analysis of genomic DNA extracted from transgenic primary generation (T_0) of *Sorghum bicolor*. (A) PCR amplification of a *uid*A gene, showing the 406 bp fragment (lanes 1-13); (B) PCR amplification of *bar* gene, showing a 407 bp fragment (lanes 1-13). M: molecular weight marker, N: control wild type sorghum plants; WC: water control; PC: plasmid positive control.

The stable integration of the *bar* gene in the transgenic events were analyzed by Southern blotting of genomic DNA digested with *Xho*I, which cuts outside the *bar* coding sequence releasing a 564 bp fragment. No hybridization signal was present in the digested DNA from the untransformed plants (Fig. 4).

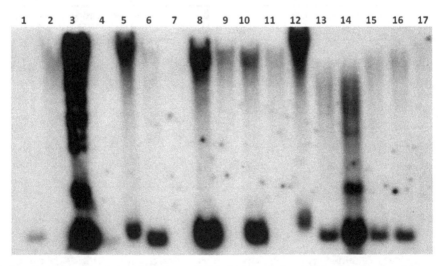

Fig. 4. Southern blot analysis of genomic DNA from transgenic *Sorghum bicolor* plants (cultivar CMSXS 102B). Genomic DNA was digested with *xho*I and hybridized with a radiolabelled 520 pb *bar* fragment. Lane 1: plasmid positive control; lanes 2-16: transformed plants ; lane 17: wild type *S. bicolor* total DNA.

5. Discussion

Immature inflorescence proved to be an excellent organ to increase considerably the quantity of tissue competent of embryogenic callus production. A large number of high quality calli is relatively easier and faster to produce from immature inflorescence.

Transient expression of anthocyanin allows us to detect in a rapid and precise manner the most efficient combination of biolistics parameters that rendered a higher transient expression. The frequency of transient activity expression as an indicator of stable transformation efficiency has already been used, successfully, by Christiansen *et al.* (2005) to optimize the transformation conditions of *Brachypodium distachyon*. They observed that treatments with a higher number of GUS spots where the ones that produced a larger number of stable transgenic events.

An important step in the transformation via biolistics is the wound suffered by the explant during the microparticle entry into the cell. Usually to minimize this type of problem and to increase the capacity for somatic embryogenesis and plant regeneration, the target cells are plasmolised by an osmotic treatment (Vain *et al.* 1993). In this study all treatments where sorghum explants were incubated in an osmotic media a few hours before bombardment produced a higher number of anthocyanin spots, confirming that plasmolysis of cells can reduce damage and increase the efficiency of bombardments.

Acceleration pressure and microcarrier flying distance are parameters that influence the ability to deliver DNA into various explants. Analyzing the transient expression of anthocyanin, it was observed that a helium gas pressure of 1000 psi combined with a distance of 3 cm rendered the higher number of anthocyanin spots. This combination of biolistic physical parameters when tested in stable transformation experiments showed an efficiency of up to 3.33% of transgenic sorghum events production. Even though, biolistic parameters should be optimized for each equipment and explant used, other authors found optimal bombardment conditions similar to our results. Casas *et al.* (1993) and Tadesse *et al.* (2003) were able to generate transgenic sorghum plants via biolistic using a macro-carrier flying distance of 6 cm and a pressure of 1100 psi with an efficiency ranging from 0,3% to 1,3%.

We introduced the *uidA* and *bar* genes under the control of the CaMV35S promoter into sorghum CMSXS102B genome. The transformed plants were determined by a combination of PCR and Southern blot analysis, together with assays demonstrating functional gene product activity, PAT and GUS. Histochemical GUS activity was absent in leaves of the T_0 plants, but could be detected in the T_1 germinating seeds of one of the events investigated in this study. The absence of GUS expression in transgenic sorghum has been reported by several investigators (Emani *et al.*, 2002; Carvalho *et al.*, 2004; Girijashankar *et al.*, 2005; Van Nguyen *et al.*, 2007). Factors such as methylation based silencing (Emani *et al.*, 2002), regulatory sequences present in the genetic cassette (Tadesse *et al.*, 2003; Carvalho *et al.*, 2004) and phenolic compounds typically present in the sorghum tissue culture (Carvalho *et al.*, 2004) might have contributed for the absence of GUS expression.

The analysis of PPT-resistance showed that the trait was expressed by all of the transgenic events recovered, probably because of the herbicide selection pressure. In addition, it was inherited by the T_1 progenies with a typical Mendelian segregation pattern in four out of six transgenic lines studied. Two lines showed a 1:1 segregation ratio; this type of transgene segregation had already been reported in wheat and maize (Cheng *et al.*, 1997; Ishida *et al.*

1996). This abnormal segregation pattern might be partially caused by gene silence or non-detectable gene expression in the transgenic plants (Cheng *et al.*, 1997; Vaucheret *et al.*, 1998).

We report a transformation methodology for calli derived from immature inflorescence of sorghum, via biolistics; these transformation conditions are already being used at Embrapa Maize and Sorghum to introduce genes of agronomical interest into the sorghum genome.

6. Acknowledgements

We thank the McKnight Foundation, Fapemig "Fundação de Apoio a Pesquisa do Estado de Minas Gerais" and CNPq "Conselho Nacional de Desenvolvimento Científico e Tecnológico"-Brazil for financial support. We also acknowledge Dr. Vicki Chandler from the Department of Plant Science, University of Arizona, Tucson, Arizona /USA for the donation of genetic cassettes p35S::C1 and p35S::*Bperu*. This publication has been funded by "Fundo MP2 Embrapa / Monsanto".

7. References

Able, J.A., Rathus, C., Godwin, I.D.: The investigation of optimal bombardment parameters for transient and stable transgene expression in sorghum. - In Vitro Cell Dev Biol. 37:341-348, 2001.

Bai, Z.L., Wang, L.Q., Zheng, L.P., Li, A.J., Wang, F.L. A study on the callus induction and plant regeneration of different sorghum explants. Acta Agric Boreali Sinica 10:60–63, 1995.

Bhat, S., Kuruvinashetti, M.S. Callus induction and plant regeneration from immature embryos of maintainer line (B) of kharif sorghum. J Maharashtra Agric Univ 20:159, 1995.

Bhat. S., Kuruvinashetti, M.S., Bhat, S. (1994) Callus induction and plantlet regeneration from immature inflorescence in some maintainer (B) lines of kharif sorghum (*Sorghum bicolor* (L.) Moench.). Karnataka J Agric Sci 7:387–390, 1994.

Brandão, R.L. Transformação genética de *Sorghum bicolor* (L. Moench) visando tolerância ao Al+3. Tese doutorado. Universidade Federal de Lavras, UFLA, Lavras, Brazil, 127 p. 2007.

Cai, T., Butler, L. G.: Plant regeneration from embryogenic callus initiated from immature inflorescences of several high-tannin sorghums. - *Plant Cell. Tissue and Organ Culture* 20: 101–110, 1990.

Carvalho, C.H.S., Zehr, U.B., Gunaratna, N., Anderson, J., Kononowicz, H.H., Hodges, T.K.D., Axtell, J.D. *Agrobacterium*-mediated transformation of sorghum: factors that affect transformation efficiency. *Genet. Mol. Biol.* 27, 259-269, 2004

Casas, A.M., Kononowicz, A.K., Zehr, U.B., Tomes D.T., Axtell, J.D., Buttler, L.G., Bressan, R. A., Hasegawa, P. M.: Transgenic sorghum plants via microprojectile bombardment. - *Proc. Natl. Acad. Sci.* 90: 11212-11216, 1993.

Casas, A.M., Kononowicz, A.K., Haan, T.G., Zhang, L., Tomes, D.T., Bressan, R.A., Hasegawa, P.M.: Transgenic sorghum plants obtained after microprojectile bombardment of immature inflorescences. - In *Vitro Cell Dev. Biol.* 33: 92-100, 1997.

Chen, C.; Meyermans, H.; Van Doorsselaere, J.; Van Montagu, M.; Boerjan, W. A gene encoding caffeoyl coenzyme A 3-*O*-methyltransferase (CCoAOMT) from *Populus trichocarpa*. Plant Physiology, v.117, p.719, 1998

Cheng, M., Fry, J.E., Pang, S., Zhou, H., Hironaka, C.M., Duncan, D. R., Conner, T. W., Wan, Y.: Genetic transformation of wheat mediated by *Agrobacterium tumefaciens*. - *Plant Physiol.* 115: 971-980, 1997.

Christiansen, P., Andersen, C.H., Didion, T., Folling, M., Nielsen, K.K.: A rapid and efficient transformation protocol for the grass *Brachypodium dist*achyon. - *Plant Cell Rep.* 23: 750-758, 2005

Devi, P., Sticklen, M.: In vitro culture and genetic transformation of sorghum by microprojectile bombardment. - *Plant Biosystems.* 137: 249-254, 2003.

Emani, C., Sunilkumar, G., Rathore, K.S.: Transgene silencing and reactivation in sorghum. - *Plant Sci.* 162: 181-192, 2002.

Ferreira, D.F.: SISVAR.exe: sistema de análise de variância. UFLA, Lavras, 2000. [SISVAR.exe: variance analysis system]

Gao, Z., Xie, X., Ling, Y., Muthukrishnan, S., Liang, G.H.: *Agrobacterium tumefaciens*-mediated sorghum transformation using mannose selection. - *Plant Biotechnol. J.* 3: 591-599, 2005.

Girijashankar, V., Sharma, H.C., Sharma, K.K., Swathisree, V., Sivarama Prasad, L., Bhat, B.V., Royer, M., Secundo, B.S., Narasu, M.L., Altosaar, I., Seetharama, N.: Development of transgenic sorghum for insect resistance against the spotted stem borer (Chilo partellus). - *Plant Cell Rep.* 24: 513-522, 2005.

Goff, S.A., Klein, T.M., Roth, A.B., Fromm, M.E., Cone, K.C., Radicella, J.P., Chandler, V.L.: Transactivation of anthocyanin biosysnthetic genes following transfer of B regulatory gene into maize tissues. - *EMBO J.* 9: 2517-2522, 1990.

Gupta, S.; Khanna, V.K.; Singh, R.; Garg, G.K. Strategies for overcoming genotypic limitations of in vitro regeneration and determination of genetic components of variability of plant regeneration traits in sorghum. Plant Cellular Tiss. Organ. Cult., v.86, p.379–388, 2006.

Gurel, S., Gurel, E., Kaur, R., Wong, J., Meng, L., Tan, H-Q., Lemaux, P.: Efficient, reproducible *Agrobacterium*-mediated transformation of sorghum using heat treatment of immature embryos. – *Plant Cell Rep*, 28:429-444, 2009.

Howe, A., Sato, S., Dweikat, I., Fromm, M., Clemente, T.: Rapid and reproducible *Agrobacterium*-mediated transformation of sorghum. - *Plant Cell Rep.*, 25: 784-791, 2006.

Ishida, Y., Saito H., Ohta, S., Hiei, Y., Komari, T., Kumashiro, T.: High efficiency transformation of maize *(Zea mays* L.) mediated by *Agrobacterium tumefaciens*. - *Nature Biotechnol.* 14: 745–750, 1996.

Jeoung, J.M., Krishanaveni, S., Muthukrishnana, S., Trick, H.N., Liang, G.H.: Optimization of sorghum transformation parameters using genes for green fluorescent protein and β-glucuronidase as visual markers. - *Hereditas*, 137: 20-28, 2002.

Jogeswar, G., Ranadheer, D., Anjaiah, V., Kishor, P.B.K.: High frequency somatic embryogenesis and regeneration in different genotypes of *Sorghum bicolor* (L.) Moench from immature inflorescence explants. - *In Vitro Cell Dev. Biol. Plant.*, 43: 159-166, 2007.

Kaeppler, H.F., Pedersen, J. F.: Evaluation of 41 elite and exotic inbred *Sorghum* genotypes for high quality callus production. *Plant Cell, Tissue and Organ Culture* 48: 71-75, 1997.

Klein T M, Wolf ED, Wu R, Sanford JC. High-velocity microprojectiles for delivering nucleic acids into living cells. Nature. 6117:70-73, 1987.

Lu, L., Wu X., Yin, X., Morrand, J., Chen, X., Folk, W.R., Zhang, Z.J. Development of marker-free transgenic sorghum [*Sorghum bicolor* (L.) Moench] using standard binary vectors with bar as a selectable marker. *Plant Cell Tiss Organ Cult.* 99:97–108, 2009.

Murashige, T., Skoog, F.A.: Revised medium for rapid growth and bioassays with tobacco tissue culture. - *Physiologia Plantarum*, 15: 473-497, 1962.

Mall, T.K., Dweikat, I., Sato, S.J., Neresian, N., Xu, K., Ge, Z., Wang, D., Elthon, T., Clemente, T. Expression of the rice CDPK-7 in sorghum: molecularand phenotypic analyses. *Plant Mol Biol* 75:467-479 2011.

Nahdi, S.; de Wet JMJ (1995). In vitro regeneration of *Sorghum bicolor* lines from shoot apeses. Internation Sorghum and Millet News letter 36:89-90.

Nguyen, T-V., Thu, T. T., Claeys, M., Angenon, G.: *Agrobacterium*-mediated transformation of sorghum (*Sorghum bicolor* (L.) Moench) using an improved in vitro regeneration system. – *Plant Cell Tiss Organ Cult.*, 91:155-164, 2007.

Nwanze, K.F., Seethrama, N., Sharma, H.C., Stenhouse, J.W., Frederiksen, R., Shantharam, S., Raman, K.: Biotechnology in pest management:improving resistance in sorghum to insect pests. Environmental impact and biosafety: issues of genetically engineered sorghum. - *Afr Crop Sci J.*, 3: 209-215, 1995.

Patil, V.M., Kuruvinashetti, M.S. Plant regeneration from leaf sheath cultures of some rabi sorghumcultivars. South Afr J Bot 64:217-219, 1998.

Rueb, S., Hensgens, L.A.M.: Improved histochemical staining for β-glucuronidase activity in monocotyledonous plants. - *Rice Genet. Newsl.* 6: 168-169, 1998.

Saghai-Maroof, M.A., Soliman, K.M., Jorgensen, R.A., Allard, R.W.: Ribosomal DNA spacer-length polymorphism in barley: Mendelian inheritance, chromosomal location, and population dynamics. - *Proc. Natl. Acad. Sci.*, 81: 8014 - 8019.

Sambrook, J., Fritsch, E.F., Maniatis, T.: *Molecular Cloning: A Laboratory Manual*. 2.nd.ed. - Cold Spring Harbor Laboratory London 1989.

Sanford, J.C.; Smith, F.D.; Russell, J.A. Optimizing the biolistic process for different biological applications. In: WU, R. (dE.). Recombinant DNA – Part H. San Diego: Academic, 1993. p.483-510 (Methods in Enzymology, 217).

Sasaki, T.; Yamamoto, Y.; Ezaki, B.; Katsuhara, M.; Ahn, S.J.; Ryan, P.R.; Delhaize, E.; Matsumoto, H. A wheat gene encoding an aluminum-activated malate transporter. Plant Journal, v.37, p.645–653, 2004.

Shyamala, D., Devi, P. Efficient regeneration of sorghum, Sorghum bicolor (L.) Moench, from shoot-tip explant. Ind J Exp Biol 41:1482–1486, 2003.

Tadesse, Y., Sági, L., Swennen, R., Jacobs, M.: Optimisation of transformation conditions and production of transgenic sorghum (*Sorghum bicolor*) via microparticle bombardment - *Rev. Plant Biotechnol. Appl. Genetics* 75: 1-18, 2003.

Taylor NJ, Fauquet CM. Particle bombardment as a tool in plant science and agricultural biotechnology. DNA Cell Biol. 21(12):963-977, 2002.

Vain, P., Mcmullen, M.D., Finer, J.J.: Osmotic treatment enhances particle bombardment-mediated transient and stable transformation of maize - *Plant Cell Rep.* 12: 84-88, 1993.

Van Nguyen, T., Thu ,T.T., Claeys, M., Angenon, G.: *Agrobacterium*-mediated transformation of sorghum (*Sorghum bicolor* (L.) Moench) using an improved in vitro regeneration system. - *Plant Cell Tiss Organ Cult.* 91: 155-164, 2007

Vaucheret, H., Béclin, C., Elmayan, T., Feuerbach, F., Godon, C., Morel, J.B., Mourrain, P., Palauqui, J.C., Vernhettes, S.: Transgene-induced gene silencing in plants - *Plant J.* 16: 651–659, 1998.

Wu L, Nandi S, Chen L, Rodriguez RL, Huang N (2002) Expression and inheritance of nine transgenes in rice. Transgenic Res.11:533-541, 2002.

Zhao, Z.Y., Cai, T., Tagliani, L., Miller, M., Wang, N., Pang, H., Rudert, M., Schroeder, S., Hondred, D., Seltzer, J., Pierce, D.: *Agrobacterium*-mediated sorghum transformation - *Plant Mol Biol*, 44: 789, 2000.

Zhu, H., Muhukrishnan, S., Krishnaveni, S., Wilde, G., Jeoung, J.M., Liang. G.H.: Biolistic transformation of sorghum using a rice chitinse gene - *Journal of Genetics & Breeding* 52: 243-252, 1998.

7

Biological Activity of *Rehmannia glutinosa* Transformed with Resveratrol Synthase Genes

Bimal Kumar Ghimire[1], Jung Dae Lim[2] and Chang Yeon Yu[3]
[1]*Department of Ethnobotany and Social Medicine Studies, Sikkim University, Sikkim*
[2]*Department of Herbal Medicine Resource, Kangwon National University, Samcheok*
[3]*Division of Applied Plant Science, Kangwon National University Chunchon,*
Kangwondo
[1]*India*
[2,3]*South Korea*

1. Introduction

The rate of increase of food crop production has decreased due different factors such as global climate change, alteration in use of land, pests, disease, salinity and drought. Food production was said to be inadequate for the increasing world population (WHO, 1996a). Therefore, it is essential to increase food production and distribution in order to meet its demand and free from hunger. Furthermore, transgenic plants can enhance yields, harvesting of crops, reduce dependency on chemical insecticides. Production of transgenic plants can address the global problems such as climate change, deficiency of food and nutrition. Development of transgenic plant involves manipulation or transfer of genes from other organisms which may improve yield, quality, herbicides, and pest or diseases resistant or environmental conditions, increased agricultural productivity and better quality foods. Modification of genetic constitute of plant by inserting transgene enhances nutritional composition of the foods and improve human health and minimizes the use of pesticides and insecticides.

A number of transgenic crop plants has been produced from a variety of crop plants to date with enhanced agronomic characteristics, for example, transgenic tomatoes with improved shelf-life, transgenic fruits and vegetables with delayed ripening time and increased length of storage. Moreover, pest and disease resistance crops have been produced, viz., papaya-ringspot-virus-resistant papaya (Gonsalves 1998), insect resistance cotton, transgenic rice plants that are resistant to rice yellow mottle virus (RYMV) (Pinto et al 1999), improved nutritional contents in the transgenic rice which exhibits an increased production of beta-carotene as a precursor to vitamin A (Ye et al 2000). In addition, technology of transgenic plant production can be used to produce vaccines and bioactive compounds in plants. For example, expression of anti-cancer antibody in rice resulted in production of vaccines against infectious disease from potato (Thanavala et al

1995). *Rehmannia glutinosa* is perennial medicinal plant belongs to the family *Scrophulariaceae*. Its root has long been used in Korea for medicinal purposes such as hemantic, robustness, cardiotonic drug, diabetes treatment, antifebriel and detoxification purposes (Choi et al., 1995). Roots of *R. glutinosa* are usually infected with various pathogens during storage and these infections cause great damage to the roots and impede the intensive farming of the crop (Lim et al 2003).

Resveratrol is found in a limited number of unrelated plants and possess antifungal activity and induction in response to pathogen infection. Resveratrol is well known for its potent antioxidant activity and health-promoting effects, cardioprotection (Ignatowicz and Baer-Dubowska 2001) and reduction of cancer risk have also been observed (Jang et al. 1997; Cal et al. 2003).It can also exert neuroprotective effects by increasing heme oxygenase activity in the brain (Zhuang et al. 2003). The expression of RS transcripts has been associated with an increasedresistance to various fungal pathogens in transgenic tobacco (Hain et al. 1993) and tomato (Thomzik et al. 1997).

The purpose of this report is to better understand the role of transgene to improve the nutrition value of important crops and to evaluate the biological activity of secondary metabolic substances such as resveratrol, SOD, phenolic compounds in *Rehmannia glutinosa* under environmental stress.

2. Transformation of *R. glutinosa* with RS gene

The peanut RS genomic DNA sequence, AhRS3 (GenBank Accession number, AF227963) a polypeptide of 389 amino acid residues, was cloned into the Xba I/Cla I sites of binary expression vector pGA643 under the CaMV35S promoter. This produced a recombinant AhRS3 expression plasmid, pMG-AhRS3. *Agrobacterium tumefaciens* strain LBA4404 harboring the binary vector pMG-AhRS3, which contains the neomycin phosphotransferase gene (npt II) directed by the nos promoter as a selectable marker, was used. A single colony of this strain was grown for 24 h at 28 ± 1 °C with shaking (150 rpm) in 20 ml of liquid Luria–Bertani (LB) medium (10 g/l tryptone, 10 g/l NaCl, 5 g/l yeast extract, pH 7) containing 100 mg/l kanamycin. The cells were centrifuged for 10 min at 7,000 x g at 4 °C and resuspended in liquid inoculation medium (MS medium with 20 g/l of sucrose) to obtain a final OD_{600} = 1.0 for use in plant infection. The surface-sterilized leaf explants were cultured for 2–3 days on MS medium containing 1 mg/l 6-benzylamino-purine (BAP), 2 mg/l thidiazuron (TDZ), 0.2 mg/l naphthalene acetic acid (NAA), Murashige and Skoog (MS) vitamin, 3% sucrose, and 0.8% agar (pH 5.2). Pre-treated explants were dipped into the Agrobacterium suspension in liquid inoculation medium for 10–15 min, blotted dry on sterile filter paper and incubated in a shoot induction medium (MS medium containing 2 mg/l BAP, 1 mg/l TDZ, 0.2 mg/l NAA, MS vitamin, 3% sucrose, and 0.8% agar at pH 5.2.) in the dark at 23±1 °C. After co-cultivation for 2 days, the explants were transferred to shoot induction medium containing 50 mg/l kanamycin and 200 mg/l timentin (mixture of ticarcillin disodium and clavulanate potassium) and were transferred to fresh selection medium every 2 weeks. Putative transgenic shoots were regenerated 6–8 weeks after the first sub-culture and were incubated in a growth chamber with a 16 h at 23±1 °C for 30 days. Putative transgenic plantlets were then transferred to pots containing autoclaved vermiculite and were grown in the glasshouse.

Transgene-positive T_0 lines were selected by PCR screening. The lines containing RS gene and *npt* II transgene sequences were chosen for the evaluation of biological activities. Transformation of transgene into the plant genome was confirmed by Southern blot analysis.

2.1 Scavenging of DPPH radicals and Inhibition of lipid peroxidation of transgenic *R. glutinosa*

Free radical-scavenging activity was evaluated using trolox as standard antioxidants. The radical-scavenging activity was measured using the stable radical 1,1-diphenyl-2-picrylhydrazyl (DPPH) as previously described (Xing et al., 1996). Various concentrations of the extracts were added to 4 ml 0.004% methanol solution of DPPH. The mixture was shaken and left for 30 min at room temperature in the dark, and the absorbance was measured with a spectrophotometer at 517 nm. The radical-scavenging activity was expressed as a percentage of the absorption of DPPH in the presence and in the absence of the compound. Calculated IC_{50} values indicate the concentration of sample required to scavenge 50% of the DPPH radical. DPPH activity was calculated as

DPPH activity (%) = $(A_{blank} - A_{sample})/A_{blank} \times 100$,

where, A_{blank} is the control reaction (containing all reagents except the test compound), and A_{sample} is the absorbance of the test compound.

Inhibition of lipid peroxidation was determined by measuring thiobarbituric acid-reactive substance production (Buege and Aust 1978).

2.1.1 Measurement of photosynthesis rate

Stomatal conductance (g_s), net CO_2 assimilation rate (A), and the ratio of internal CO_2 concentration (C_i) were measured according to the method described by Ivan et al. (2001) using a portable photosynthesis system (LCA-4; Analytical Development Co., Hoddesdon, Herts, UK) at an air temperature 28 ± 1 °C. For this measurement, ten fully expanded young leaves were selected from the transgenic lines and control plants at the maximum tillering stage. All measurements were performed three times on sunny days between 1000 and 1400 hours on the surface of leaves from August to September under a saturating photosynthetic photo flux density (PPFD) of 1,500 μmol/m/s. Measurements were made at the center of the leaf surface immediately after the CO_2 concentration decrease was stable. Each leaf was allowed to stabilize for 4–6 min before measuring the g_s, C_i, and A.

2.1.2 HPLC analysis

The HPLC analysis was applied using the modified method of Banwart et al. 1985. The mobile phase consisted of solvent A and B. Solvent A contained 98% water and 2% glacial acetic acid in 0.018M ammonium acetate. Solvent B was 70% solvent A and 30% organic solution, the latter being composed of 82% methanol, 16% n-butanol and 2% glacial acetic acid in 0.018M ammonium acetate. Following injection of 20μL of the sample, the flow rate of the mobile phases was maintained at 1mL min−1. A linear HPLC gradient was employed. The HPLC system consisted of a Young-Lin M930 liquid chromatograph pump and an M720 detector (Young-Lin Instruments Co., Ltd). The column for quantitative analysis was a YMC-Pack ODS-AM-303 (250×4.6mm I.D.), and the UV absorption was measured at 280 nm.

2.1.3 SOD activity

SOD activity of root *R. glutinosa* was measured by the nitro blue tetrazolium (NBT) reduction method (Beyer and Fridovich, 1987). Test tubes containing reaction solution with 3mL of assay buffer, 60 _L of crude enzyme and 30_L of riboflavin were illuminated for 7 min in an aluminium foil lined box containing two 20-W Slyvania Groiux Fluorescent lamps at 25 ∘C. After reaction, the absorbance of the blank solution and reaction solution was measured with a spectrophotometer (Hitachi Ltd., Tokyo, Japan) at 560 nm. SOD activities were calculated as a following equation:

SOD activity (%) = $(1 - A/B) \times 100$

A: absorbance of sample; B: absorbance of blank.

2.1.4 Paper disc diffusion assay

Bacterial pathogens and fungal strains were grown in liquid medium (micrococcus, nutrient, and YM media) for 20 h to a final concentration of 10^6–10^7 CFU/ml. Aliquots of 0.1 ml of the test microorganisms were spread over the surface of agar plates. Sterilized filter-paper discs (Whatman No. 1, 6 mm) were saturated with 50 μl of the methanol extract at 10,000 ppm and left to dry in a laminar flow cabinet. The soaked, dried discs were then placed in the middle of the plates and incubated for 24 h. Antimicrobial activity was measured as the diameter (mm) of the clear zone of growth inhibition. Negative controls were prepared using the same solvents employed to dissolve the plant extracts.

In order to evaluate morphological and agronomic performance of transgenic *R. glutinosa*, plants of each PCR positive T1 transgenic lines and seed derived control plants were transferred to field containing bed soil and evaluated for morphological characters.

2.2 Results and discussion

2.2.1 Biological activities of transgenic *R. glutinosa*

2.2.1.1 Scavenging of DPPH radicals of transgenic *R. glutinosa*

The free radical scavenging activities of non-transgenic control and transgenic R. glutinosa extracts, α-tocopherol, are presented in Fig.1. A solution of each extract at a concentration of 1.0 mg/ml was prepared. The activities of transgenic sample extracts were between 16.00 and 20.00 μg/ml at 1.0 mg/ml. Most of the transgenic leaves samples showed high antioxidant activity using DPPH as compared to non-transgenic control plants. With regard to RC50 values (the concentration of antioxidant required to achieve absorbance equal to 50% that of a control containing no antioxidants), RS3 transgenic lines showed highest radical-scavenging abilities (RC50 = 16.00 ± 2.00 μm). The DPPH free radical scavenging and

LDL peroxidation activities of trans-3'-H-Rglu and trans-resveratrol isolated from transgenic R. glutinosa evaluated (Fig. 2 & 3). DPPH activity of trans-resveratrol were significantly higher (72 ± 4.5 μm) than trans-3'-H-Rglu (198 ± 6.8 μm). This could be attributed to the higher level of accumulation of resveratrol compounds in the transgenic R. glutinosa (Fig. 4).

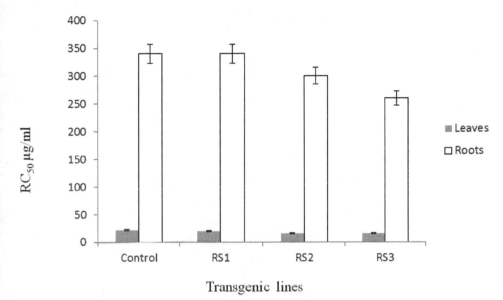

Fig. 1. DPPH free radical scavenging activity of extract in transgenic and non-transgenic *Rehmannia glutinosa*

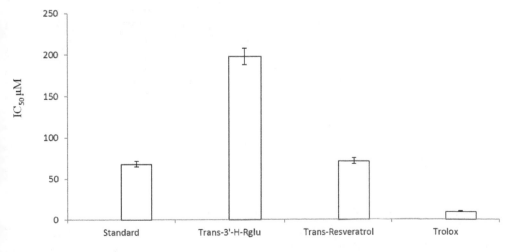

Fig. 2. DPPH scavenging activities of stilbenes isolated from transgenic *Rehmannia glutinosa*.

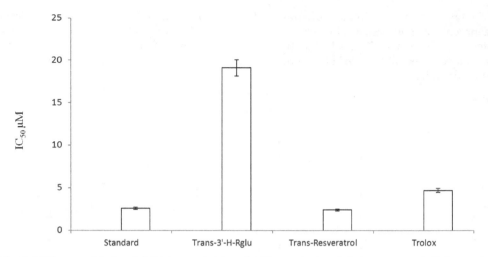

Fig. 3. LDL peroxidation inhibition activities of stilbenes isolated from transgenic *Rehmannia glutinosa*.

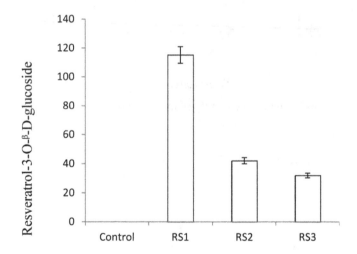

Transgenic lines

Fig. 4. Accumulation of Resveratrol-3-O-β-D-glucoside levels observed in leaves of RS3-transformed transgenic *Rehmannia glutinosa*.

2.2.2 Superoxide Dismutase (SOD) activity of transgenic *R. glutinosa*

The SOD activities non transgenic plant and transgenic plants (without water stress) were 13.81 and 11.23% respectively. In contrast, the SOD activities non transgenic plant and transgenic plants (with water stress) were 24.59 and 3.8 % respectively (Fig. 5).

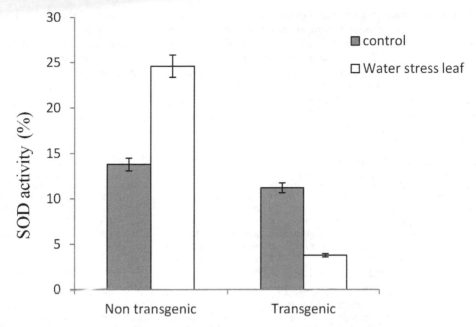

Fig. 5. SOD activities of stilbenes isolated from transgenic *Rehmannia glutinosa*.

2.2.3 Phenolic compound analysis of transgenic *R. glutinosa*

The quantitative analysis of phenolic compounds of non-transgenic and transgenic *R. glutinosa* extract performed using HPLC is given in Table 1. We found quantitative differences in total phenolic compounds between transgenic and control plants stem and root samples under hyper irrigation treatment (Table 2).

Lines	Plant parts	Hyd[1]	Chl[2]	Cat[3]	Caf[4]	Syr[5]	Sal[6]	Cou[7]	Fer[8]	Hes[9]	Nar[10]	Hyr[11]	Cin[12]	Que[13]	Nan[14]	Total
						ug/g										
Control	S	9.12	37.3	1.33	13.25	25.68	0.25	97.19	67.02	54.11	52.02	19.07	82.77	0	0	412.69
	R	4.13	0	0	0	2.19	0	7.21	12.35	6.71	0	53.45	98.53	26.39	0	210.94
RS1	S	3.71	34.13	1.14	0.25	20.81	3.35	90.84	76.28	47.66	55.17	16.99	185.22	0	0	497.72
	R	1.48	0	0	0	10.21	0	18.04	8.53	3.23	0	14.42	198.69	14.57	0	269.18
RS2	S	0.25	29.77	0.25	0.25	11.26	0.25	74.26	83.06	34.41	176.93	73.53	83.69	0	0	537.88
	R	9.48	0	0	0	4.84	0	7.75	7.41	0	10.96	23.88	114.2	19.99	0	189.01
RS3	S	0.25	43.08	0.25	0.25	12.48	0.25	117.69	60.2	48.63	84.61	43	100.91	0	0	468.27
	R	14.61	6.47	0	0	10.02	0	23.32	17.95	0.77	0	24.18	264.23	19.57	0	360.01

S: shoot, R: root, Hyd[1]: *p*-hydroxybenzoic acid, Chl[2]: Chlorogenic acid, Cat[3]: Cathechin, Caf[4]: Caffeic acid, Syr[5]: Syringic acid, Sal[6]: Salicylic acid, Cou[7]: *p*-coumeric acid, Fer[8]: Ferulic acid, Hes[9]: Hesperidin, Nar[10]: Narigen, Hyr[11]: Hyricetin, Cin[12]: trans-cinnamic acid, Que[13]: Quercitin, Nan[14]: Narigenin.

Table 1. Distribution of major phenolic compounds in control and transgenic plants transformed by resveratrol synthase in *R. glutinosa* under hyper irrigation.

The average total concentrations of phenolic compounds in control plant stem and roots were 412.69 and 210.94 µg/g dry weight (DW), respectively; in comparison, transgenic stem and root samples had higher concentrations of 468.27–537.88 and 189.01–360.01µg/g DW, respectively. The phenolic compounds that increased in the transgenic lines were p-hydroybenzoic acid, p-coumaric acid, ferulic acid, narigenin, trans-cinnamic acid, chlorogenic acid. Similarly, we found quantitative differences in total phenolic compounds between transgenic and control plants stem and root samples under pathogen treatment (Table 2). The average total concentrations of phenolic compounds in control plant stem and roots were 364.58 and 181.20 µg/g DW, respectively; in comparison, transgenic stem and root samples had higher concentrations of 555.00–919.16 and 312.70–677.26 µg/g DW, respectively.

Lines		Hyd[1]	Chl[2]	Cat[3]	Caf[4]	Syr[5]	Sal[6]	Cou[7]	Fer[8]	Hes[9]	Nar[10]	Hyr[11]	Cin[12]	Que[13]	Nan1[14]	Total
						ug/g										
Control	S	7.08	5.75	0	0	0	0.25	0.25	22.8	0	293.04	26.14	0	0	22.11	364.58
	R	21.4	14.28	0	9.82	0	7.78	5.73	12.53	0	0	33.23	94.47	17.66	0	181.2
RS1	S	27.98	39.58	2.57	0	53.45	22.22	28.97	195.6	0	195.35	56.84	0	0	0	555
	R	8.97	10.64	0	0	21.45	0	10.21	6.93	0	3.35	33.91	211.4	20.16	0	318.05
RS2	S	35.27	78.8	1.18	0.25	60.42	240.9	38.61	137.3	15.97	245.74	38.84	6.87	0	133.01	919.16
	R	0	0	0	0	36.87	0	9.04	4.25	8.37	0	8.35	223.25	22.58	0	312.7
RS3	S	0	0	0	0	0	0	153.91	353.5	25.04	97.34	47.42	0	0	0	677.26
	R	30.54	14.61	0	7.01	0	6.02	46.33	21.62	9.93	0	26.56	464.94	19.7	0	602.11

S: shoot, R: root, Hyd[1]: p-hydroxybenzoic acid, Chl[2]: Chlorogenic acid, Cat[3]: Cathechin, Caf[4]: Caffeic acid, Syr[5]: Syringic acid, Sal[6]: Salicylic acid, Cou[7]: p-coumeric acid, Fer[8]: Ferulic acid, Hes[9]: Hesperidin, Nar[10]: Narigen, Hyr[11]: Hyricetin, Cin[12]: trans-cinnamic acid, Que[13]: Quercitin, Nan[14]: Narigenin.

Table 2. Distribution of major phenolic compounds in control and transgenic plants transformed by resveratrol synthase in *R. glutinosa* under infected pathogen (*Fusarium oxysporum*).

2.2.4 Antimicrobial activities transgenic *R. glutinosa*

Antimicrobial activities of the non-transgenic and transgenic plants were assessed by a paper disc diffusion assay. The results indicated variation in the antimicrobial properties of the resveratrol-3-O-B-D glucoside and resveratrol extracted from the transgenic *R. glutinosa* (Table 3). In general, the resveratrol was more effective than the resveratrol-3-O-B-D glucoside against all the microbes tested. The strongest inhibitory effect was against *E. coli* and *S. typhimurium* at concentration of 1 mg/ml. Antimicrobial activity in plant extracts depends not only on the presence of phenolic compounds, but also on the presence of various secondary metabolites (Gordana et al., 2007). These observations suggest that the antimicrobial activity of transgenic *R. glutinosa* enhanced by the RS3 genes transformed into the *R. glutinosa* genome. However, other phenolic acid-like phenols are thought to contribute to plant defences against pests and pathogens (Awika & Rooney, 2004).

Compounds	Conc. (ppm)	Clear zone (mm)						
		P. jadinii	*C. albicans*	*S. aureus*	*B. subtilis*	*K. pneumonia*	*E. coli*	*S. typhimurium*
Resveratrol -3-O-B-D- glucoside	20000	11	10.8	10.6	10.1	9.8	12.4	12.7
Resveratrol	20000	13.7	12.8	12.9	14.2	11.5	19.8	18.6

Table 3. Antimicrobial activity of stilbenes compounds isolated from transgenic *R. glutinosa*.

2.2.5 Morphological characterization of transgenic *R. glutinosa*

Phenotypic differences were observed within the different transgenic lines and between the transgenic and non-transgenic control plants (Table 4). However, there were no apparent differences in terms of root length and root diameter. Significant differences in root weight were observed between transgenic and non-transgenic lines and showed reduced weight over control plants.

Line	Root length (cm)	root diameter	Root weight
Control	24.3	20	330.1
RS1	20.3	14	142
RS2	23.6	18	226.3
RS3	21.9	15	159.5
RS4	21.7	17	273.1

Table 4. Growth characteristics of transgenic *R. glutinosa*.

2.2.6 Analysis of catapol content of transgenic *R. glutinosa*

The catapol contents and composition in subterranean parts of the transgenic lines and non-transformed plants were investigated using HPLC (Fig. 6). Overexpression of RS3 gene significantly increased the catapol, compared to that of wild type *R. glutinosa*.

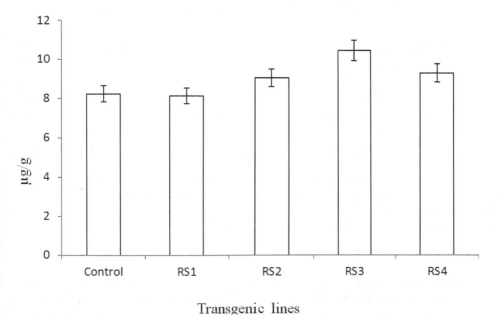

Fig. 6. Content of catapol in root of *R. glutinosa*

2.2.7 Effect of the photosynthesis rate in transgenic *R. glutinosa*

To compare the effect of RS3 gene overexpression on the photosynthesis rate and yield of transgenic and control plants, we measured stomatal conductance (gs), CO_2 concentration (CI), and photosynthesis rate (A) and found significant differences in these factors between transgenic and control plants (Table 5). The photosynthesis rate increased progressively with increasing CO_2 concentration. Photosynthesis rate of both non-transgenic and transgenic plants reduced by the increased duration of dry stress, being much lower at 15 days. Comparatively, transgenic lines showed higher photosynthetic control plants. Therefore, it is very possible that the higher level of RS3 gene in the transgenic plant is responsible for its enhanced photosynthetic performance.

		Photosynthetic rate					
Treatment Days		Non-transgenic plant			Transgenic plant		
		A (μmol m^{-2}s^{-1})	gs (μmol m^{-2}s^{-1})	Ci (ppm)	A (μmol m^{-2}s^{-1})	Gs (μmol m-2s-1)	Ci (ppm)
Control-		14.63 ± 0.1	0.29 ± 0.0	241.67 ± 0.85	17.42 ± 0.14	0.27 ± 0.02	212.47 ± 0.17
	3D	9.68 ± 0.06	0.11 ± 0.01	195.8 ± 3.14	17.1 ± 0.06	0.32 ± 0.0	223.98 ± 1.88
Dry stress	9D	3.63 ± 0.14	0.02 ± 0.0	45.52 ± 5.92	14.39 ± 0.04	0.23 ± 0.0	220.13 ± 1.24
	15D	1.13 ± 0.01	0.0	215.82 ± 0.78	1.69 ± 0.31	0.0	194.13 ± 4.21

A: Photosynthetic rate; gs: Stomatal conductance; Ci: Intercellular; CO_2concentration.

Table 5. Photosynthetic rate of non-trangenic and transgenic *R. glutinosa* under water stress.

It can be concluded that introduction of RS3 gene into the *R. glutinosa* genome resulted into increased production of stilbenes compounds that enhances the biological activity of plants. Increased in the resveratrol compounds further enhanced the disease resistance capacity of plant. This may cause beneficial effects on human and plant defence system.

3. References

Anwart WL, Porter PM, Granato TC, Hassett JJ (1985) HPLC separation and wavelength area ratios of more than 50 phenolics and flavonoids. J. Chem. Ecol. 11: 383–395.

Awika J M and Rooney LW (2004). Sorghum phytochemical and their potential impact on human health. Phytochemistry, 65: 1199–1221.

Beyer Jr, WE, Fridovich I, (1987) Assaying for superoxide dismutase activity: some large consequences of minor changes in conditions. Anal. Biochem. 161: 559–566.

Buege JA, Aust SD (1978) Microsomal lipid peroxidation. Methods Enzymol. 52: 302–310.

Cal C, Garban H, Jazirehi A, Yeh C, Mizutani Y and Bonavida B (2003) Resveratrol and cancer: chemoprevention, apoptosis, and chemo-immunosensitizing activities. Curr. Med. Chem. Anti-Canc. Agents. 3: 77–93.

Gonsalves D (1998) Control of papaya ringspot virus in papaya: A case study. *Annual Review of Phytopathology*, 36: 415-437.

Gordana S C, Jasna MC, Sonja MD, Tumbas VT, Markov SL, and Dragoljub DC (2007) Antioxidant potential, lipid peroxidation inhibition and antimicrobial activities of Satureja montana L. Subsp. kitaibelii extracts. International Journal of Molecular Sciences, 8: 1013-1027.

Hain R, Reif HJ, Krause E, Langebartels R, Kind H, Vornam B, Wiese W, Schmelzer E, Schreier PH, Stocker RH and Stenzel K (1993) Disease resistance results from phytoalexin expression in a novel plant. Nature 361: 153–156.

Ivan Y, Tsonko T, Violeta V, Katya G, Peter I, Nikolai T, Tatyana P (2001) Changes in CO_2 ssimilation, transpiration and stomatal resistance of different wheat cultivars experiencing drought under field conditions. Bulg J Plant Physiol 27:20–33

Ignatowicz E and Baer-Dubowska W (2001) Resveratrol, a natural chemopreventive agent against degenerative diseases. Pol. J. Pharmacol. 53: 557–569.

Jang M, Cai L, Udean GO, Slowing KV, Thomas CF, Beecher CWW, Fong HHS, Farnsworth NR, Kinghorn AD, Mehta RG, Moon RC and Pezzuto JM (1997) Cancer chemopreventive activity of resveratrol, a natural product derived from grapes. Science 275: 218–220.

Lim JD, Yun SJ, Lee SJ, Chung IM, Kim MJ, Heo K and Yu CY (2004) Comparison of resveratrol contents in medicinal plants. Korean J. Medicinal Crop Sci. 12: 163–170.

Pinto Y M, Kok RA and Baulcombe DC (1999) Resistance to rice yellow mottle virus (RYMV) in cultivated African rice varieties containing RYMV transgenes. *Nature Biotechnology* 17:702–707.

Thanavala Y, Yang Y, Lyons P, Mason HS, Arntzen C (1995). Immunogenicity of Transgenic Plant-derived Hepatitis B Surface Antigen. Proc. Nat. Acad. Sci. USA. 92: 3358-3361.

Thomzik JE, Stenzel K, Stocker R, Schreier PH, Hain R and Stahl DJ (1997) Synthesis of a grapevine phytoalexin in transgenic tomatoes (Lycopersicon esculentum Mill.) conditions resistance against Phytophthora infestations. Physiol. Mol. Plant Pathol. 51: 265–278.

WHO.: 1996a, Micronutrient Malnutrition: Half theWorld's Population Affected,World Health Organization, 13 Nov. 1996 no. 78, pp. 1–4.

Xing Q, Kadota S, Tadata T, Namba T (1996) Antioxidant effect of phenylethanoids from *Cistanche deserticola*. Biological and Pharmaceutical Bulletin. 19: 1580-1585.

Ye X, et al (2000) Engineering the provitamin A (beta carotene) biosynthesis pathway into (carotenoid-free) rice endosperm. Science, 287, 303-305.

Zhuang H, Kim YS, Koehler RC and Dore S (2003) Potential mechanism by which
    resveratrol, a red wine constituent, protects neurons. Ann. New York Acad. Sci.
    993: 276–286.

8

Phytoremediation of Bis-Phenol A via Secretory Fungal Peroxidases Produced by Transgenic Plants

Tomonori Sonoki[1], Yosuke Iimura[2] and Shinya Kajita[3]

[1]*Faculty of Agriculture and Life Science, Hirosaki University,*
[2]*National Institute of Advanced Industrial Science and Technology,*
[3]*Graduate School of Bio-Applications and Systems Engineering,*
Tokyo University of Agriculture and Technology,
Japan

1. Introduction

The fungal lignin-degrading enzymes lignin peroxidase (LiP, E.C. 1.11.1.14), Mn-dependent peroxidase (MnP, E.C. 1.11.1.13), and phenol oxidase (laccase) (Lac, E.C. 1.10.3.2) can degrade or polymerize organic pollutants such as polychlorophenols, polycyclic aromatic hydrocarbons, and chlorinated hydrocarbons (Fernando and Aust, 1994; Hammel, 1989; Hirano et al., 2000; Levin et al., 2003; Lin et al., 1990; Lovley et al., 1994; Mohn and Tiedje, 1992; Reddyy et al., 1998). However, to maintain such fungal lignin-degrading enzymes at adequate levels for degradation or detoxification (bioremediation), appropriate additions of both microorganisms and nutrients are essential over long periods of time. Recently, phytoremediation technology has gained attention for its potential as an ecological remediation tool of contaminated soil and water, as plants can grow autotrophically. Establishment of effective phytoremediation technology is a suitable strategy for the long-term remediation of contaminated areas. Phytoremediation includes some processes based on the plant functions as follows; phytostabilization, which is accumulation of pollutants in the rhizosphere by absorption on the root surface, precipitation, and complexation of pollutants; rhizodegradation, which is degradation of pollutants by interaction with rhizosphere microorganisms; phytoaccumulation (phytoextraction), which is uptake and accumulation of pollutants by plants; phytodegradation (phytotransformation), which is uptake and degradation of pollutants by plants; and phytovolatilization, which is uptake and volatilization of pollutants by transpiration from contaminated area. To widely apply the benefit of phytoremediation, improvement and reinforcement of the abilities for uptake, accumulation and degradation of pollutants using genetic engineering are one of the important development subjects.

There have been many reports of phytoremediation using transgenic plants. For example, glutathione S transferase and cytochrome P450 expression showed high resistance to pesticides (Gullner et al., 2001; Doty et al., 2000), the overexpression of bacterial mercury reductase showed high resistance to organic mercury (Bizilly et al., 2003) and effective volatilization of ionic mercury (Haque et al., 2010), pentaerythritol tetranitrate reductase-expressing plants were able to degrade glycerol trinitrate and 2,4,6-trinitrotoluene (French et al., 1999), introduction of bacterial genes involved in polychlorinated biphenyl (PCB) degradation in plants showed removal of PCB from a contaminated area (Novakova et al., 2009), the bacterial arsenite S-adenosylmethyltransferase expression induced arsenic methylation and volatilization (Xiang-Yan et al., 2011), the expression of gamma-glutamylcysteine synthetase and the genes involved in phytochelatin synthesis in plant showed more resistance and accumulation of cadmium (Zhu et al, 1999, Wawrzyński et al, 2006), and the yeast metallothionein expressing tobacco showed effective copper uptake (Thomas et al, 2003).

Recently, attempts are carried out to enhance the environmental remediation in contaminated area by using appropriate genetically modified plants with usage of fungal peroxidases. This chapter mainly focused on the removal of bis-phenol A (BPA; 2,2-bis(4-hydroxyphenyl)propane), which is one of the major chemicals used in plastics and resins and is well known to disrupt endocrine systems in humans and other animals, from contaminated areas with usage of transgenic technology. Although many organisms can degrade and metabolize BPA, which can lead to a reduction of the estrogenicity and toxicity of BPA (Kang et al., 2006), lignin-degrading basidiomycete fungi are particularly powerful degraders of organic pollutants including BPA. These fungi produce oxidative enzymes, such as LiP, Lac, and MnP, which can degrade and polymerize BPA both *in vivo* and *in vitro*. Therefore, an overview of our recent results regarding the phytoremediation of BPA with fungal peroxidase-expressing transgenic plants by lignin peroxidase (LiP), laccase (Lac), and manganese peroxidase (MnP) were presented together with the other potential uses of these transgenic plants in this chapter.

2. LiP-expressing transgenic tobacco

cDNA (Accession no. AB158478.1) encoding LiP from the reverse transcription (RT) products of total RNA prepared from mycelia of *Trametes versicolor* IFO1030 was isolated. The cloned cDNA was ligated into binary vector pBI121 (Brasileiro et al., 1991) with double cauliflower mosaic virus (CaMV) 35S promoter sequence (Figure 1), and was introduced into the genome of the tobacco (*Nicotiana tabacum* Samsun NN) by the leaf-disk method via *Agrobacterium tumefacience* LBA4404 (Liang et al., 1989).

Integration of the cDNA into the genome of tobacco was confirmed by polymerase chain reaction (PCR) upon 10 independent transgenic lines. Two of the lines showed growth inhibition and thus were excluded from further analysis. Western blot analysis with root extracts of transgenic tobaccos and antiserum raised against LiP protein were performed to confirm the production of LiP protein in roots of transgenic lines. . To prepare the antiserum against LiP of *T. versicolor* IFO1030, we synthesized one peptide, whose sequence was [240]CNGTTFPGTGDNQG[254]E, and conjugated it with keyhole limpet hemocyanin (KLH). The resultant peptide-KLH conjugant was injected into a 10-wk-old rabbit. After four injections, antiserum was collected and used for Western blot analysis. The expected signal was observed in the cell-free extracts of roots from LiP transgenic tobaccos (Figure 2).

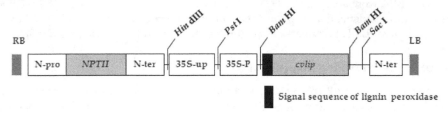

Fig. 1. Gene construct of T-DNA region of Ti plasmid.

RB, Right border of T-DNA; N-pro, promoter region of nopaline synthase gene; *NPTII*, neomycin phosphotransferase gene; N-ter, terminator region of nopaline synthase gene; 35S-up, upstream region of cauliflower mosaic virus (CaMV) 35S promoter sequence; 35S-P, CaMV 35S promoter sequence; *cvlip*, cDNA encoding LiP of *T. versicolor* IFO1030 plus signal sequence; LB, left border of T-DNA.

Fig. 2. Western blot analysis of LiP in cell-free extracts of roots of LiP transgenic lines.

Lanes; 1, LiP transgenic line (FLP)-1; 2, FLP-2; 3, FLP-3; 4, FLP-4; 5, FLP-5; 6, FLP-8; 7, control plant.

To test the ability of BPA removal by LiP-expressing transgenic plants, we transferred two-month-old transgenic lines on MS medium (Murashige and Skoog, 1962) to fresh MS liquid medium containing 3 g/L of glucose and 100 µg/L of kanamycin. After one week of incubation at 25°C, BPA was added to the medium at the final concentration of 100 µM and the medium was hydroponically incubated for another week. The six LiP-expressing transgenic lines showed 2- to 4-fold higher BPA removal ability than that of control plants during aqueous cultivation (Figure 3). LiP is a well-known enzyme that carries out direct and indirect oxidation of a number of environmental pollutants. Our confirmation that transgenic plants could express LiP in their roots and remove BPA will help us to establish improved methods for phytoremediation of contaminated environments.

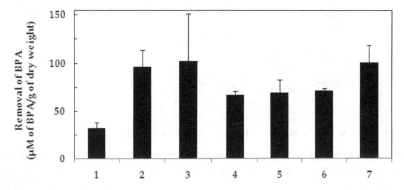

Fig. 3. Removal of BPA by LiP-expressing transgenic lines.

The levels of BPA were analyzed by HPLC (λ=278 nm). The values shown are the average of results from three independent experiments. Lanes; 1, control; 2, FLP-1; 3, FLP-2; 4, FLP-3; 5, FLP-4; 6, FLP-5; 7, FLP-8. Error bars on the graph indicate standard deviations (N=3).

3. Lac-expressing transgenic tobacco

Lac is a member of the multicopper oxidase family found in a wide range of organisms such as animals, plants, bacteria, and fungi. The reduction of oxygen to water is accompanied by the oxidation of substrate by laccase.

cDNA encoding Lac (Accession no. D13372.1) from the reverse transcription products of total RNA prepared from mycelia of *T. versicolor* IFO1030 was cloned. The cDNA under the control of double CaMV 35S promoter was introduced into the genome of *N. tabacum* Samsun NN by the leaf-disk method via *A. tumefacience* LBA4404 (Figure 4).

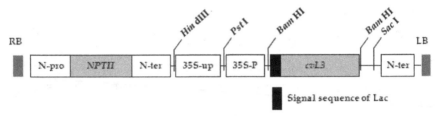

Fig. 4. Gene construct of T-DNA region of Ti plasmid.

cvL3, cDNA encoding Lac of *T. versicolor* IFO1030 plus signal sequence. Other abbreviations are listed in Figure 1.

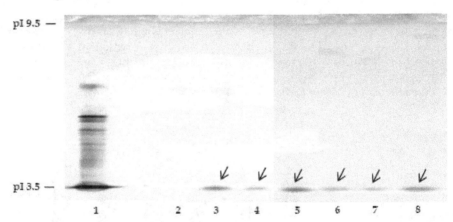

Fig. 5. Active staining of secreted Lac from the roots of transgenic lines.

Concentrated 60 µg of crude extracellular protein was analyzed by IEF and active staining using 4-chloro-1-naphthol. Lanes, 1, Concentrated aqueous cultivation medium of *T. versicolor* IFO1030; 2, control; 3, Lac transgenic line (FL)-4; 4, FL-5; 5, FL-9; 6, FL-20; 7, FL-22; 8, FL-23.

Two-month-old transgenic lines, which were incubated on MS medium, were transferred to fresh MS liquid medium and subjected to further incubation. After two weeks, to confirm the expression of Lac protein and secretion from the roots of each transgenic line into the rhizosphere, we concentrated the aqueous culture medium and analyzed it by iso-electric focusing electrophoresis (IEF) and active staining using 4-chloro-1-naphtol (Figure 5). Six independent transgenic lines apparently secreted active Lac protein into their rhizosphere, and we tested four of those to determine their ability to remove BPA. As described above, four independent transgenic lines were cultivated hydroponically. After one week of incubation, BPA was added to the medium at the final concentration of 100 μM and hydroponic incubation was done for another week. The ability to remove BPA of these Lac-expressing transgenic tobaccos was more than 5-fold that of the control line during hydroponic cultivation (Figure 6).

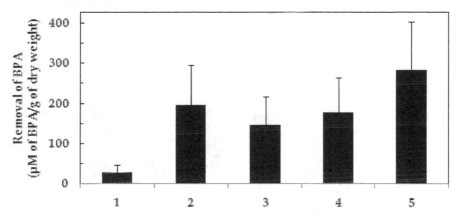

Fig. 6. BPA removal ability of Lac transgenic lines.

The levels of BPA were analyzed by HPLC (λ=278 nm). The results shown are the average of three independent experiments. Lanes; 1, control; 2, FL-4; 3, FL-9; 4, FL-20; 5, FL-22. Error bars on the graph indicate standard deviations (N=3).

All of these Lac-expressing transgenic tobaccos were somewhat shorter than control plants at the flowering stage, and most of the transgenic anthers failed to dehisce after blooming, while the anthers of control plants were normally dehiscent (Figure 7). In addition, the nondehiscent anthers were brown in contrast to the greenish control lines. Brown pigmentation and rough epidermis were observed on the surface of transgenic anthers. Greater Lac activity was detected in the cell-free extracts of transgenic anthers than in the controls; however, there was no correlation with lignin contents in transgenic anthers (Figure 8). Histochemical analysis of anther tissues revealed apparent deformation of the stomium in transgenic plants (Figure 9). Beals reported that the stomium in anther tissue plays a crucial role in the dehiscence of anthers in tobacco (Beals, 1997), indicating that such deformation of stomium observed in the transgenic anther tissue might affect the appearance of the nondehiscent phenotype. The expression of Lac could promote the efficient removal of BPA, but it also influences some aspects of flower development.

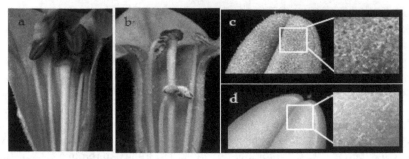

Fig. 7. Phenotypes of anthers of Lac transgenic lines.

Transgenic and control tobaccos were cultivated at 24°C. a, Transgenic flower with nondehiscent anthers. b, Control flower with normal anthers. c, Stereomicroscopic view of a transgenic anther. d, Stereomicroscopic view of a normal anther.

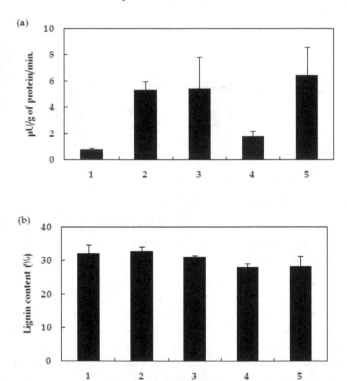

Fig. 8. Laccase activity and lignin content in anther tissues.

Transgenic and control tobaccos were cultivated at 24°C. a, Laccase activity. Cell-free extracts were prepared from both transgenic and control anthers before they dehisced. Laccase activity was calculated using the extinction coefficient (6400 $M^{-1}cm^{-1}$) of oxidized guaiacol ($\lambda=436nm$), and activity was expressed as definitive units (1 unit = 1 mol guaiacol oxidized per min) (Eggert et al, 1996). b, Lignin content. Lignin was quantified by the Klason method. The results shown are the average of three independent experiments. Error bars on the graph indicate standard deviations (N=3). Lanes; 1, control; 2, FL4; 3, FL9; 4, FL20; 5, FL22

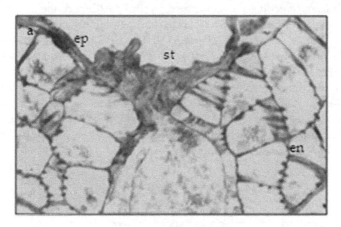

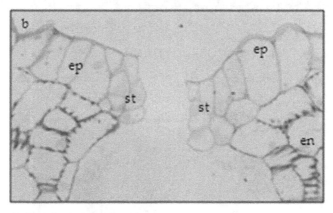

Fig. 9. Histochemical analysis of anther tissues.

Safranin-stained thin sections of a mature anther from a transgenic (a) and a control plant (b). ep, epidermis; st, stomium; en, endothecium cell.

4. MnP-expressing transgenic hybrid aspen

MnP is a heme peroxidase that can oxidize phenolic compounds in the presence of Mn (II) and hydrogen peroxide. Mn (II) is oxidized to Mn (III) by MnP; the resultant Mn (III) makes a chelating compound with an organic acid, and then organic compounds such as BPA are oxidized by the chelating compound. Previously, we isolated a cDNA (Accession no. AR429405) encoding MnP from *T. versicolor* and introduced it into the genome of *N. tabacum* Samsun NN. The transgenic tobacco could express MnP and produce Mn (III) as a result of Mn (II) oxidation in the rhizosphere during hydroponic cultivation (Iimura et al., 2002).

Moreover, isolated cDNA was also introduced into the genome of hybrid aspen Y63 (*Populus seiboldii* x *Populus gradientata*) under the control of double CaMV 35S promoter (Figure 10), as described previously (Kajita et al., 2004). Integration of the T-DNA into the genome of each transgenic line was confirmed by PCR. Although the expression of cDNA encoding MnP was confirmed by RT-PCR in six independent transgenic lines, MnP activity was detected in four of the six lines (Figure 11). The BPA-removing activities of the four MnP-expressing transgenic hybrid aspens were more than twice that of the control lines (Figure 12). Interestingly, the expression of the MnP gene showed no phenotypical differences between the MnP-expressing and control plants, unlike the expressions of the LiP and Lac genes. The lack of negative effects of MnP expression on growth and development will be advantageous when it is used in phytoremediation. Our results showed that the transgenic plants could express MnP in their roots and contribute to the effective removal of BPA from a hydroponic medium.

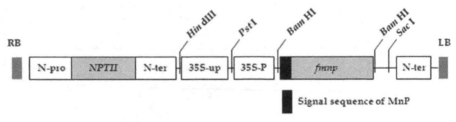

Fig. 10. Gene construct of T-DNA region of Ti plasmid.

fmnp, cDNA encoding MnP of *T. versicolor* IFO1030 plus signal sequence. Other abbreviations are listed in Figure 1.

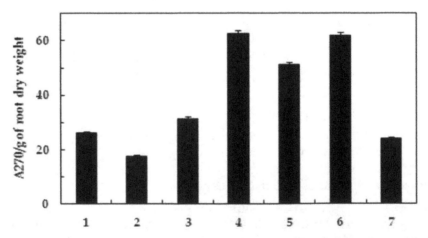

Fig. 11. MnP activity in root exudates of transgenic lines. Undamaged root tissues were dipped in 50 mM malonate buffer (pH 4.5) containing 1 mM manganese sulfate. After incubation for 24 hrs at 37°C, the absorbance of supernatant was measured at 270 nm. Lanes; 1, control plant; 2, MnP transgenic line (FM)-1; 3, FM-2; 4, FM-3; 5, FM-4; 6, FM-7; 7, FM-8. Error bars on the graph indicate standard deviations (N=3).

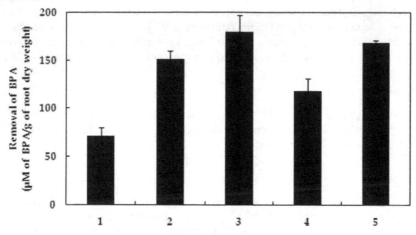

Fig. 12. BPA removal ability of MnP transgenic lines.

The levels of BPA were analyzed by HPLC (λ=278 nm). The values shown are the average of results from three independent experiments. Lanes; 1, control; 2, FM-2; 3, FM-3; 4, FM-4; 5, FM-7. Error bars on the graph indicate standard deviations (N=3).

As described above, fungal peroxidase (LiP, Lac, and MnP)-expressing transgenic plants showed effective BPA removal ability, but no reaction products of BPA conversions by these fungal peroxidase-expressing transgenic plants were detected under our analytical conditions. The enzymatic reaction of fungal peroxidases is non-specific and free radical-based, so it is difficult to detect the reaction products. BPA might be degraded or polymerized, as reported in some previous studies of lignolytic enzymes (Hirano et al., 2000; Fukuda et al., 2001; Tsutsumi et al., 2001; Uchida et al., 2001). The increase of BPA removal efficiency by the fungal peroxidase expression in plants would contribute to the development of remediation systems for the cleanup of contaminated areas.

5. Conclusions

Plants can metabolize BPA. Cultured cells of plants were able to glucosylate BPA (Nakajima et al., 2002; Hamada et al., 2002), and, in seedlings, BPA was absorbed from roots and translocated to leaves after glucosylation (Nakajima et al., 2002). In addition, some glycosylated forms of BPA showed less estrogenic activity than that of non-glycosylated BPA (Morohoshi et al., 2003), and oxidative enzymes in plants such as peroxidases stimulated the degradation and polymerization of BPA (Sakuyama et al., 2003). Although the ability of plants to detoxify might be useful for remediation of soil and water contaminated with BPA, the expressions of fungal peroxidases in plants by genetic engineering, as reviewed above, reinforces their ability with respect to the detoxification of BPA. Furthermore it is worth noting that the MnP- and Lac-expressing transgenic plants could remove pentachlorophenol effectively from contaminated areas during hydroponic cultivation (Iimura et al., 2002; Sonoki et al., 2005). Plants could secrete Lac and generate Mn (III) in the rhizosphere, and then the Lac and Mn (III) might be able to affect hydrophobic substrates, such as pentachlorophenol, which is difficult for plant roots to absorb.

Plants producing fungal secretory peroxidases would provide us useful tools for the remediation of areas contaminated with environmental pollutants. Further studies on the effective expression and the secretion of introduced enzymes and the application with other substrates will play an important role in the development of phytoremediation technology.

6. Acknowledgement

This work was supported in part by Grant-in-Aid for Scientific Research A (no. 21248037) from Japan Society for the Promotion of Science (JSPS) and Hirosaki University Grant for Exploratory Research by Young Scientists.

7. References

Beals, T. P., Goldberg, R. B. 1997. A novel cell ablation strategy blocks tobacco anther dehiscence. *Plant Cell* Vol. 9, No. 9, pp. 1527–1545, ISSN 1040-4651.

Bizilly, S. P., Kim, T., Kandasamy, M. K., Meagher, R. B. 2003. Subcellular targeting of methylmercury lyase enhances its specific activity for organic mercury detoxification in plants. *Plant Physiol.* Vol. 131, No. 2, pp. 463-471, ISSN 0032-0889.

Brasileiro, A. C., Leple, J. C., Muzzin, J., Ounnoughi, D., Michel, M. F., Jouanin, L. 1991. An alternative approach for gene transfer in trees using wild-type *Agrobacterium* strains. *Plant Mol. Biol.* Vol. 17, No. 3, pp. 441-452, ISSN 0167-4412.

Doty, S. L., Shang, T. Q., Wilson, A. M., Tangen, J., Westergreen, A. D., Newman, L. A., Strand, S. E., Gordon, M. P. 2000. Enhanced metabolism of halogenated hydrocarbons in transgenic plants containing mammalian cytochrome P450 2E1. *Proc. Natl. Acad. Sci. USA* Vo. 97, No. 12, pp. 6287–6291, ISSN 0027-8424.

Eggert, C., Temp, U., Eriksson, K. E. 1996. The ligninolytic system of the white rot fungus *Pycnoporus cinnabarinus*: purification and characterization of the laccase. *Appl. Environ. Microbiol.* Vol. 62, No. 4, pp. 1151–1158, ISSN 0099-2240.

Fernando, T., Aust, S. D. 1994. Biodegradation of toxic chemicals by white-rot fungi, In: *Biological degradation and bioremediation of the toxic chemicals.* Chaudhry, G. R., pp. 386-402, Chapman & Hall, ISBN 978-0-412-62290-8, London.

Fukuda, T., Uchida, H., Takashima, Y., Uwajima, T., Kawabata, T., Suzuki, M. 2001. Degradation of bisphenol A by purified laccase from *Trametes villosa. Biochem, Biophys. Res. Commun.* Vol. 284, No. 3, pp. 704-706, ISSN 0006291x.

Gullner, G., Komives, T., Rennenberg, H. 2001. Enhanced tolerance of transgenic poplar plants overexpressing gamma-glutamylcysteine synthetase towards chloroacetanilide herbicides. *J. Exp. Bot.* Vol. 52, No. 358, pp, 971–979, ISSN 0022-0957.

Hamada, H., Tomi, R., Asdada, Y., Furuya, T. 2002. Phytoremediation of bisphenol A by cultured suspension cells of *Eucalyptus perriniana*-regioselective hydroxylation and glycosylation. *Tetrahedron Lett.*, Vo. 43, No. 22, pp. 4087-4089, ISSN 0040-4039.

Hammel, K. E. 1989. Organopollutant degradation by ligninolytic fungi. *Enzyme Microb. Technol.* Vo. 11, No. 11, pp. 776-777, ISSN 0141-0229.

Haque, S., Zeyaullh, M., Nabi, G., Srivastava, P. S., Ali, A. 2010. Transgenic tobacco plant expressing environmental *E. coli merA* gene for enhanced volatilization of ionic mercury. *J. Microbiol. Biotechnol.* Vol. 20, No. 5, pp. 917-924, ISSN 1017-7825.

Hirano, T., Honda, Y., Watanabe, T., Kuwahara, M. 2000. Degradation of bisphenol A by the lignin-degrading enzymes, manganese peroxidase, produced by the white-rot basidomycete, *Pleurotus ostreatus*. *Biosci. Biotechnol. Biochem.* Vol. 64, No. 9, pp. 1958-1962, ISSN 0916-8451.

Iimura, Y., Ikeda, S., Sonoki, T., Hayakawa, T., Kajita, S., Kimbara, K., Tatsumi, K., Katayama, Y. 2002. Expression of a gene for Mn-peroxidase from *Coriolus versicolor* in transgenic tobacco generates potential tools for phytoremediation. *Appl. Microbiol. Biotechnol.* Vo. 59, No. 2-3, pp. 246-251, ISSN 0175-7598.

Kajita, S., Honaga, F., Uesugi, M., Iimura, Y., Masai, E., Kawai, S., Fukuda, M., Morohoshi, N., Katayama, Y. 2004. Generation of transgenic hybrid aspen that express a bacterial gene for feruloyl-CoA hydratase/lyase (FerB), which is involved in lignin degradation in *Sphingomonas paucimobilis* SYK-6. *J. Wood Sci.* Vo. 50, No. 3, pp. 275-280, ISSN 1435-0211.

Kang, J. H., Katayama, Y., Kondo, F. 2006. Biodegradation or metabolism of bisphenol A: from microorganisms to mammals. *Toxicology* Vol. 217, No. 2-3, pp. 81–90, ISSN 0030-483X.

Levin, L., Viale, A., Forchiassin, A. 2003. Degradation of organic pollutants by the white rot basidomycete *Trametes trogii*. *Int. Biodeterior. Biodegrad.* Vo. 52, No. 1, pp. 1-5, ISSN 0964-8305.

Liang, X. W., Dron, M., Schmid, J., Dixon, R. A., Lamb, C. J. 1989. Developmental and environmental regulation of a phenylalanine ammonia-lyase-beta-glucronidase gene fusion in transgenic tobacco plants. *Proc. Natl. Acad. Sci. USA* Vo. 86, No. 23, pp. 9284-9288, ISSN 0027-8424.

Lin, J.E., Wang, H. Y., Hickey, R. F. 1990. Degradation kinetics of pentachlorophenol by *Phanerochaete chrysosporium*. *Biotech. Bioeng.* Vol. 35, No. 11, pp. 1125–1134, ISSN 0006-3592.

Lovley, D. R., Woodward, J. C., Chapelle, F. H. 1994. Characterization of the mnp2 gene encoding manganese peroxidase isozyme 2 from the basidomycete *Phanerochaete chrysosporium*. *Gene* Vol. 142, pp. 231-235, ISSN 0378-1119.

Mohn, W. W. and Tiedje, J. M. 1992. Microbial reductive dehalogenation. *Microbiol. Rev.* Vol. 56, No. 3, pp. 482-507, ISSN 1098-5557.

Morohoshi, K., Shiraishi, F., Oshima, Y., Koda, T., Nakajima, N., Edmonds, J.S., Morita, M. 2003. Synthesis and estrogenic activity of bisphenol A mono- and Di-beta-D-glucopyranosides, plant metabolites of bisphenol A. *Environ. Toxicol. Chem.* Vol. 22, No. 10, pp. 2275-2279, ISSN 0730-7268.

Murashige, T., Skoog, F. 1962. A revised medium for the rapid growth and bioassay with tobacco tissue cultures. *Plant Physiol.* Vol. 15, pp. 473-497, ISSN 0032-0889.

Nakajima, N., Ohshima, Y., Serizawa, S., Kouda, T., Edmonds, J. S., Shiraishi, F., Aono, M., Kubo, A., Tamaoki M., Saji, H., Morita, M. 2002. Processing of bisphenol A by plant tissues: glucosylation by cultured BY-2 cells and glucosylation/translocation by plants of *Nicotiana tobacum*. *Plant Cell Physiol.* Vol. 43, No. 9, pp. 1036-1042, ISSN 0032-0781.

Novakova, M., Mackova, M., Chrastilova, Z., Viktorova, J., Szekeres, M., Demnerova, K., Macek, T. 2009. Cloning the bacterial *bphC* gene into *Nicotiana tabacum* to improve the efficiency of PCB phytoremediation. *Biotechnol. Bioeng.* Vol. 102, No. 1, pp. 29-37, ISSN 0006-3592.

Reddyy, G. V. B., Gelpke, M. D. S., Gold, M. H. 1998. Degradation of 2,4,6-trichlorophenol by *Phanerochaete crysosporium*: Involvement of reductive dechlorination. *J. Bacteriol.* Vol. 180, No. 19, pp. 5159-5164, ISSN 0021-9193.

Sakuyama, H., Endo, Y., Fujimoto, K., Hatano, Y. 2003. Oxidative degradation of alkylphenols by horseradish peroxidase. *J. Biosci. Bioeng.* Vol. 96, No. 3, pp. 227-231, ISSN 1389-1723.

Sonoki, T., Kajita, S., Ikeda, S., Uesugi, M., Tatsumi, K., Katayama, Y., Iimura, Y. 2005. Transgenic tobacco expressing fungal laccase promotes the detoxification of environmental pollutants. *Appl. Microbiol. Biotechnol.* Vol. 67, No. 1, pp. 138-142, ISSN 0175-7598.

Thomas, J. C., Davies, E. C., Malick, F. K., Endreszi, C., Williams, C. R., Abbas, M., Petrella, S., Swisher, K., Perron, M., Edwards, R., Osenkowski, P., Urbanczyk, N., Wiesend, W. N., Murray, K. S. 2003. Yeast metallothionein in transgenic tobacco promotes copper uptake from contaminated soils. Biotechnol. Prog. Vol. 19, No. 2, pp. 273-280, Online ISSN 1520-6033.

Tsutsumi, Y., Haneda, T., Nishida, T. 2001. Removal of estrogenic activities of bisphenol A and nonylphenol by oxidative enzymes from lignin degrading basidomycetes. *Chemosphere* Vol. 42, No. 3, pp. 271-276, ISSN 0045-6535.

Uchida, H., Fukuda, T., Miyamoto, H., Kawabata, T., Suzuki, M., Uwajima, T. 2001. Polymerization of bisphenol A by purified laccase from *Trametes villosa*. *Biochem. Biophys. Res. Commun.* Vol. 287, No. 2, pp. 355-358, ISSN 0006-291X.

Wawrzyński, A., Kopera, E., Wawrzyńska, A., Kamińska, J., Bal, W., Sirko, A. 2006. Effects of simultaneous expression of heterologous genes involved in phytochelatin biosynthesis on thiol content and cadmium accumulation in tobacco plants. J. Exp. Bot. Vol. 57, No. 10, pp. 2173-2182, ISSN 0022-0957.

Xiang-Yan, M., Jie, Q., Li-Hong, W., Gui-Lan, D., Guo-Xin, S., Hui-Lan, W., Cheng-Cai, C., Hong-Qing, L., Barry, P. R., Yong-Guan, Z. 2011. Arsenic biotransformation and volatilization in transgenic rice. *New Phytologist* Vol. 191, No. 1, pp. 49-56, ISSN 0028-646X.

Zhu, Y. L., Pilon-Smits, E. A., Tarun, A. S., Weber, S. U., Jouanin, L., Terry, N. 1999. Cadmium tolerance and accumulation in Indian mustard is enhanced by overexpressing gamma-glutamylcysteine synthetase. *Plant Physiol.* Vol. 121, No. 4, pp. 1169-1178, ISSN 0032-0889.

9

Methods to Transfer Foreign Genes to Plants

Yoshihiro Narusaka, Mari Narusaka,
Satoshi Yamasaki and Masaki Iwabuchi
Research Institute for Biological Sciences (RIBS), Okayama
Japan

1. Introduction

Genome sequencing of several organisms has resulted in the rapid progress of genomic studies. Genetic transformation is a powerful tool and an important technique for the study of plant functional genomics, i.e., gene discovery, new insights into gene function, and investigation of genetically controlled characteristics. In addition, the function of genes isolated using map-based cloning of mutant alleles has been confirmed by functional complementation using genetic transformation. Furthermore, genetic transformation enables the introduction of foreign genes into crop plants, expeditiously creating new genetically modified organisms. Gene transformation and genetic engineering contribute to an overall increase in crop productivity (Sinclair et al., 2004).

This review outlines general methods for plant transformation and focuses on the development of the *Arabidopsis* transformation system.

2. Plant transformation methods

Plant transformation was first described in tobacco in 1984 (De Block et al., 1984; Horsch et al., 1984; Paszkowski et al., 1984). Since that time, rapid developments in transformation technology have resulted in the genetic modification of many plant species. Methods for introducing diverse genes into plant cells include *Agrobacterium tumefaciens*-mediated transformation (De la Riva, 1998; Hooykaas & Schilperoort, 1992; Sun et al., 2006; Tepfer, 1990; Zupan & Zambryski, 1995), recently reclassified as *Rhizobium radiobacter*, direct gene transfer into protoplasts (Gad et al., 1990; Karesch et al., 1991; Negrutiu et al., 1990; Neuhaus & Spangenberg, 1990), and particle bombardment (Birch & Franks, 1991; Christou, 1992; Seki et al., 1991; Takeuchi et al., 1992, 1995; Yao et al., 2006).

2.1 Gene transformation

Several gene transformation techniques utilize DNA uptake into isolated protoplasts mediated by chemical procedures, electroporation, or the use of high-velocity particles (particle bombardment). Direct DNA uptake is useful for both stable transformation and transient gene expression. However, the frequency of stable transformation is low, and it takes a long time to regenerate whole transgenic plants.

2.1.1 Chemical procedures

Plant protoplasts treated with polyethylene glycol more readily take up DNA from their surrounding medium, and this DNA can be stably integrated into the plant's chromosomal DNA (Mathur & Koncz, 1997). Protoplasts are then cultured under conditions that allowed them to grow cell walls, start dividing to form a callus, develop shoots and roots, and regenerate whole plants.

2.1.2 Electroporation

Plant cell electroporation generally utilizes the protoplast because thick plant cell walls restrict macromolecule movement (Bates, 1999). Electrical pulses are applied to a suspension of protoplasts with DNA placed between electrodes in an electroporation cuvette. Short high-voltage electrical pulses induce the formation of transient micropores in cell membranes allowing DNA to enter the cell and then the nucleus.

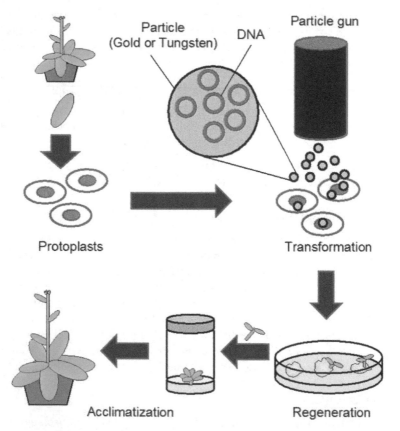

Fig. 1. Plant transformation process using particle bombardment includes the following steps: (1) Isolate protoplasts from leaf tissues. (2) Inject DNA-coated particles into the protoplasts using particle gun. (3) Regenerate into whole plants. (4) Acclimate the transgenic plants in a greenhouse.

2.1.3 Particle (microprojectile) bombardment

Particle bombardment is a technique used to introduce foreign DNA into plant cells (Birch & Franks, 1991; Christou, 1992, 1995; Gan, 1989; Takeuchi et al., 1992; Yao et al., 2006) (Figure 1). Gold or tungsten particles (1–2 µm) are coated with the DNA to be used for transformation. The coated particles are loaded into a particle gun and accelerated to high speed either by the electrostatic energy released from a droplet of water exposed to high voltage or using pressurized helium gas; the target could be plant cell suspensions, callus cultures, or tissues. The projectiles penetrate the plant cell walls and membranes. As the microprojectiles enter the cells, transgenes are released from the particle surface for subsequent incorporation into the plant's chromosomal DNA.

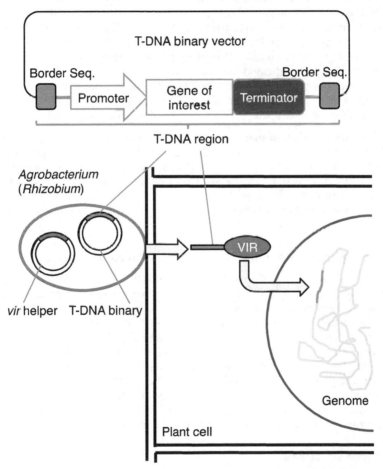

Fig. 2. The *Agrobacterium*-mediated transformation process includes the following steps: (1) Isolate genes of interest from the source organism. (2) Insert the transgene into the Ti-plasmid. (3) Introduce the T-DNA containing-plasmid into *Agrobacterium*. (4) Attach the bacterium to the host cell. (5) Excise the T-strand from the T-DNA region. (6) Transfer and integrate T-DNA into the plant genome.

2.2 Using *Agrobacterium* for plant transformation

Agrobacterium-mediated transformation is the most commonly used method for plant genetic engineering (Bartlett et al., 2008; Leplé et al., 1992; May et al., 1995; Sun et al., 2006; Tsai et al., 1994; Tzfira et al., 1997). The pathogenic soil bacteria *Agrobacterium tumefaciens* that causes crown gall disease has the ability to introduce part of its plasmid DNA (called transfer DNA or T-DNA) into the nuclear genome of infected plant cells (Figure 2) (Binns & Thomashaw, 1988; Gelvin, 2000; Nester et al., 1984; Tzfira et al., 2004; Zupan & Zambryski, 1995).

2.3 Transforming *Arabidopsis thaliana*

Arabidopsis thaliana, a small flowering plant, is a model organism widely used in plant molecular biology. The first *in planta* transformation of *Arabidopsis* included the use of tissue culture and plant regeneration (Feldmann & Marks, 1987). The *Agrobacterium* vacuum (Bechtold et al., 1993) and floral dipping (Clough & Bent, 1998) are efficient methods to generate transgenic plants. They allow for plant transformation without the need for tissue culture. The floral dipping method markedly advanced the ease of creating *Arabidopsis* transformants, and it is the most widely used transformation method. These methods were later simplified and substantially improved (Davis et al., 2009; Zhang et al., 2006), significantly reduced the required labor, cost, and time, as compared with earlier procedures.

However, these transformation methods have some problems. The floral dipping method involves dipping *Arabidopsis* flower buds into an *Agrobacterium* cell suspension, requiring large volumes of bacterial culture grown in liquid media. The large shakers and centrifuges, necessary to house the media, require sufficient experimental space. These factors limit transformation quantities. Here, we describe an improved method for *Agrobacterium*-mediated transformation that does not require the large volumes of liquid culture necessary for floral dipping.

2.3.1 Improved method for *Agrobacterium*-mediated transformation

A. thaliana can be stably transformed with high efficiency using T-DNA transfer by *Agrobacterium tumefaciens*. *Agrobacterium*-mediated transformation using the floral dipping method is the most widely used method for transforming *Arabidopsis*. We have showed that *A. thaliana* can be transformed by inoculating flower buds with 5 μl of *Agrobacterium* cell suspension, thus avoiding the use of large volumes of *Agrobacterium* culture (Narusaka et al., 2010). Using this floral inoculating method, we obtained 15–50 transgenic plants per three transformed *A. thaliana* plants. The floral inoculating method can be satisfactorily used in subsequent analyses. This simplified method, without floral dipping, offers an equally efficient transformation as previously reported methods. This method reduces overall labor, cost, time, and space. Another important aspect of this modified method is that it allows many independent transformations to be performed at once.

2.3.2 *Agrobacterium* strains

The *Agrobacterium* strain GV3101 (C58 derivative) is frequently used to transform many binary vectors, e.g., pBI121, pGPTV, pCB301, pCAMBIA, and pGreen, into *Arabidopsis*. It carries rifampicin resistance (10 mg l⁻¹) on the chromosome (Koncz & Schell, 1986). On the other hand, LBA4404 is a popular strain for tobacco transformation but is less effective for *Arabidopsis*.

2.3.3 *Agrobacterium* transformation—freeze/thaw and electroporation procedures

Agrobacterium can be transformed with plasmid DNA using the freeze/thaw (Höfgen & Willmitzer, 1998; Holsters et al., 1978) and electroporation (den Dulk-Ras & Hooykaas, 1995; Mersereau et al., 1990; Shen & Forde, 1989) procedures. The freeze/thaw procedure is very simple and does not require special equipment.

Reagents

- *Agrobacterium* strain
- 20 mM CaCl$_2$
- Liquid nitrogen
- Luria–Bertani (LB) agar plate
- Liquid LB medium

Equipments

- Microcentrifuge
- Water bath
- Eppendorf tube (1.5 ml)
1. Pellet 1.5 ml of overnight-grown *Agrobacterium* (GV3101) cells by centrifugation in an Eppendorf tube at 14,000 rpm for 1 min at 4°C.
2. Resuspend in 1 ml of ice-cold 20 mM CaCl$_2$.
3. Recentrifuge at 14,000 rpm for 1 min at 4°C.
4. Resuspend in 200 µl of ice-cold 20 mM CaCl$_2$.
5. Add binary vector DNA (500 ng or 5–10 µl from an alkaline lysis miniprep) to the suspension. Mix by pipetting.
6. Freeze the Eppendorf tube in liquid nitrogen for 5 min and thaw at 37°C in a water bath for 5 min. Repeat two times.
7. Cool on ice.
8. Add 1 ml LB liquid medium to the Eppendorf tube and incubate at 28°C for 2–5 hrs with gentle agitation (150 rpm; water bath).
9. Spread 50–200 µl of the cells onto LB agar medium containing appropriate antibiotics and incubate at 28°C for two days.

2.3.4 Selecting transformed *Agrobacterium* using polymerase chain reaction (PCR)

This method is designed to quickly screen for plasmid inserts directly from *Agrobacterium* colonies. Alternatively, the insert presence can be determined by DNA sequencing.

Reagents

- PCR components (one reaction):
  Autoclaved, distilled water - 11.625 µl
  10× PCR buffer - 1.5 µl
  2.5 mM dNTPs - 1.2 µl
  10 pmol µl^{-1} Primer #1 - 0.3 µl
  10 pmol µl^{-1} Primer #2 - 0.3 µl
  Taq DNA polymerase (5 U/µl) - 0.075 µl

  Total PCR master mix volume - 15.0 µl

- *Taq* DNA polymerase: Takara EX *Taq* (Takara, Otsu, Japan) (recommended)
- TBE (Tris/Borate/EDTA) buffer

Equipments

- PCR tubes (0.2 ml)
- Thermocycler
- Electrophoresis system
1. Prepare sufficient PCR master mix for the number of samples tested.
2. Add 15 µl of PCR master mix to each PCR tube.
3. Select *Agrobacterium* colonies from the plate using a sterile toothpick or pipette tip.
4. Insert selected colony sample into the PCR master mix and mix with a sterile toothpick or pipette tip.
   (Note: Sufficient mixing results in complete cell lysis and high yields.)
5. Briefly centrifuge tubes to collect all liquid and insert them into the PCR.
6. Set the thermocycler conditions and start PCR.
   Conditions: Preliminary denaturation at 95°C for 3 min then 40 cycles at 95°C for 20 sec, 55°C for 30 sec, and 72°C for 30 sec.
   (Note: Preliminary denaturation is very important for initial cell breakage.)
7. Run 8–10 µl of each PCR sample on 1.0% agarose gel in 1× TBE buffer at 100 V for 30 min to visualize the PCR results. Stain gels according to your lab method.

2.3.5 Simplified *Arabidopsis* transformation: Floral inoculating method

Until now, a limited number of constructs could be transformed into *Arabidopsis* because of difficulty growing large volumes of *Agrobacterium*. Therefore, we focused on improvements to the floral dipping method (Figure 3) (Narusaka et al., 2010). The problem of space and volume can be solved by using a small culture volume. Each plant is transformed using only 30–50 µl of bacteria grown in 2 ml of liquid culture. Our present method, as described below, is a simple modification of the method reported by Clough & Bent (1998).

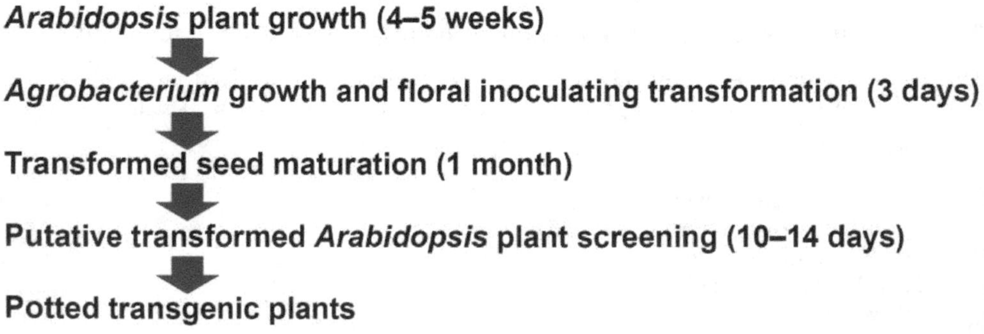

Arabidopsis plant growth (4–5 weeks)
⬇
Agrobacterium growth and floral inoculating transformation (3 days)
⬇
Transformed seed maturation (1 month)
⬇
Putative transformed *Arabidopsis* plant screening (10–14 days)
⬇
Potted transgenic plants

Fig. 3. Transformation using *Agrobacterium* and the floral inoculating method

Recent papers (Liu et al., 2008; Zhang et al., 2006) illustrate the floral dipping process. Clough and Bent (1998) reported that neither Murashige and Skoog (MS) salts and hormones nor optical density (OD) makes a difference in transformation efficiency. An *Agrobacterium* cell suspension containing 0.01–0.05% Silwet L-77 (vol/vol) was used in the uptake of *Agrobacterium* into female gametes, instead of vacuum-aided infiltration of inflorescences.

Reagents

- *A. thaliana*: There are marked differences in transformation efficiency between various ecotypes. For floral dipping transformation, efficiency in the Landsberg *erecta* (L*er*-0) ecotype is lower than that in the Columbia (Col-0) ecotype. Transformation efficiency in Wassilewskija (Ws-0) is very high among *Arabidopsis* ecotypes.
- *Agrobacterium* strain: GV3101 (Koncz & Schell, 1986) (recommended) or others.
- 0.1% (wt/vol) agar solution
- 70% (vol/vol) ethanol
- Sodium hypochlorite solution containing 1% available chlorine and 0.02% (vol/vol) Tween 20
- Distilled water
- MS medium: 1× MS plant salt mixture (Wako Pure Chemical Industries, Osaka, Japan), 1× Gamborg's vitamin solution (Sigma-Aldrich, St. Louis, MO, USA), 1% (wt/vol) sucrose, 0.05% (wt/vol) MES, and pH 5.7 adjusted with 1 N KOH
- Bacto agar (Becton, Dickinson and Company, Franklin Lakes, NJ, USA) (recommended)
- LB agar plate
- Liquid LB
- Glycerol
- Transformation buffer: 1/2× MS plant salt mixture, 1× Gamborg's vitamin solution, 5% (wt/vol) sucrose, and pH 5.7, adjusted with 1 N KOH
- 5% (wt/vol) sucrose solution
- Silwet L-77
- 6-Benzylaminopurine (BAP) (final concentration 0.01 µg ml[-1])
- Claforan (Aventis Pharma AG, Zürich, Switzerland) (final concentration 2 mg ml[−1])
- Kanamycin (final concentration 30 µg ml[-1])
- Hygromycin (final concentration 20 µg ml[-1])
- Bialaphos (final concentration 7.5 µg ml[-1])
- Peat moss (Soil Mix, Sakata Seed Corp., Yokohama, Japan)
- Expanded vermiculite granules

Equipments

- Growth chamber
- Plant pot (3-inch)
- Conical tube (15 ml)
- Eppendorf tubes (2 ml)

1. Grow *A. thaliana* plants. (Note: Plant health is an important factor. Healthy *A. thaliana* plants should be grown until they are flowering.) There are two different procedures: standard (A) and quick (B) (Zhang et al., 2006). We generally use the quick procedure, which is useful for rare seeds and seeds with low germination frequency. It is also used to retransform a transgenic line with a second construct.

Fig. 4. Part 1.

Fig. 4. part 2. Floral inoculating transformation of *Arabidopsis thaliana*. (A) Clipping primary bolts. (B, C, and D) Using a micropipette, inoculate flower buds with 5 µl of *Agrobacterium* when plants have just started to flower after clipping primary bolts. (E) Place inoculated plants under a dome or cover for 16–24 hrs to maintain high humidity. (F) Remove the cover and grow the plants in a greenhouse or growth chamber until maturity. (G, H) Screening of putative transformed *Arabidopsis* plants. G: 10 days, H: 21 days. Arrows indicate putative transformed *Arabidopsis* plants.

1.a. Standard procedure (A): Suspend seeds in 0.1% (wt/vol) agar solution and keep in darkness for 2–4 days at 4°C to break dormancy. Spread seeds on wet soil (a mixture of peat moss and expanded vermiculite granules at a 1:2 ratio) in a 3-inch pot and grow under long-day conditions (16-hr light/8-hr dark) at 22°C. Thin to three seedlings per pot. Do not cover with a bridal veil, window screen, or cheesecloth.

1.b. Quick procedure (B): Sterilize seeds by treatment with 70% (vol/vol) ethanol for 1 min then immerse in sodium hypochlorite solution containing 1% available chlorine and 0.02% (vol/vol) Tween 20 for 7 min. Wash seeds five times with sterile distilled water. Place seeds on MS medium containing 0.8% (wt/vol) Bacto agar. Keep seeds in darkness for 2–4 days at 4°C to break dormancy. Grow under long-day conditions (16-hr light/8-hr dark) for 3 weeks at 22°C. Transfer to pots per Step 1a. Do not cover with a bridal veil, window screen, or cheesecloth.

2. Clip primary bolts to encourage proliferation of secondary bolts (Figure 4A). Plants will be ready approximately 4–6 days after clipping.

3. Prepare the *Agrobacterium* strain carrying the gene of interest. Spread a single *Agrobacterium* colony on an LB agar plate with suitable antibiotics. Incubate the culture at 28°C for two days.

4. Use feeder culture to inoculate a 2-ml liquid culture in LB with suitable antibiotics to select for the binary plasmid in a 15-ml Conical tube at 28°C for 16–24 hrs. Mid-log cells or a freshly saturated culture (Clough and Bent 1998) can be used. (Optional: If needed, keep 500 µl of *Agrobacterium* culture in a 25% (vol/vol) glycerol stock at -80°C.)

5. Spin down 1.5 ml of the *Agrobacterium* cell suspension in 2-ml Eppendorf tubes and resuspend in 1 ml transformation buffer. OD_{600} value adjustment is not required. Each small pot containing three plants requires approximately 150 µl of culture. (Optional: 5% (wt/vol) sucrose solution may be used instead of transformation buffer.)

    Just before inoculation, add Silwet L-77 to a concentration of 0.02% (vol/vol) and immediately mix well. (Optional: If using transformation buffer, add 0.01 µg ml-1 BAP just before transformation.)

6. Apply 5 µl of *Agrobacterium* inoculum to the flower buds (Figures 4B, C, and D), inoculating each plant with a total of 30–50 µl of inoculum.

7. Place inoculated plants under a dome or cover for 16–24 hrs to maintain high humidity (Figure 4E). Avoid excessive exposure to light. (Optional: For higher rates of transformation, inoculate newly forming flower buds with *Agrobacterium* 2–3 times at 7-day intervals.)

8. Water and grow plants normally, tying up loose bolts with wax paper, tape, stakes, twist-ties, or other means. Stop watering as seeds become mature (Figure 4F).

9. Harvest dry seeds. Though transformants are usually independent, independence can be guaranteed if seeds come from separate plants.

10. Surface-sterilize seeds by immersion in 70% (vol/vol) ethanol for 1 min, followed by immersion in sodium hypochlorite solution containing 1% available chlorine and 0.02% (vol/vol) Tween 20 for 10 min. Then, wash seeds five times with sterile distilled water.

    To select for transformed plants, resuspend liquid-sterilized seeds in approximately 8 ml of 0.1% (wt/vol) agar solution containing 2 mg ml-1 Claforan. Sow seeds per Step 1b in MS medium containing 0.8% Bacto agar and appropriate antibiotics or herbicide selective markers at the following concentrations: kanamycin (final concentration 30 µg ml-1), hygromycin (20 µg ml-1), and bialaphos (7.5 µg ml-1). Claforan is necessary for *Agrobacterium* decontamination (Figures 4G and H).

11. Transplant putative transformants to soil per Step 1a. Grow, test, and use.

2.3.6 Screening of transgenic plants by PCR

Transgenes can be detected by plant genome DNA analysis with PCR (Figure 5). Although transgenes can be distinguished from their surrounding host plant genome, their presence should be determined by DNA sequencing.

PCR-based transgene detection is a simple and highly sensitive process. Subsequent PCR tests are assessed by agarose gel electrophoresis, and results are visualized by the presence or absence of the appropriately sized DNA fragment. If PCR shows a positive result, the transgene may be present. Transgene presence is confirmed by incorporating it into the genome by DNA sequencing. In contrast, a negative PCR result implies that the transgene is not present.

Simplified DNA isolation method

A small plant leaf disc (3–4 mm diameter) can be directly used as a PCR template. *Arabidopsis*, tomato, Chinese cabbage, Komatsuna (*Brassica rapa*), and tobacco leaf discs are good template candidates.

Reagents

Buffer A: 100 mM Tris-HCl (pH 9.5), 1 M KCl, 10 mM EDTA (ethylenediaminetetraacetic acid)

Equipments

- Cork borer (3–4 mm diameter)
- Disposable blade
- Eppendorf tube (1.5 ml)
- PCR tube (0.2 ml)
- Heat block
1. Cut each plant leaf disc using a cork borer (3–4 mm diameter) or disposable blade (leaf piece should be approximately 3 mm × 3 mm).
2. Place the leaf disc into an Eppendorf tube.
3. Add 100 µl of Buffer A.
4. Incubate for 10 min at 95°C.
5. Vortex thoroughly.
6. Transfer 0.5 µl of the template DNA supernatant to a PCR tube.

PCR detection method

Reagents

- PCR components (one reaction):

  Autoclaved, distilled water - 3.9 µl
  2× PCR buffer for KOD FX - 10.0 µl
  2 mM dNTPs - 4.0 µl
  10 pmol µl^{-1} Primer #1 - 0.6 µl
  10 pmol µl^{-1} Primer #2 - 0.6 µl
  KOD FX (1.0 U/µl) - 0.4 µl
  Total PCR master mix volume - 19.5 µl

  Add template DNA - 0.5 µl
  Total reaction volume - 20.0 µl

- DNA polymerase: KOD FX (Toyobo Co., Ltd, Osaka, Japan) (required)
- TBE buffer

Equipments

- PCR tubes (0.2 ml)
- Thermocycler
- Electrophoresis system
1. Prepare a PCR master mix for the number of samples tested.
2. Add 19.5 µl of PCR master mix to the template DNA and gently mix by pipetting.
3. Briefly centrifuge tubes to collect all liquid and insert into the PCR.
4. Set the Thermocycler condition and start PCR. Conditions: Preliminary denaturation step at 94°C for 2 min, followed by 40 cycles at 98°C for 10 sec, 55°C for 15 sec, and 68°C for 30 sec.
5. Run 8–10 PCR samples on 1.0% agarose gel in 1× TBE buffer at 100 V for 30 min to visualize the PCR results. Stain gels according to your lab method.

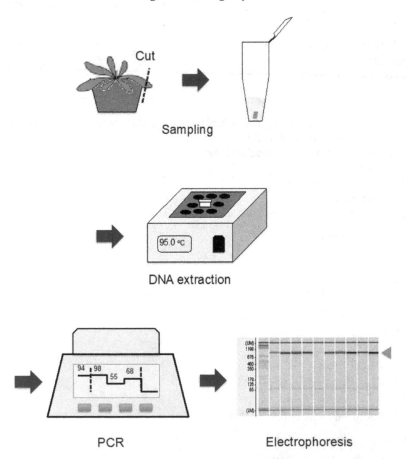

Fig. 5. Screening regimen for transgenic plants by PCR.

3. Conclusion

The floral inoculating method resulted in 15–50 transgenic plants per three transformed *A. thaliana* plants (Table 1). The method can be satisfactorily used for subsequent analyses. This simplified method does not utilize plant inversion or floral dipping, which requires large volumes of *Agrobacterium* culture. It offers equally efficient transformation as previously reported methods with the added benefit of reduced labor, cost, time, and space. Of further importance, this modified method allows many independent transformations to be performed at once.

Vector	Antibiotic marker (final concentration)	Ecotype	%Transformation[a]
pBI101	kanamycin (30 µg ml⁻¹)	Columbia (Col-0)	0.32 ± 0.02
		Wassilewskija (Ws-0)	0.86 ± 0.12
pGWB1[b]	kanamycin (30 µg ml⁻¹) hygromycin (20 µg ml⁻¹)	Wassilewskija (Ws-0)	0.31 ± 0.05
[a] Values are mean ± SE. [b] Refer to Nakagawa et al. (2007).			

Table 1. Transformation efficiency using floral inoculating method

4. Acknowledgment

We thank Ms. Mariko Miyashita for her excellent technical assistance. This work was supported in part by the Industrial Technology Research Grant Program in 2004 and 2009 from the New Energy and Industrial Technology Development Organization (NEDO) of Japan, by the Programme for Promotion of Basic and Applied Researches for Innovations in Bio-oriented Industry to Y.N., and by a Grant-in-Aid for Scientific Research (KAKENHI) (21580060 to Y.N. and 21780038 to M.N.).

5. References

Bartlett, J.G.; Alves, S.C.; Smedley, M.; John W.; Snape, J.W. & Harwood, W.A. (2008). High-throughput *Agrobacterium*-mediated barley transformation. *Plant Methods*, Vol.4, No.22, pp. 1-12, ISSN 1746-4811

Bates, G.W. (1999). Plant transformation via protoplast electroporation. *Methods in Molecular Biology*, Vol.111, pp. 359-366, ISSN 1064-3745

Bechtold, N.; Ellis, J. & Pelletier, G. (1993). *In planta Agrobacterium* mediated gene transfer by infiltration of adult *Arabidopsis thaliana* plants. *Comptes rendus de l'Académie des Sciences*, Vol.316, pp. 1194-1199, ISSN 0764- 4469

Binns, A.N. & Thomashow, M.F. (1988). Cell biology of *Agrobacterium* infection and transformation of plants. *Annual Reviews in Microbiology*, Vol.42, pp. 575-606, ISSN 0066-4227

Birch, R.G. & Franks, T. (1991). Development and optimization of micropojectile systems for plant genetic transformation. *Australian Journal of Plant Physiology*, Vol.18, pp.453-469, ISSN 0310-7841

Christou, P. (1992). Genetic transformation of crop plants using microprojectile bombardment. *The Plant Journal*, Vol.2, pp. 275-281, ISSN 0960-7412

Christou, P. (1995). Strategies for variety-independent genetic transformation of important cereals, legumes and woody species utilizing particle bombardment. *Euphytica*, Vol.85, pp. 13-27, ISSN 0014-2336

Clough, S.J. & Bent, A.F. (1998). Floral dip: a simplified method for *Agrobacterium*-mediated transformation of *Arabidopsis thaliana*. *The Plant Journal*, Vol.16, pp. 735-743, ISSN 0960-7412

Davis, A.M.; Hall, A.; Millar, A.J.; Darrah, C. & Davis, S.J. (2009). Protocol: Streamlined sub-protocols for floral-dip transformation and selection of transformants in *Arabidopsis thaliana*. *Plant Methods*, Vol.5, pp. 1-7, ISSN 1746-4811

De Block, M.; Herrera-Estrella, L.; van Montagu, M.; Schell, J. & Zambryski, P. (1984). Expression of foreign genes in regenerated plants and their progeny. *EMBO Journal*, Vol.3, pp. 1681-1689, ISSN 0261-4189

De la Riva, G.A.; González-Cabrera, J.; Vázquez-Padrón, R. & Ayra-Pardo, C. (1998). *Agrobacterium tumefaciens*: a natural tool for plant transformation. *Electronic Journal of Biotechnology*, Vol.1, No.3, pp. 118-133, ISSN 0717-3458

Den Dulk-Ras, A. & Hooykaas, P.J.J. (1995). Electroporation of *Agrobacterium tumefaciens*. *Methods in Molecular Biology*, Vol.55, pp. 63-72, ISSN 1064-3745

Feldmann, K.A. & Marks, M.D. (1987). *Agrobacterium*-mediated transformation of germinating seeds of *Arabidopsis thaliana*: a non-tissue culture approach. *Molecular and General Genetics*, Vol.208, pp. 1-9, ISSN 0026-8925

Gad, A.E.; Rosenberg, N. & Altman, A. (1990). Liposome-mediated gene delivery into plant cells. *Physiologia Plantarum*, Vol.79, pp. 177-183, ISSN 0031-9317

Gan, C. (1989). Gene gun accelerates DNA-coated particles to transform intact cells. *The Scientist*, Vol.3, No.18, p. 25, ISSN 0890-3670

Gelvin, S.B. (2000). *Agrobacterium* and plant genes involved in T-DNA transfer and integration. *Annual Review of Plant Physiology and Plant Molecular Biology*, Vol. 51, pp. 223-256, ISSN 1040-2519

Höfgen, R. & Willmitzer, L. (1998). Storage of competent cells for *Agrobacterium* transformation. *Nucleic Acids Research*, Vol.16, p. 9877, ISSN 0305-1048

Holsters, M.; de Waele, D.; Depicker A.; Messens E.; van Montagu, M. & Schell, J. (1978). Transfection and transformation of *Agrobacterium tumefaciens*. *Molecular and General Genetics*, Vol.163, pp. 181-187, ISSN

Hooykaas, P.J.J. & Schilperoort, R.A. (1992). *Agrobacterium* and plant genetic engineering. *Plant Molecular Biology*, Vol.19, pp. 15-38, ISSN 0167-4412

Horsch, R.B.; Fraley, R.T.; Rogers, S.G.; Sanders P.R.; Lloyd, A. & Hoffmann, N. (1984). Inheritance of functional foreign genes in plants. *Science*, Vol.223, pp. 496-498, ISSN 00368075

Karesch, H.; Bilang, R.; Scheid, O.M. & Potrykus, I. (1991). Direct gene transfer to protoplasts of *Arabidopsis thaliana*. *Plant Cell Reports*, Vol.9, pp. 571-574, ISSN 0721-7714

Koncz, C. & Schell, J. (1986). The promoter of the T_L-DNA gene 5 controls the tissue-specific expression of chimeric genes carried by a novel type of *Agrobacterium* binary vector. *Molecular and General Genetics*, Vol.204, pp. 383-396, ISSN 0026-8925

Leplé, J.C.; Brasileiro, A.C.M.; Michel, M.F.; Delmotte, F. & Jouanin, L. (1992). Transgenic poplars: expression of chimeric genes using four different constructs. *Plant Cell Reports*, Vol.11, pp. 137-141, ISSN 0721-7714

Liu, N.Y.; Zhang, Z.F. & Yang, W.C. (2008). Isolation of embryo-specific mutants in *Arabidopsis*: plant transformation. *Methods in Molecular Biology*, Vol.427, pp. 91-100, ISSN 1064-3745

Mathur, J. & Koncz, C. (1997). PEG-mediated protoplast transformation with naked DNA. *Methods in Molecular Biology*, Vol.82, pp.267-276, ISSN 1064-3745

May, G.D.; Afza, R.; Mason, H.S.; Wiecko, A.; Novak, F.J. & Arntzen, C.J. (1995). Generation of transgenic banana (*Musa acuminata*) plants via *Agrobacterium*-mediated transformation. *Nature Biotechnology*, Vol.13, pp. 486-492, ISSN 1087-0156

Mersereau, M.; Pazour, G.J. & Das, A. (1990) Efficient transformation of *Agrobacterium tumefaciens* by electroporation. *Gene*, Vol.90, pp.149-151, ISSN 0378-1119

Nakagawa, T.; Kurose, T.; Hino, T.; Tanaka, K.; Kawamukai, M.; Niwa, Y.; Toyooka, K.; Matsuoka, K.; Jinbo, T. & Kimura, T. (2007). Development of series of gateway binary vectors, pGWBs, for realizing efficient construction of fusion genes for plant transformation. *Journal of Bioscience and Bioengineering*, Vol.104, pp. 34-41, ISSN 1389-1723

Narusaka, M.; Shiraishi, T.; Iwabuchi, M. & Narusaka, Y. (2010). The floral inoculating protocol: a simplified *Arabidopsis thaliana* transformation method modified from floral dipping. *Plant Biotechnology*, Vol.27, pp. 349-351, ISSN 1467-7644

Negrutiu, I.; Dewulf, J.; Pietrzak, M.; Botterman, J.; Rietveld, E.; Wurzer-Figurelli, E.M.; Ye, De & Jacobs, M. (1990). Hybrid genes in the analysis of transformation conditions. II. Transient expression vs stable transformation: analysis of parameters influencing gene expression levels and transformation efficiency. *Physiologia Plantarum*, Vol.79, pp. 197-205, ISSN 0031-9317

Nester, E.W.; Gordon, M.P.; Amasino, R.M. & Yanofsky, M.F. (1984). Crown gall: a molecular and physiological analysis. *Annual Review of Plant Physiology*, Vol.35, pp. 387-413, ISSN 0066-4294

Neuhaus, G. & Spangenberg, G. (1990). Plant transformation by microinjection techniques. *Physiologia Plantarum*, Vol.79, pp. 213-217, ISSN 0031-9317

Paszkowski, J.; Shillito, R.D.; Saul, M.; Mandak, V.; Hohn, T. & Potrykus, I. (1984). Direct gene transfer to plants. *EMBO Journal*, Vol.3, No.12, pp. 2717-2722, ISSN 0261-4189

Seki, M.; Shigemoto, N.; Komeda, Y.; Imamura, J.; Yamada, Y. & Morikawa, H. (1991). Transgenic *Arabidopsis thaliana* plants obtained by particle-bombardment-mediated transformation. *Applied Microbiology and Biotechnology*, Vol.36, pp. 228-230, ISSN 0175-7598

Shen, W.J. & Forde, B.G. (1989). Efficient transformation of *Agrobacterium* spp. by high voltage electroporation. *Nucleic Acids Research*, Vol.17, p. 885, ISSN 0305-1048

Sinclair, T.R.; Purcell, L.C. & Sneller, C.H. (2004). Crop transformation and the challenge to increase yield potential. *TRENDS in Plant Science*, Vol.9, No.2, pp. 70-75, ISSN 1360-1385

Sun, H.J.; Uchii, S.; Watanabe, S. & Ezura, H. (2006). A highly efficient transformation protocol for micro-Tom, a model cultivar for tomato functional genomics. *Plant and Cell Physiology*, Vol.47, pp. 426-431, ISSN 0032-0781

Takeuchi, Y.; Dotson, M. & Keen, N.T. (1992). Plant transformation: a simple particle bombardment device based on flowing helium. *Plant Molecular Biology*, Vol.18, No.4, pp. 835-839, ISSN 0167-4412

Tepfer, D. (1990). Genetic transformation using *Agrobacterium rhizogenes*. *Physiologia Plantarum*, Vol.79, pp. 140-146, ISSN 0031-9317

Tsai, C.J.; Podila, G.K. & Chiang, V.L. (1994). *Agrobacterium*-mediated transformation of quaking aspen (*Populus tremuloides*) and regeneration of transgenic plants. *Plant Cell Reports*, Vol.14, pp. 94-97, ISSN 0721-7714

Tzfira, T.; Jensen, C.S.; Wang, W.; Zuker, A.; Vincour, B.; Altman, A. & Vainstein, A. (1997). Transgenic *Populus tremula*: a step-by-step protocol for its *Agrobacterium*-mediated transformation. *Plant Molecular Biology Reporter*, Vol.15, pp. 219-235, ISSN 0735-9640

Tzfira, T.; Li, J.; Lacroix, B. & Citovsky, V. (2004). *Agrobacterium* T-DNA integration: molecules and models. *TRENDS in Genetics*, Vol.20, pp. 375-383, ISSN 0168-9525

Yao, Q.; Cong, L.; Chang, J.L.; Li, K.X.; Yang, G.X. & He, G.Y. (2006). Low copy number gene transfer and stable expression in a commercial wheat cultivar via particle bombardment. *Journal of Experimental Botany*, Vol.57, No.14, pp. 3737-3746, ISSN 0022-0957

Zhang, X.; Henriques, R.; Lin, S.S.; Niu, Q.W. & Chua, N.H. (2006). *Agrobacterium*-mediated transformation of *Arabidopsis thaliana* using the floral dip method. *Nature Protocols*, Vol.1, pp. 1-6, ISSN 1754-2189

Zupan, J.R. & Zambryski, P. (1995). Transfer of T-DNA from *Agrobacterium* to the plant cell. *Plant Physiology*, Vol.107, pp. 1041-1047, ISSN 0032-0889

Part 2

Crop Improvement

10

Stability of Transgenic Resistance Against Plant Viruses

Nikon Vassilakos
Benaki Phytopathological Institute
Greece

1. Introduction

Plant viruses constitute one of the main problems of the agricultural production worldwide (Kang et al., 2005). To date, there are not therapeutical measures available for the control of plant-virus diseases in the field and the main control strategy used in practice is based on prevention measures. Genetic resistance is by far the most effective way to control plant viruses. However, 'traditional' genetic sources of resistance to viruses are rare (Lecoq et al., 2004) and due to the high rate of mutation of the viral genomes this resistance even when applicable, is frequently broken under field conditions. The era of *Agrobacterium*-mediated genetic transformation of plants which started at the 80s (Thomashow et al., 1980; Zambryskiet al., 1980) offered new promising prospects for engineered genetic resistance to viruses with numerous following studies reporting a successful use of the transgenic technology against almost all genera of plant viruses or even viroids (Lin et al., 2007; Prins et al., 2008; Ritzenthaler, 2005; Schwind et al., 2009). However, mainly due to public concerns for the safety of using transgenic plants in agriculture only in a relatively small number of virus diseases transgenic technology has been used in the field and in these cases it was proved an efficient and safe way of control (Fuchs et al., 2007). The mechanism of resistance in the vast majority of the applications of transgenic-plant strategy is based on RNA-silencing. RNA-silencing is a sequence specific RNA degradation mechanism, highly conserved between kingdoms, which in plants, among other functions, operates as a natural antiviral defense system (Eamens et al., 2008). The role of RNA-silencing as an antiviral weapon has been further supported by the fact that almost every known plant virus species encodes for at least one protein with RNA-silencing suppression activity (Diaz-Pendon & Ding, 2008). This knowledge raised the first concerns regarding the efficiency of RNA-silencing based resistance against viruses under field conditions. As silencing is sequence specific, the resistance of transgenic plants engineered to be resistant to typically one virus could be broken by a different, heterologous virus that could infect the plants in the field. The hypothesis was that the heterologous virus through its silencing suppressor protein(s) could repress the RNA silencing machinery of the plant as a whole, resulting in the loss of the initially engineered resistance. In addition, the extensive research on RNA-silencing that is going on for over a decade has revealed a number of environmental and plant physiological factors that can influence the silencing mechanism and consequently the effectiveness of RNA-silencing based transgenic resistance to viruses under field conditions.

This review summarizes a fair amount of data that have been produced during the last decade in studies that have examined the role of heterologous viruses, the effect of temperature, the influence of the developmental stage of the plants in the stability of the transgenic resistance to viruses as well as recent findings for a direct effect of light intensity on the RNA silencing machinery. Moreover, new approaches for the implementation of RNA silencing in transgenic plant virus resistance are discussed as possible ways to overcome constrains of the current applications.

2. Strategies for engineering resistance to plant viruses

After the revolutionary work that was carried on *Agrobacterium* as a vector for plant transformation, the breakthrough for the creation of transgenic resistance to plant viruses came by Beachy's group which showed that the expression of the coat protein gene of *Tobacco mosaic virus* (TMV) in transgenic plants is conferring resistance to TMV (Abel et al., 1986). This discovery led the way for the production of an enormous number of transgenic plants resistant to viruses, using most types of viral genes. This genetically engineered resistance, referred to as pathogen-derived resistance (PDR) (Sanford & Johnston, 1985), mechanistically was divided into two categories; protein mediated and RNA-mediated. In protein mediated resistance the transformation cassette is designed in such a manner that the introduced viral gene, most commonly either of the coat protein, the replicase or a defective movement protein gene, would be able to be translated and expressed into the plant and somehow interfere with the disassembly, the replication or the movement respectively, of the intruding virus. However, this division is rather simplistic as in most cases of resistance which were designed to be protein mediated, it was proved that multiple mechanisms were involved, most frequently the RNA-mediated one (Lin et al., 2007; Prins et al., 2008; Ritzenthaler, 2005). RNA-mediated resistance is related to RNA-silencing which is probably the most important and common strategy for engineered resistance to plant viruses and will be discussed more extensively below.

Besides the PDR strategy, alternative biotechnological approaches for the manufacturing of plants resistant to viruses include the expression of plant virus-resistance genes in other plants than those from which they were isolated (Farnham, 2006; Seo et al., 2006; Spassova et al., 2001) and the expression of peptides (Lopez-Ochoa et al., 2006; Rudolph et al., 2003; Uhrig, 2003) or antibodies. After the first successful application of the later strategy in 1993 by Tavladoraki and co-workers, with antibodies that reduced the susceptibility to *Artichoke mottle crinkle virus* using a single-chain variable fragment (scFv) directed against the CP of the virus, technical difficulties hampered a wider application of this methodology. Nevertheless, several studies have reported the creation of plants resistant to viruses by expressing scFvs targeting structural as well as non-structural viral proteins (Binz & Plückthun, 2005; Prins et al., 1995; Prins et al., 2005; Ziegler & Torrance, 2002). The mechanisms of protein mediated resistance and of alternative methodologies are out of the scope of this review and will not be discussed further.

3. RNA-silencing based transgenic resistance against plant viruses

RNA silencing constitutes a vital element of the innate antiviral 'immune' response in plants. It uses cytoplasm-associated small interfering RNAs (siRNAs) to specifically target and inactivate invading nucleic acids. Besides siRNAs, a vast population of small RNAs (sRNAs) accumulates in plant tissues, which includes microRNAs (miRNAs), *trans*-acting siRNAs (ta-siRNAs), heterochromatin-associated siRNAs (also referred to as *cis*-acting siRNAs that are linked to transcriptional gene silencing) and natural antisense transcript siRNAs. These sRNAs through RNA silencing mediate repressive gene regulation and play important role in developmental control, preservation of genome integrity and plant responses to adverse environmental conditions, including biotic stress (Brodersen & Voinnet, 2006; Chapman & Carrington, 2007; Pasquinelli et al., 2005; Ruiz-Ferrer & Voinnet, 2009; Vaucheret, 2006). To date, it has primarily been the cytoplasmic siRNA silencing pathway (also referred to as post transcriptional gene silencing, PTGS) that has been exploited by genetic engineering to confer resistance to plant viruses (Mlotshwa et al., 2008; Tenllado et al., 2004).

RNA silencing, is activated as a response to double-stranded RNA (dsRNA). Viruses, as well as transgenes, arranged as inverted repeats (IR), can directly produce dsRNA (which at a subsequent stage will give rise to primary siRNAs), whereas highly transcribed, sense orientated, single copy transgenes produce aberrant transcripts that serve as a substrate for producing dsRNA (subsequently processed to secondary siRNAs). In the latter case dsRNA is synthesized by one member of a family of cellular RNA-depended RNA polymerases (RdRPs) which counts six members in *Arabidopsis* (RDR1-6). Subsequently, the dsRNA can be targeted by a member of a group of Dicer-like ribonucleases (DCL1-4 in *Arabidopsis*) with each of them being involved in specific sRNA pathway(s) and generating specific size of sRNA duplexes (18-25nt in length). All four *Arabidopsis* DCL enzymes appear to be involved – directly or indirectly – in the production of siRNAs from DNA plant viruses, whereas the activities of DCL-4 and DCL-2 are mainly related to the production of siRNAs from single stranded RNA (ssRNA) viruses (Blevins et al., 2006; Ruiz-Ferrer & Voinnet, 2009; Vaucheret, 2006, and references therein). dsRNA cleavage is facilitated by another group of dsRNA-binding proteins (HYPONASTIC 1or HYL 1 and DRB2-5 in *Arabidopsis*). Then, siRNAs are stabilized by 2'O-methylation in their overhanging 3'ends and exported to cytoplasm for PTGS. One selected sRNA strand together with one member of the ARGONAUTE (AGO) family of proteins form the core of a nuclease complex (RNA induced silencing complex, RISC) that targets and cleaves sequence-specifically homologous ssRNA (Ronemus et al., 2006; Ruiz-Ferrer & Voinnet, 2009). The AGO family in *Arabidopsis* is predicted to contain ten members and for some of them a RNA slicer activity has been verified (Brodersen & Voinnet, 2006; Chapman & Carrington, 2007, and references therein). Many excellent reviews cover the functions of sRNAs and their role in RNA-silencing pathways in plants in great detail (Brodersen & Voinnet, 2006; Chapman & Carrington, 2007; Pasquinelli et al., 2005; Ruiz-Ferrer & Voinnet, 2009; Vaucheret, 2006; Mlotshwa et al., 2008).

RNA silencing impedes viral multiplication in plants by two major ways. First it degrades the dsRNA intermediates of virus replication themselves as well as the cognate mRNAs (referred to as cell-autonomous silencing), a procedure that leads to the increase of accumulation of the respective siRNAs. Second, it generates a mobile signal that triggers the degradation of homologous mRNAs in distant cells (systemic silencing). This systemic branch of antiviral RNA silencing is related to siRNA population or their dsRNA precursors that move between neighboring cells through plasmodesmata and over long distances through the phloem (Kalantidis et al., 2008).

RNA-silencing based resistance against viruses was first reported by Lindbo et al. (1993) and was shown to be related to the previously observed co-suppression mechanism (Napoli et al., 1990; Van der Krol et al., 1990). The following years, engineering of transgenic plants to harbor single-stranded sense and to a less extend antisense viral sequences became a common strategy to pre-activate the silencing machinery and obtain resistance against the homologous virus from which the introduced sequence has derived (Ritzenthaler, 2005). Further exploiting this knowledge led to constructing IR transgenes from which long double-stranded (ds) RNA precursors of siRNAs were directly generated. The utilization of such IR transgene constructs has become the method of choice for providing genetically engineered resistance to viruses because a single copy is sufficient to provide immunity, there is no expression of viral proteins, short genome incomplete sequences can be used and efficiencies of up to 90% of all transgenic plants produced to be resistant to the homologous virus were achieved (Lin et al., 2007; Tenllado et al., 2004; Ritzenthaler, 2005). In contrast to the situation with RNA viruses, the use of RNA silencing against DNA viruses most often resulted in delays in symptom development and did not always prevent virus replication (Lin et al., 2007). However, immune lines against *Tomato yellow leaf curl virus* (TYLCV) have been reported by Yang and co-workers (2004), and Fuentes and associates (2006).

In order to overcome the weakness of RNA-silencing based resistance [ineffective against viruses whose sequence differs from that of the transgene by more than 10% (Bau et al., 2003; Jones et al., 1998)], Bucher et al. (2006) fused 150-nt fragments of viral sequences of four tospoviruses in a single small chimeric IR construct. This strategy resulted in a high frequency of produced resistant plants. A most recent approach used modified plant miRNA cistrons to produce a range of antiviral artificial miRNAs (amiRNAs) (Niu et al., 2006; Qu et al., 2007; Schwab et al., 2006; Simon-Mateo & Antonio Garcia, 2007; Zhang et al, 2011).

4. Factors that influence the RNA-silencing based transgenic resistance

4.1 Heterologous viruses

Since 1998 where the first viral suppressor of silencing was discovered it has been established that most known virus species carry at least one RNA silencing suppressor (Diaz-Pendon & Ding, 2008; Ding & Voinnet, 2007). The awareness of this viral counter-defensive strategy against the innate antiviral defense system of plants guided several groups to investigate the effect that could invoke on transgenic resistance of plants that were immune to a virus, the infection with a different virus carrying a strong silencing suppressor.

The first studies were presented in 2001 by Savenkov and Valkonen, and Mitter and co-workers. Savenkov and Valkonen produced transgenic tobacco plants resistant to *Potato virus A* (PVA, genus *Potyvirus*) and examined whether the resistance to PVA was affected by infection of the transgenic plants with *Potato virus Y* (PVY), another potyvirus that was known to suppress RNA silencing through its HC-Pro protein (Diaz-Pendon & Ding, 2008; Ding & Voinnet, 2007). The PVY infection resulted in increased steady-state levels of the transgene mRNA in the transgenic plants. PVA challenge was followed 15 days after inoculation with PVY. In contrast to healthy (non-PVY inoculated) transgenic plants, in which no detectable infection with PVA was observed following challenge with PVA, all the PVY-infected transgenic plants were readily systemically infected by PVA. Moreover, in all PVA-infected plants, new leaves continued to display the severe symptoms, indicating no recovery from disease up to 90 days post inoculation. It was concluded that RNA-silencing mediated resistance in transgenic plants against viruses may be suppressed by infection of the plants with heterologous viruses that encode suppressors of gene silencing (Savenkov & Valkonen, 2001). Not equally definite was the outcome from the studies of Mitter et al. (2001; 2003) which showed that in transgenic tobacco plants, infection with *Cucumber mosaic virus* (CMV, genus *Cucumovirus*) expressing the silencing suppressor 2b protein could transiently suppress the silencing mediated immunity to PVY but solely in new leaves that emerged after CMV inoculation and for a limited period of time. The experiments were carried out for six months and different time intervals were examined between the two virus inoculations. It was shown that longer periods of time between CMV inoculation and challenge of transgenic plants with PVY led to a larger proportion of PVY-susceptible plants. Nevertheless, in these plants the relative PVY titers tended to be lower as compared with untransformed control plants and the movement of PVY in the transgenic plants was restricted relatively to the controls. Most importantly, CMV infection supported only a transient PVY infection and did not prevent recovery of the transgenic plants. Moreover, re-inoculation with PVY of the recovered plants or of plants that had been infected with CMV nine weeks earlier, failed to establish a PVY infection. Finally, although CMV infection resulted in increased transgene-derived mRNA levels in the leaves where breakdown of immunity had been recorded, the transgene-specific siRNAs levels were left unaffected.

Simon-Meteo et al. (2003) performed similar experiments on *Nicotiana benthamiana* plants that displayed RNA-silencing based resistance and were regenerated from recovered tissue of plants which showed a delayed resistance to *Plum pox virus* (PPV, genus *Potyvirus*). They used two heterologous viruses with distinct silencing suppressors, CMV and *Tobacco vein mottling virus* (TVMV, genus *Potyvirus* carrying an HC-Pro silencing suppressor). Each heterologous virus and PPV were inoculated either simultaneously or sequentially with an interval of two to four weeks onto transgenic plants. Both viruses, when applied sequentially, were able to reactivate transgene expression, but surprisingly, only the silencing suppression caused by CMV and not that originating from TVMV, was able to revert the transgenic resistant plants to a PPV-susceptible phenotype.

Taking into consideration these first studies several of the numerous succeeding reports (Fuentes et al, 2006; Germundsson & Valkonen, 2006; Praveen et al, 2010; Kawazu et al, 2009; Yang et al, 2004) of engineered transgenic resistance to plant viruses have examined the possible effect of heterologous virus infection in the resistance. However, not always an influence on resistance was observed. Missiou et al. (2004) in transgenic potato plants resistant to PVY examined the effect on the resistance of *Potato virus X* (PVX, genus *Potexvirus*, carrying the P25 silencing suppressor) infection simultaneously with PVY or one week prior to the challenge with PVY. In either of the two variations, infections with PVX occurred without a PVY infection to be detected. Similarly, resistance of transgenic cucumbers incorporating the 54K polymerase domain of *Cucumber fruit mottle mosaic virus* (CFMMV) was not influenced by infection with the potyviruses *Zucchini green mottle mosaic virus* (ZYMV), *Zucchini fleck mosaic virus* (ZFMV), the ipomovirus *Cucumber vein yellowing virus* (CVYV) or CMV (Gal-On et al., 2005). In a different work, Lennefors et al. (2007) tested whether the high levels of RNA silencing-based resistance to *Beet necrotic yellow vein virus* (BNYVV) in transgenic sugar beet roots could be reduced by co-infection with common soil-borne and aphid-borne beet viruses. The plants were first inoculated with the aphid transmitted *Beet mild yellowing virus* (BMYV), *Beet yellows virus* (BYV), or both viruses. Four weeks later, the plants were transplanted to soil infested with BNYVV, *Beet soil borne virus* (BSBV) and *Beet virus Q* (BVQ) and their fungal vector, *Polymyxa betae*. The effectiveness of the resistance was not detectably compromised even following co-infection with all five viruses. Most recently, transgenic tobacco plants were produced, transformed with an IR construct corresponding to sequences of the TMV movement protein gene and the exhibited resistance to TMV was not affected by infection with CMV regardless of the order that the latter was inoculated (prior to or simultaneously with TMV) (Hu et al., 2011). In a different approach, amiRNAs expressed in tomato plants against CMV coding sequences resulted in resistance against the virus which was not noticeably affected by infection with TMV or TYLCV (Zhang et al., 2011). Moreover, the stability of transgenic resistance of tobacco plants against *Tobacco rattle virus* (TRV) (Vassilakos et al., 2008) remained largely unaffected by infection with CMV, PVY or *Tomato spotted wilt virus* (TSWV) (Vassilakos, unpublished results).

In contrast, in *N. benthamiana* plants expressing a *Grapevine virus A* (GVA) minireplicon and displaying high resistance to GVA, infection with *Grapevine virus B* (GVB, genus *Vitivirus*, carrying a P10 silencing suppressor) or PVY resulted in suppression of the GVA-specific defense (Brumin et al., 2009). Interestingly, in these tests GVA and GVB or PVY inocula were applied simultaneously as a mixture of saps derived from plants infected with the respective viruses, unlike previous studies, in which only sequential inoculations with the heterologous viruses resulted in reduced resistance. Finally, sweetpotato transgenic plants transformed with an IR construct targeting the replicase encoding sequences of *Sweetpotato chlorotic stunt virus* (SPCSV, genus *Crinivirus*) and *Sweetpotato feathery mottle virus* (SPFMV, genus *Potyvirus*) exhibited mild or no symptoms and virus accumulation was significantly reduced following SPCSV infection. However, development of severe sweetpotato virus disease symptoms (attributed to infection by both viruses) occurred in transgenic plants infected with a SPFMV isolate with a limited sequence similarity to the sequence used in the transgene (Kreuze et al., 2008).

The results from the studies that examined the effect of heterologous virus infection on the silencing-based transgenic resistance indicated that this kind of resistance, despite the immunity that can confer to the plants against a specific virus, could be compromised to some degree if applied in the field where mixed virus infections occur frequently. However, it became evident that the outcome of the interference between the heterologous viruses and the silencing machinery of the plant is not so easily predictable (Table 1).

The reasons for the discrepancies are unclear, but could be related to the mode of action of the viral suppressor proteins of the different virus tested. Viral silencing suppressors are highly diverse in sequence, structure and activity, and could target multiple points in RNA silencing pathways whereas viruses with large genomes may encode several functionally distinct proteins to achieve silencing suppression (Diaz-Pendon & Ding, 2008; Ding &

Voinnet, 2007). It is considered that suppressor proteins interfere either with siRNAs biogenesis or siRNA function without a multifunctional nature to be excluded. For instance, most studies agree that the potyviral HC-Pro probably specifically blocks accumulation of secondary siRNAs and leaves primary siRNA accumulation unimpaired, whereas P25 blocks accumulation of primary siRNAs (Diaz-Pendon & Ding, 2008). In contrast, the 2b protein of cucumoviruses directly sequestrate siRNAs duplexes using a pair of hook-like structures that interact more promiscuously with long and short dsRNA (Diaz-Pendon & Ding, 2008; Ding & Voinnet, 2007; Ruiz-Ferrer & Voinnet, 2009). Additionally, it binds AGO1 and blocks slicing without interfering with sRNA loading *in vitro*. Although apparently contradictory, these two anti-silencing 2b activities are reconcilable, because 2b's affinity for dsRNA is weak and its interaction with AGO1 could increase 2b local concentrations and enhance specific binding to siRNAs (Ruiz-Ferrer & Voinnet, 2009). Besides, Buchmann et al. (2009) reported that geminivirus AL2 and L2 proteins act as inhibitors of transcriptional gene silencing, which is the branch of silencing that targets DNA viruses.

Additional antiviral plant defense pathways could also be involved in the interference between the heterologous virus infection and the transgenic resistance or as yet unknown factors involved in specific virus species interactions. Thus, the CMV 2b protein has been shown also to block silencing indirectly by interfering with the salicylic acid mediated defense pathway (Li & Ding, 2001). Moreover, *N. benthamiana* plants transformed with an IR construct containing partial *N* gene sequences from five tospoviruses [TSWV, *Groundnut ring spot virus* (GRSV), *Tomato chlorotic spot virus* (TCSV), *Watermelon silver mottle virus* (WSMoV) *Tomato yellow ring virus* (TYRV-t)] displayed resistance against all five viruses. However, co-infection of one of the tospoviruses with a genetically distant strain of the same species (TYRV-s), resulted in specific intraspecies breakdown of resistance through a procedure that involved complementation of the silencing suppressors of the two viruses (Hassani-Mehraban et al., 2009) (Table 1).

4.2 Temperature

It has been well known to plant virologists that temperature strongly influences plant-virus interactions. In high temperature, symptoms are frequently attenuated and virus titers in infected plants are decreased. In contrast, outbreaks of virus diseases are frequently associated with low temperatures (Hull, 2002).

Kalantidis and co-workers (2002) examined the influence of elevated temperature on siRNAs in CMV-resistant transgenic tobacco plants. Two transgenic lines, one expressing very high and the other very low levels of siRNAs, were tested for siRNAs concentration at 25°C and 32°C and at two time points, 20 and 30 days post-germination. At the early time point, transgene derived siRNAs could be detected only in the first line at 25°C and in both lines at 32°C. However, in the first line transgene specific siRNAs were at 32°C in a significantly higher concentration compared to that of 25°C. The analysis of samples taken at the second time point revealed the presence of transgene derived siRNAs in both lines at 25°C. However, at 32°C, siRNAs were detected in both plant lines at a higher concentration. Apparently, in these experiments, except for temperature the developmental stage of the plants also influenced the siRNA concentration (discussed further below).

Factor		Transgenic Plant	Engineered resistance against	Effect on the resistance	Reference
Heterologous viruses	PVY	N. tabacum	PVA	Suppressed	Savenkov & Valkonen, 2001
	CMV	N. tabacum	PVY	Reduced	Mitter et al., 2001; 2003
	CMV TVMV	N. benthamiana	PPV	Suppressed Unaltered	Simón-Mateo et al., 2003
	PVX	Potato	PVY	Unaltered	Missiou et al., 2004
	ZYMV ZFMV CVYV CMV	Cucumber	CFMMV	Unaltered	Gal-On et al., 2005
	BMYV BYV BSBV BVQ	Sugar beet	BNYVV	Unaltered	Lennefors et al., 2007
	SPFMV-C	Sweetpotato	SPCSV SPFMV-Uganda	Suppressed	Kreuze et al., 2008
	GVB, PVY	N. benthamiana	GVA	Suppressed	Brumin et al., 2009
	TYRV-s	N. benthamiana	TSWV GRSV TCSV WSMoV TYRV-t	Suppressed	Hassani-Mehraban et al., 2009
	CMV	N. tabacum	TMV	Unaltered	Hu et al, 2011
	TMV TYLCV	Tomato	CMV	Unaltered	Zhang et al, 2011
	CMV, PVY TSWV	N. tabacum	TRV	Unaltered	Vassilakos (unpublished)
Temperature	32⁰C	N. tabacum	CMV	n/t	Kalantidis et al., 2002
	15⁰C	N. benthamiana	CymRSV	Suppressed	Szittya et al., 2003
		N. tabacum	TMV CMV	Unaltered	Hu et al, 2011
		N. tabacum	TRV	Suppressed locally	Vassilakos (unpublished)
Light	High/Low Intensity	N. benthamiana	PPV	n/t	Kotakis et al., 2010
Early developmental stage		N. benthamiana	PMMoV	Reduced	Tenllado & Díaz-Ruíz, 1999
		Squash	SqMV	Suppressed	Jan at al., 2000
		Papaya	PRSV	Suppressed	Tennant at al., 2001
		N. tabacum	CMV	n/t	Kalantidis et al., 2002
		N. tabacum	TRV	Reduced	Vassilakos et al., 2008

Table 1. Synopsis of the studies described in the text that involved experiments with transgenic plants resistant to viruses and the influence to the resistance of the various factors examined; n/t, not tested.

Szittya and associates (2003) provided further insight into the mechanism that is involved in these observations. Through a set of delicate experiments they demonstrated that RNA silencing induced by viruses or transgenes is inhibited at low temperatures and enhanced with rising temperatures. They used wild type *Cymbidium ringspot virus* (CymRSV) encoding a p19 viral suppressor and a mutated one unable to express p19 (Cym19stop). In virus transfected *N. benthamiana* protoplasts, virus derived siRNA were undetectable at 15°C and gradually increased with temperature from 21 to 27°C indicating that virus-induced cell-autonomous silencing is temperature dependent. The effect of temperature on virus-induced systemic RNA silencing was also examined. *N.benthamiana* plants were inoculated with CymRSV and Cym19stop and grown at different temperatures. CymRSV infected plants died within 2 weeks at 15, 21 and 24°C whereas CymRSV symptoms were attenuated at 27°C and associated with reduced virus level. Confirming the role of p19 as a suppressor of systemic silencing, plants infected with the Cym19stop showed a recovery phenotype at 21 and 24°C. At 27°C, the mutant virus was unable to infect the plants, while at 15°C, Cym19stop-infected plants displayed strong viral symptoms demonstrating that at low temperature, RNA silencing failed to protect the plants even when the virus lacked the silencing suppressor. In addition, using a strain of *Agrobacterium tumefaciens* carrying a green fluorescent protein (GFP) gene construct which was infiltrated sole or together with p19, to wt *N.benthamiana* or *N.benthamiana* plants expressing GFP, it was shown that transgene-induced silencing is also temperature dependent. The stability of RNA silencing mediated transgenic virus resistance at different temperatures was examined using transgenic *N.benthamiana* plants expressing a CymRSV-derived RNA. After inoculation with CymRSV the plants displayed strong resistance at 24°C whereas at 15°C, severe symptoms were developed and CymRSV RNA accumulated to a high level demonstrating that the transgene-mediated virus resistance was broken at low temperature. A temperature effect was also observed on the antisense-mediated endogen gene inactivation of *Arabidopsis* and potato plants, in which antisense inhibition of genes involved in carbohydrate metabolism is broadly used. Interestingly, in contrast to siRNA production, miR157, miR169 and miR171 RNAs accumulated to equal levels at 15, 21 and 24°C in arabidopsis suggesting that accumulation of miRNAs is not affected by temperature.

Chellappan and co-workers (2005) expanding the above findings quantified gemini virus-derived siRNAs at different temperatures and evaluated their distribution along the virus genome for isolates of five species of cassava geminiviruses, consisting of recovery and non-recovery types. In cassava plants, geminivirus-induced RNA silencing increased by raising the temperature from 25°C to 30°C and the appearance of symptoms in newly developed leaves was reduced, irrespectively of the nature of the virus. Consequently, high temperature rendered non-recovery type geminiviruses to recovery-type viruses. The distribution of virus derived siRNAs on the respective virus genome at three temperatures (25°C, 25°C-30°C and 30°C) remained unaltered only for recovery-type viruses. siRNAs derived from recovery-type viruses accumulated at moderately higher levels during virus-induced silencing at higher temperatures. However, siRNAs from non-recovery-type viruses accumulated six times higher than those observed for infections with recovery-type viruses at high temperature. Thus, the decreased symptom severity and virus concentration that were recorded at higher temperature indicate a similar effect of temperature on ssDNA and RNA viruses although there was a differential effect of temperature on the level of virus-derived siRNAs between recovery and non-recovery types of ssDNA viruses.

As with the effect of heterologous viruses, inhibition of RNA silencing or decreasing of siRNAs concentration in low temperature has not always been observed. Thus, transgene anti-sense induced RNA silencing was not inhibited in potato plants at low temperature (Sos-Hegedus et al., 2005). Moreover, tomato plants carrying an IR construct derived from *Potato spindle tuber viroid* (PSTVd) sequences and exhibiting resistance to PSTVd infection, did not show an elevated IR-siRNA accumulation at 31°C in comparison to 21°C (Schwind et al., 2009). In a more recent study, transgenic tobacco plants transformed separately with IR constructs corresponding to sequences of TMV movement protein gene or CMV replication protein gene, exhibited at both 15°C and 24°C similar high levels of resistance to TMV or CMV, respectively (Hu et al, 2011). In addition, the resistance against TRV of transgenic tobacco plants (Vassilakos et al., 2008) grown at 15°C was influenced only in the inoculated leaves but not systemically (Vassilakos, unpublished results).

In summary (Table 1), the well-known temperature effect on the development of viral diseases is closely associated to the RNA silencing antiviral pathway and consequently influences the efficiency of silencing-based transgenic resistance. However, it appears that the low temperature effect on the transgenic resistance depends on additional factors that remain to be identified, fact supported by inconsistencies in the results of the diverse studies described here. Importantly, although at low temperature the siRNA-based silencing machinery is partially inactivated as an adaptive response of plants to adverse conditions, the miRNA-mediated, which is essential for regulatory functions, continues to operate ensuring plant growth (Szittya et al., 2003).

4.3 Light

Studies on the effect of light on transgenic resistance to viruses are not available, however light has been implicated as one of the factors that affect RNA silencing initiation and maintenance in several studies. Although in most of them light effect on silencing was not clearly isolated from that of temperature (Nethra et al., 2006; Vaucheret et al., 1997) recently, Kotakis et al. (2010) investigated solely the role of light intensity in physiological ranges on RNA silencing. They used as a system *N. benthamiana* transgenic lines engineered to express GFP, which exhibited spontaneously silencing at different frequencies and of different spreading intensities. The authors demonstrated that high light intensity increased the frequency of plants displaying both short range and systemic silencing. In contrast, plants grown under low light conditions, showed lower silencing frequencies. In addition, increased light intensity positively affected siRNA levels corresponding to the GFP transgene (sense) transcript. In a different set of experiments, *N. benthamiana* plants were used, incorporating an IR structure derived from the NIb gene of *Plum pox virus* (PPV) and it was shown that levels of all distinguishable siRNA classes corresponding to the IR transcript were also positively affected by high light intensity (Table 1). Although in the latter case, the effect of light intensity on virus resistance was not tested, the authors proposed that light conditions comprise an additional environmental factor that should be taken under consideration when transgenic technology against viral infections applies on the field.

4.4 Plant developmental stage

Quite a few studies with plants carrying sense transgenes and displaying RNA-silencing mediated resistance have suggested an influence of plant developmental stage on the degree of the expressed resistance. Tenllado and Diaz-Ruiz (1999) reported that a higher percentage of transgenic *N. benthamiana* plants, transformed with the 54K read-through domain of the replicase gene of *Pepper mild mottle virus* (PMMoV), displayed complete virus resistance at maturity than at an earlier stage of development. Subsequently, Jan et al (2000) demonstrated that a recovery type of resistance, in squash genetically transformed with the coat protein genes of *Squash mosaic virus* (SqMV), was due to RNA silencing that was activated at a later developmental stage, independently of virus infection. However, a different phenotype of complete resistance was not altered after SqMV inoculation at early developmental stages. Moreover, analysis of crosses between lines exhibiting complete resistance, recovery and susceptible phenotypes revealed that the time of activation of silencing, besides the developmental stage, is affected by the interaction of transgene inserts. Similarly, transgenic papaya plants were susceptible to *Papaya ringspot virus* (PRSV) at a younger stage but resistant when inoculated at an older stage (Tennant et al., 2001).

As mentioned already, Kalantidis and associates (2002) showed that siRNA accumulation in transgenic tobacco, incorporating an IR construct carrying CMV sequences, was higher at later developmental stages. No significant differences in the siRNA concentration were observed between leaves of different age from a single plant or from the seven-leaf stage on, while the siRNA concentration reached a plateau that remained stable in the course of further development.

In a more recent work, *N. tabacum* plants were transformed with the 57-kDa read-through domain of the replicase gene of TRV and were highly resistant to homologous (to the transgene sequence) TRV isolates and moderately resistant to the genetically distinct TRV-GR. Very young transgenic plants with detectable levels of transgene transcript were resistant only systemically to homologous isolates and were susceptible to TRV-GR. Conversely, older plants (at a five-leaf stage) containing a low steady state level of transcripts were immune to homologous isolates and displayed moderate resistance against TRV-GR (Vassilakos et al., 2008).

In conclusion (Table 1), most studies agree that younger transgenic plants accumulate reduced amounts of transgene specific siRNAs compared to older ones, or correspondingly accumulate higher amount of transgene specific transcripts suggesting a reduced efficiency of transgenic resistance against plant viruses. However, the resistance phenotype was not always affected in younger plants, possibly due to reasons associated with the type of the transgene construct used, its integration into the plant genome or the viral sequences that are targeted.

5. Conclusion

A great deal of progress has been made towards comprehension of plant virus biology and the ways in which plants defend themselves against these pathogens. RNA silencing has provided a promising potential for generating virus-resistant transgenic plants and this potential is certainly not cancelled by the awareness of factors that may affect under specific conditions the acquired resistance. However, as with any other pathogen control strategy, RNA silencing does not constitute a panacea and a number of issues should be taken into consideration before being applied in the field. Noticeably, silencing based transgenic resistance is not influenced solely by the factors that were presented in this review. However, planting into areas where endemic virus diseases occur and mixed virus infections are expected especially during early stages of the vegetation period, time intervals of low air temperature and greenhouse or open field cultivation practices could affect the stability of transgenic resistance against plant viruses.

Further exploitation of our knowledge on RNA-silencing pathways is essential to improve the efficiency of the existing strategies or for the development of potential new strategies which will hopefully lead to a better reception by the public. Recent advances like the construction of chimeric IR constructs incorporating sequences derived from different virus species if combined with epidemiological data and pest risk analyses could reduce the effect of mixed virus infections on the resistance (Bucher et al., 2006; Dafny-Yelin & Tzfira, 2007; Kung et al., 2009). Recently, virus resistance was achieved through expression of amiRNAS against viral coding sequences (Ding & Voinnet, 2007; Duan et al., 2008; Niu et al., 2006; Qu et al., 2007; Simon-Mateo & Antonio Garcia, 2006; Zhang et al, 2011). Although there was evidence that amiRNA-mediated virus resistance may not be inhibited by low temperature (Niu et al., 2006) this possibly depends on the plant species examined (Qu et al., 2007). Moreover, the durability of this approach, which resulted in relatively few antiviral small RNAs compared with those of the long dsRNA approach, needs to be further demonstrated (Duan et al., 2008; Simon-Mateo & Antonio Garcia, 2006).

6. Acknowledgement

The author wishes to thank Christina Varveri for helpful discussions and for critically reading the manuscript, and to express his apologies to all those, whose papers were not cited.

7. References

Abel, P.P.; Nelson, R.S.; De, B.; Hoffmann, N.; Rogers, S.G.; Fraley, R.T. & Beachy. R.N. (1986). Delay of disease development in transgenic plants that express the Tobacco mosaic virus coat protein gene. *Science*, Vol. 232, pp. 738–43

Bau, H.J.; Cheng, Y.H.; Yu, T.A.; Yang, J.S. & Yeh, S.D. (2003), Broad-spectrum resistance to different geographic strains of *Papaya ringspot virus* in coat protein gene transgenic papaya. *Phytopathology*, Vol. 93, pp. 112-120

Binz, H.K. & Plückthun, A. (2005). Engineered proteins as specific binding reagents. *Current Opinon in Biotechnology*, Vol.16, pp. 459–69

Blevins, T,; Rajeswaran, R.; Shivaprasad, P.V.; Beknazariants, D.; Si-Ammour, A.; Park, H-S.; Vazquez, F.; Robertson, D.; Meins, Jr F.; Hohn, T. & Pooggin M.M. (2006). Four plant Dicers mediate viral small RNA biogenesis and DNA virus induced silencing. *Nucleic Acids Research*, Vol. 34, pp. 6233–46

Brodersen, P. & Voinnet, O. (2006). The diversity of RNA silencing pathways in plants. *Trends in Genetics*, Vol. 22, pp. 268–80

Brumin, M.; Stukalov, S.; Haviv, S.; Muruganantham, M.; Moskovitz, Y.; Batuman, O.; Fenigstein, A. & Mawassi, M. (2009). Post-transcriptional gene silencing and virus resistance in *Nicotiana benthamiana* expressing a *Grapevine virus A* minireplicon. *Transgenic Research*, Vol.18, pp. 331-45

Bucher, E.; Lohuis, D.; van Popple, P.M.J.A.; Geerts-Dimitriadou, C.; Goldbach, R. & Prins, M. (2006). Multiple virus resistance at a high frequency using a single transgene construct. *Journal of General Virology*, Vol. 87, pp. 3697-3701

Buchmann, R.C.; Asad, S.; Wolf, J.N.; Mohannath, G. & Bisaro, D.M. (2009) Geminivirus AL2 and L2 Proteins Suppress Transcriptional Gene Silencing and Cause Genome-Wide Reductions in Cytosine Methylation. *Journal of Virology*, Vol. 83, pp. 5005-13

Chapman, E.J. & Carrington, J.C. (2007). Specialization and evolution of endogenous small RNA pathways. *Nature Reviews Genetics*, Vol. 8, pp. 884–96

Chellappan, P.; Vanitharani, R.; Ogbe, F. & Fauquet, C.M. (2005). Effect of Temperature on Geminivirus-Induced RNA Silencing in Plants. *Plant Physiology*, Vol.138, pp. 1828–41.

Dafny-Yelin, M. & Tzfira, T. (2007). Delivery of Multiple Transgenes to Plant Cells. *Plant Physiology*, Vol.145, pp. 1118-28

Diaz-Pendon, J.A. & Ding, S-W.(2008). Direct and Indirect Roles of Viral Suppressors of RNA Silencing in Pathogenesis. *Annual Review of Phytopathology*, Vol.46, pp. 303–26

Ding, S.W. & Voinnet, O. (2007). Antiviral immunity directed by small RNAs. *Cell*, Vol. 130, pp. 413–26.

Duan, C-G.; Wang, C-H. ; Fang, R-X. & Guo, H-S. (2008). Artificial MicroRNAs Highly Accessible to Targets Confer Efficient Virus Resistance in Plants. *Journal of Virology,*Vol. 82, pp. 11084-95

Eamens, A.; Wang, M-B.; Smith, N.A. &, Waterhouse, P.M. (2008). RNA Silencing in Plants: Yesterday, Today, and Tomorrow. *Plant Physiology*, Vol. 147, pp. 456–68

Farnham, G. & Baulcombe, D.C. (2006). Artificial evolution extends the spectrum of viruses that are targeted by a disease-resistance gene from potato. *Proceedings of National Academy of Science of USA*, Vol.103, pp. 18828–33

Fuchs, M. & Gonsalves, D. (2007). Safety of virus-resistant transgenic plants two decades after their introduction: lessons from realistic field risk assessment studies. *Annual Review of Phytopathology*, Vol.45, pp. 173-202

Fuentes, A.; Ramos, P.L.; Fiallo, E.; Callard, D.; Sanchez, Y.; Peral, R.; Rodriguez, R. & Pujol, M. (2006). Intron–hairpin RNA derived from replication associated protein C1 gene confers immunity to *Tomato yellow leaf curl virus* infection in transgenic tomato plants. *Transgenic Research*, Vol. 15, pp. 291-304

Gal-On, A.; Wolf, D.; Antignus, Y.; Patlis, L.; Ryu, K.H.; Min, B.E.; Pearlsman, M.; Lachman, O.; Gaba, V.; Wang, Y.; Shiboleth, Y.M.; Yang, J. & Zelcer, A. (2005). Transgenic cucumbers harboring the 54-kDa putative gene of *Cucumber fruit mottle mosaic tobamovirus* are highly resistant to viral infection and protect non-transgenic scions from soil infection. *Transgenic Research*, Vol. 14, pp. 81-93

Germundsson, A. & Valkonen J.P.T. (2006). P1- and VPg-transgenic plants show similar resistance to Potato virus A and may compromise long distance movement of the virus in plant sections expressing RNA silencing-based resistance. *Virus Research*, Vol. 116, pp. 208–213

Hassani-Mehraban, A.; Brenkman, A.B.; van den Broek, N.J.F.; Goldbach, R. & Kormelink, R. (2009). RNAi-Mediated Transgenic Tospovirus Resistance Broken by Intraspecies Silencing Suppressor Protein Complementation. *Molecular Plant Microbe Interactions*, Vol.22, pp. 1250–7

Hu, Qiong.; Niu, Yanbing.; Zhang, Kai.; Liu, Yong. & Zhou, Xueping. (2011). Virus-derived transgenes expressing hairpin RNA give immunity to *Tobacco mosaic virus* and *Cucumber mosaic virus*. *Virology Journal*, Vol. 8, pp.:41

Hull, R. (2002) *Matthews' Plant Virology*. Academic Press, ISBN 0-12-361160-1, San Diego, USA

Jan, F-J.; Pang, S-Z.; Tricoli, D.M. & Gonsalves, D. (2000). Evidence that resistance in squash mosaic comovirus coat protein-transgenic plants is affected by plant developmental stage and enhanced by combination of transgenes from different lines. *Journal of General Virology*, Vol. 81, pp. 2299–2306

Ji, L.H. & Ding S.W. (2001). The suppressor of transgene RNA silencing encoded by *Cucumber mosaic virus* interferes with salicylic acid-mediated virus resistance. *Molecular Plant Microbe Interactions*, Vol.14, pp. 715–24

Jones, A.L.; Johansen, I.E.; Bean, S.J.; Bach, I. & Maule, A.J. (1998). Specificity of resistance to pea seed-borne mosaic potyvirus in transgenic peas expressing the viral replicase (NIb) gene. *Journal of General Virology*, Vol. 79, pp. 3129-3137

Kalantidis, K.; Psaradakis, S.; Tabler, M. & Tsagris, M. (2002). The occurrence of CMV-specific short RNAs in transgenic tobacco expressing virus-derived double-stranded RNA is indicative of resistance to the virus. *Molecular Plant Microbe Interactions*, Vol. 15, pp. 826–33

Kalantidis, K.; Schumacher, H.T.; Alexiadis, T. & Helm, J.M. (2008). RNA silencing movement in plants. *Biology of the Cell* Vol. 100, pp. 13–26

Kang, B.C.; Yeam, I. & Jahn, M.M. (2005). Genetics of plant virus resistance. *Annual Review of Phytopathology*, Vol.43, pp. 581–621

Kawazu, Y.; Fujiyama, R. & Noguchi Y. (2009). Transgenic resistance to *Mirafiori lettuce virus* in lettuce carrying inverted repeats of the viral coat protein gene. *Transgenic Research*, Vol. 18, pp. 113–120

Kotakis, C.; Vrettos, N.; Kotsis , D.; Tsagris, M.; Kotzabasis K. & Kalantidis, K.(2010). Light intensity affects RNA silencing of a transgene in *Nicotiana benthamiana* plants. *BMC Plant Biology*, Vol. 10, pp. 220

Kreuze, J.F.; Klein, I.S.; Lazaro, M.U.; Chuquiyuri, W.J.C.; Morgan, G.L.; Mejía, P.G.C.; Ghislain, M.& Valkonen, J.P.T. (2008). RNA silencing-mediated resistance to a crinivirus (Closteroviridae) in cultivated sweetpotato (*Ipomoea batatas* L.) and development of sweetpotato virus disease following co-infection with a potyvirus. *Molecular Plant Pathology*, Vol. 9, pp. 589-98

Kung, Y.-J.; Bau, H.-J.; Wu, Y.-L.; Huang, C.-H.; Chen, T.-M. & Yeh, S.-D. (2009). Generation of transgenic papaya with double resistance to *Papaya ringspot virus* and *Papaya leaf-distortion mosaic virus*. *Phytopathology* Vol. 99, pp.1312-1320

Lecoq, H.; Moury, B.; Desbiez, C.; Palloix, A. & Pitrat, M. (2004). Durable virus resistance in plants through conventional approaches: a challenge. *Virus Research*, Vol. 100, pp. 31-9

Lennefors, B-L.; van Roggen, P.M.; Yndgaard, F.; Savenkov, E.I. & Valkonen J.P.T. (2007). Efficient dsRNA-mediated transgenic resistance to *Beet necrotic yellow vein virus* in sugar beets is not affected by other soilborne and aphid-transmitted viruses. *Transgenic Research*, Vol. 17, pp. 219-28

Lin, S-S.; Henriques, R.; Wu, H-W.; Niu. Q-W.; Yeh, S-D. & Chua, N-H. (2007). Strategies and mechanisms of plant virus resistance. *Plant Biotechnology Reports*, Vol. 1, pp. 125–34

Lindbo, J.A.; Silva-Rosales, L.; Proebsting, W.M. & Dougherty, W.G. (1993). Induction of a highly specific antiviral state in transgenic plants: implications for regulation of gene expression and virus resistance. *Plant Cell*, Vol. 5, pp. 1749–59

Lopez-Ochoa, L.; Ramirez-Prado, J. & Hanley-Bowdoin, L. (2006). Peptide aptamers that bind to a geminivirus replication protein interfere with viral replication in plant cells. *Journal of Virology*, Vol. 80, pp. 5841–53

Missiou, A.; Kalantidis, K.; Boutla., A.; Tzortzakaki, S.; Tabler, M. & Tsagris, M. (2004). Generation of transgenic potato plants highly resistant to *Potato virus Y* (PVY) through RNA silencing. *Molecular Breeding*, Vol. 14, pp. 185–97

Mitter, N.; Sulistyowati, E. & Dietzgen, R.G. (2003). *Cucumber mosaic virus* infection transiently breaks ds-RNA-induced immunity to *Potato virus Y* in tobacco. *Molecular Plant Microbe Interactions*, Vol. 16, pp. 936-44

Mitter, N.; Sulistyowati, E.; Graham, M.W. & Dietzgen, R.G. (2001). Suppression of gene silencing: a threat to virus-resistant transgenic plants? *Trends in Plant Science*, Vol. 6, pp. 246-7

Mlotshwa, S.; Pruss, G.J. & Vance, V. (2008). Small RNAs in viral infection and host defense. *Trends in Plant Science*, Vol. 13, pp. 375-82

Napoli, C.A.; Lemieux, C.; & Jorgensen, R.A. (1990). Introduction of a Chimeric Chalcone Synthase Gene into Petunia results in reversible co-suppression of homologous genes in trans. *Plant Cell*, Vol.2, pp. 279–89

Nethra, P.; Nataraja, K.N.; Rama, N. & Udayakumar, M. (2006). Standardization of environmental conditions for induction and retention of posttranscriptional gene silencing using tobacco rattle virus vector. *Current Science*, Vol. 90, pp. 431-435

Niu, Q-W.; Lin, S-S.; Reyes, J.L.; Chen, K-C.; Wu, H-W.; Yeh, S-D. & Chua, N-H. (2006). Expression of artificial microRNAs in transgenic *Arabidopsis thaliana* confers virus resistance. *Nature Biotechnology*, Vol. 24, pp. 1420–8

Pasquinelli, A.E.; Hunter, S. & Bracht, J. (2005). MicroRNAs: a developing story. *Current Opinion in Genetics & Development*, Vol. 15, pp. 200–5

Praveen S.; Ramesh, S.V.; A. Z.; Mishra, A.K.; Koundal, V. & Palukaitis, P. (2010). Silencing potential of viral derived RNAi constructs in *Tomato leaf curl virus*-AC4 gene suppression in tomato. *Transgenic Research*, Vol. 19, pp. 45–55

Prins, M.; De Haan, P.; Luyten, R.; Van Veller, M.; Van Grinsven, M.Q.J.M. & Goldbach, R. (1995). Broad resistance to tospoviruses in transgenic plants by expressing three tospoviral nucleoprotein gene sequences. *Molecular Plant Microbe Interactions*, Vol. 8, pp. 85–91

Prins, M.; Laimer, M.; Noris, E.; Schubert, J.; Wassenegger, M. & Tepfer, M. (2008). Strategies for antiviral resistance in transgenic plants. *Molecular Plant Pathology*, Vol. 1, pp. 73–83

Prins, M.; Lohuis, D.; Schots, A. & Goldbach, R. (2005). Phage display selected single-chain antibodies confer high levels of resistance against *Tomato spotted wilt virus*. *Journal of General Virology*, Vol. 86, pp. 2107–13

Qu, J.; Ye, J. & Fang, R. (2007). Artificial microRNA-mediated virus resistance in plants. *Journal of Virology*, Vol.81, pp. 6690–9

Ritzenthaler, C. (2005). Resistance to plant viruses: old issue, news answers? *Current Opinion in Biotechnology*, Vol. 16, pp. 118–22

Ronemus, M.; Vaughn, M.W. & Martienssen, R.A. (2006). MicroRNA-Targeted and Small Interfering RNA–Mediated mRNA Degradation Is Regulated by Argonaute, Dicer, and RNA-Dependent RNA Polymerase in Arabidopsis. *Plant Cell*, Vol. 18, pp. 1559–74

Rudolph, C.; Schreier, P.H. & Uhrig J.F. (2003). Peptide-mediated broad-spectrum plant resistance to tospoviruses. *Procceedings of National Academy of Science of USA*, Vol. 100, pp.4429–34

Ruiz-Ferrer, V. & Voinnet, O. (2009). Roles of Plant Small RNAs in Biotic Stress Responses. Annual Review of Plant Biology, Vol. 60, pp. 485–510

Sanford, J.C. & Johnston, S.A. (1985). The concept of pathogen derived resistance. *Journal of Theoretical Biology*, Vol. 113, pp. 395–405

Savenkov, E.I. & Valkonen, J.P.T. (2001). Coat protein gene-mediated resistance to *Potato virus A* in transgenic plants is suppressed following infection with another potyvirus. *Journal of General Virology*, Vol.82, pp. 2275-278

Schwab, R.; Ossowski, S.; Riester, M.; Warthmann, N. & Weigel, D. (2006). Highly Specific Gene Silencing by Artificial MicroRNAs in Arabidopsis. *Plant Cell*, Vol. 18, pp. 1121–33

Schwind, N.; Zwiebel, M.; Itaya, A.; Ding, B.; Wang, M-B.; Krczal, G. & Wassenegger, M. (2009). RNAi-mediated resistance to *Potato spindle tuber viroid* in transgenic tomato expressing a viroid hairpin RNA construct. *Molecular Plant Pathology*, Vol. 10, pp. 459-69

Seo, Y-S.; Rojas, M.R.; Lee, J-Y.; Lee, S-W.; Jeon, J-S.; Ronald, P.; Lucas, W.J. & Gilbertson, R.L. (2006). A viral resistance gene from common bean functions across plant families and is up-regulated in a non-virus-specific manner. *Procceedings of National Academy of Science of USA*, Vol. 103, pp. 11856–61

Simon-Mateo, C. & Antonio Garcıa, J. (2006). MicroRNA-Guided Processing Impairs *Plum Pox Virus* Replication, but the Virus Readily Evolves To Escape This Silencing Mechanism. *Journal of Virology*, Vol. 80, pp. 2429-36

Simón-Mateo, C.; López-Moya, J.J.; Shan Guo, H.; González, E. & García J.A. (2003). Suppressor activity of potyviral and cucumoviral infections in potyvirus-induced transgene silencing. *Journal of General Virology*, Vol. 84, pp. 2877-83

Sos-Hegedus, A.; Lovas, A.; Kondrak, M.; Kovacs, G. & Banfalvi, Z. (2005). Active RNA silencing at low temperature indicates distinct pathways for antisense-mediated gene-silencing in potato. *Plant Molecular Biology*, Vol. 59, pp. 595–602

Spassova, M.I.; Prins, T.W.; Folkertsma, R.T.; Klein-Lankhorst, R.M.; Hille, J.; Goldbach R.W. & Prins, M. (2001). The tomato gene Sw-5 is a member of the coiled coil, nucleotide binding, leucine-rich repeat class of plant resistance genes and confers resistance to TSWV in tobacco. *Molecular Breeding*, Vol.7, pp. 151–61

Szittya, G.; Silhavy, D.; Molnar, A.; Havelda, Z.; Lovas, A.; Lakatos, L.; Banfalvi, Z. & Burgyan, J. (2003). Low temperature inhibits RNA silencing-mediated defence by the control of siRNA generation. *EMBO Journal*, V. 22, pp. 633–40

Tavladoraki, P.; Benvenuto, E.; Trinca, S.; Demartinis, D.; Cattaneao, A. & Galeffi, P. (1993). Transgenic plants expressing a functional single chain Fv antibody are specifically protected from virus attack. *Nature*, Vol. 366, pp. 469–72

Tenllado, F. & Diaz-Ruiz J.R. (1999). Complete resistance to pepper mild mottle tobamovirus mediated by viral replicase sequences partially depends on transgene homozygosity and is based on a gene silencing mechanism. *Transgenic Research*, Vol. 8, pp. 83–93

Tenllado, F.; Llave, C. & Diaz-Ruiz J.R. (2004). RNA interference as a new biotechnological tool for the control of virus diseases in plants. *Virus Research*, Vol. 102, pp. 85–96

Tennant, P.; Fermin, G.; Fitch, M.M.; Manshardt, R.M.; Slightom, J.L. & Gonsalves, D. (2001). *Papaya ringspot virus* resistance of transgenic Rainbow and SunUp is affected by gene dosage, plant development, and coat protein homology. *European Journal of Plant Pathology*, Vol. 107, pp. 645–53

Thomashow, M.F., Nutter, R.; Montoya, A.L.; Gordon, M.P. & Nester, E.W. (1980). Integration and organization of Ti plasmid sequences in crown gall tumors. *Cell*, Vol. 19, pp. 729-39

Uhrig, J.F. (2003). Response to Prins: broad virus resistance in transgenic plants. *Trends in Biotechnology*, Vol. 21, pp. 376–7

Van der Krol, A.R.; Mur, L.A.; Beld, M.; Mol, J.N.M. & Stuitje, A. (1990). Flavonoid genes in petunia: addition of a limited number of gene copies may lead to a suppression of gene expression. *Plant Cell*, Vol.2, pp. 291–9

Vassilakos, N.; Bem, F.; Tzima, A.; Barker, H.; Reavy, B.; Karanastasi, E. & Robinson, D.J. (2008). Resistance of transgenic tobacco plants incorporating the putative 57-kDa polymerase read-through gene of *Tobacco rattle virus* against rub-inoculated and nematode-transmitted virus. *Transgenic Research*, Vol. 17, pp. 929-41

Vaucheret, H. (2006). Post-transcriptional small RNA pathways in plants: mechanisms and regulations. *Genes & Development*, Vol. 20, pp. 759-71

Vaucheret, H.; Nussaume, L.; Palauqui, J.C. & Quillere, I. & Elmayan, T. (1997). A transcriptionally active state is required for post-transcriptional silencing (cosuppression) of nitrate reductase host genes and transgenes. *Plant Cell*, Vol. 9, pp. 1495-1504

Yang, Y.; Sherwood, T.A.; Patte, C.P.; Hiebert, E. & Polston, J.E. (2004). Use of *Tomato yellow leaf curl virus* Rep gene sequences to engineer TYLCV resistance in tomato. *Phytopathology*, Vol. 94, pp. 490–6

Zambryski, P.; Holsters, M.; Kruger, K.; Depicker, A.; Schell, J.; Van Montagu, M. & Goodman, H.M. (1980). Tumor DNA structure in plant cells transformed by *A. tumefaciens*. *Science*, Vol. 209, pp. 1385-91

Zhang, X.; Li, H.; Zhang, J.; Zhang, C.; Gong, P.; Ziaf, K.; Xiao, F. & Ye Z. (2011). Expression of artificial microRNAs in tomato confers efficient and stable virus resistance in a cell-autonomous manner. *Transgenic Research*, Vol. 20, pp. 569–581

Ziegler, A. &, Torrance, L. (2002). Applications of recombinant antibodies in plant pathology. *Molecular Plant Pathology*, Vol.3, pp. 401–7

11

Genetic Enhancement of Grain Quality-Related Traits in Maize

H. Hartings, M. Fracassetti and M. Motto
CRA-Unità di Ricerca per la Maiscoltura, Bergamo,
Italy

1. Introduction

Maize (*Zea mays*) is a major food and animal feed worldwide and occupies a relevant place in the world economy and trade as an industrial grain crop (White & Johnson, 2003). Currently more than 70% of maize production is used for food and feed; therefore, knowledge of genes involved in grain structure and chemical is important for improving the nutritional and food-making properties of maize.

Although, plant breeding has been extremely successful at improving the yield of maize, quality has received less attention. However, important advances were made by breeders in this area as well, resulting in maize with a wide range of compositions. In fact, by exploiting genetic variation, the composition of the kernel was altered for both the quantity and quality (structure and chemical diversity) of starch, protein, and oil throughout kernel development. Furthermore, the ability of plant scientists to use existing genetic variation and to identify and manipulate commercially important genes will open new avenues to design novel variation in grain composition. This will provide the basis for the development of the next generation of speciality in maize and of new products to meet future needs.

This chapter focuses on gene discovery, exploitation, and genetic variation known to affect the development and chemical composition of maize kernel. Throughout the chapter we have attempted to summarize the current status in these areas with a particular reference to deposition of storage proteins, starches, lipids, and carotenoids, and research pertinent to enhance kernel quality-related traits. Finally, we provide a brief outlook on future developments in this field and the resultant opportunities and application of conventional and molecular breeding for the development of new maize products better suited to its various end uses.

2. Kernel growth and development

The great economical and nutritional value of the maize kernel is mainly due to its high starch content, as it represents approximately 75% of the mature seed weight. However, the protein complement (ca. 10% of the mature seed weight), mainly found in the form of zeins (storage proteins) is essential for human and animal nutrition. Yet the question remains of why the selection for higher starch level irremediably results in less protein content, as illustrated by the Illinois Long-Term Selection Experiment, which is spanning over more than 100 generations of classical breeding (Mooses et al., 2004).

As a typical angiosperm, the maize kernel comprise two zygotic tissues, namely the embryo (germ) and the endosperm, that are embedded in the testa (or seed coat) and the pericarp (or fruit wall), which fuse into a thin protective envelope. The endosperm is the main storage site of starches and proteins, whereas the embryo reserves mainly lipids. However, the economical and nutritional value of the kernel is mostly derived from the endosperm, a starch-rich tissue, that supports the embryo at germination.

In maize, endosperm makes up the majority of kernel dry matter (70-90%) and is the predominant sink of photosynthates and other assimilates during reproductive growth; therefore, factors that mediate endosperm development to a large extent also determine grain yield. Furthermore, the endosperm of seed can serve as a valuable system to address fundamental questions related to the improvement of seed size in crops.

High-throughput genomics and post-genomics approaches are now providing new tools for a better understanding of the genetic and biochemical networks operating during kernel development. Recently, large databases of maize gene expressed sequence tags (ESTs) have been made available (i.e http://www.maizegdb.org), and transcriptome analyses aimed at identifying genes involved in endosperm development and metabolism have been published, along with computer software to systematically characterize them, has made possible to analyze gene expression in developing maize endosperm more thoroughly to identify tissue-specific genes involved in endosperm development and metabolism (Lai et al., 2004; Verza et al., 2005; Liu et al., 2008; Prioul et al., 2008). These studies have shown that in maize, at least 5000 different genes could be expressed during development. However, about 35% of them are orphan genes, whose functions remain enigmatic, possibly corresponding to endosperm specific genes (Liu et al., 2008), as also observed in wheat (Wan et al., 2008). Furthermore, Mèchin and co-workers (2004) have established a proteome reference map for the maize endosperm. They found that metabolic processes, protein destination and synthesis, cell rescue, defence, cell death and ageing were the most abundant functional categories detected in the maize endosperm.

Collectively, the transcriptome and proteome maps constitute a powerful tool for physiological studies and are the first step for investigating maize endosperm development and metabolism. Although, mRNAs are the primary products of gene expression and their levels are often weakly correlated to corresponding protein levels (Gygi et al., 1999), the analyses of the changes in the transcription profiles of endosperm mutants may allow to formulate predictions regarding the biological role of these loci in endosperm development and metabolism. This information is useful for identifying distinctive, previously uncharacterised, endosperm-specific genes; in addition, it provides both further research material for academic laboratories, and material for plant breeders and food processors to include in their respective research or product pipelines.

3. Accumulation of storage products

The structure and biochemical properties of seed storage compounds have been widely investigated over the past 30 years due to their abundance, complexity, and impact on the overall nutritional value of the maize seed. A great deal is now known about the compounds that are made and stored in seeds, as well as how they are hydrolyzed and absorbed by the embryo. For more detailed reviews describing the nature and biochemistry of maize endosperm and embryo storage products, we refer the reader to a number of recent reviews (i.e. Hannah, 2007; Holding & Larkins, 2009; Motto et al., 2009; Val et al. 2009).

3.1 Storage protein

The primary storage proteins in the maize grain are prolamines called "zeins". Specifically, the zeins are the most abundant protein storage component (>60%) in developing endosperm tissues and are constituted by alcohol-soluble compounds with a characteristic amino acid composition, being rich in glutamine, proline, alanine, and leucine, and almost completely devoid of lysine and tryptophan (Gibbon & Larkins, 2005). From a nutritional point of view, the exceedingly large proportion of codons for hydrophobic amino acids in α-zeins is mostly responsible for the imbalance of maize protein reserves. Therefore, the reduction in α-zein protein accumulation with biased amino acid content could provide a correction to this imbalance. Zeins have also unique functional and biochemical properties that make them suitable for a variety of food, pharmaceutical, and manufactured goods (Lawton, 2002).

Based on their evolutionary relationships, zeins are divided into four protein subfamily of α- (19 and 22-kDa), β- (15 kDa), γ- (16-, 27-, and 50-kDa), and δ-zeins (10- and 18-kDa), that are encoded by distinct classes of structural genes (Holding & Larkins, 2009). Miclaus et al. (2011) have recently reported that α-zein genes have evolved from a common ancestral copy, located on the short arm of chromosome 1, to become a 41-member gene family in the reference maize genome, B73. According to these workers once genes are copied, expression of donor genes is reduced relative to new copies. In particular, epigenetic processes that modify the information content of the genome without changing the DNA sequence, seems to contribute to silencing older copies: some of them can be reactivated when endosperm is maintained as cultured cells, indicating that copy number variation might contribute to a reserve of gene copies.

The proper deposition of zeins inside subcellular structures called protein bodies (PBs) confers the normal vitreous phenotype to the endosperm. PBs are specialized endosperm organelles that form as an extension of the membrane of the rough endoplasmic reticulum (RER), into which zeins are secreted as the signal peptide is processed. After being secreted into the RER, the β- and γ-zeins form a matrix, which is penetrated by the α- and δ-zeins, enlarging the PB and making it a spherical structure of 1-2 μm (Lending & Larkins 1989). Alterations in size, shape or number of PBs generally determine the opaque phenotype (Holding and Larkins 2009), the sole exception being *floury1* (*fl1*), an opaque mutant with no alterations in PB size or shape (Holding et al. 2008). Recently, maize storage protein mutants created through RNAi showed that γ- zein RNA interference (RNAi) maize mutant lines exhibited slightly altered PB body formation and that a more drastic effect was observed in the β-γ- combined mutant, where protein bodies showed an irregular shape, particularly in their periphery (Wu & Messing, 2010). Further studies reported by Llop-Touset et al., (2010) have indicated that the N-terminal proline-rich domain of γ-zein plays an important role in PB formation. To gain a deeper insight into the relationship between RNA and protein localization in plants, Washida et al. (2009) have identified that the cis-localization elements of the 10-kDa δ-zein are responsible for PB-ER targeting. Their results indicate that there is a close relationship between RNA and protein localization in plant cells and that RNA localization may be an important process in mediating the deposition of storage protein in the endomembrane system in plants.

3.1.1 Endosperm mutants altering storage protein synthesis

As highlighted before, endosperm growth and development is a complex phenomenon that may be driven by the coordinate expression of numerous genes. Strategies using spontaneous and induced mutants allow the characterization of the complex underlying gene expression system integrating carbohydrate, amino acid, and storage protein metabolisms and operating during endosperm growth and development. In this respect several endosperm mutants altering the timing and the rate of zein synthesis have been described (reviewed by Motto et al., 2009). The mutants altering the rate of zein synthesis exhibit a more or less defective endosperm and have a lower than normal zein content at maturity. Many of these genes have been mapped to chromosomes and their effect on zein synthesis has been described (Table 1). All mutants confer an opaque phenotype to the endosperm, and, as zein synthesis is reduced, the overall lysine content is elevated, giving potential for use in the development of "high-lysine" maize.

Genotype	Inheritance	Effect on zein accumulation	Molecular bases
Opaque-2 (o2)	Recessive	22-kDa elimination, 20-kDa reduction,	Transcriptional activator
Opaque-5 (o5)	Recessive	No reduction	MGD1
Opaque-6 (o6)	Recessive	General reduction	
Opaque-7 (o7)	Recessive	General reduction 20 and 22-kDa	ACS-like protein
Opaque-15 (o15)	Recessive	27-kDa reduction, reduction γ-zein	
Opaque-2 modifiers	Semidominant	27-kDa overproduction	
Floury-1 (fl1)	Semidominant	General reduction	Transmembrane protein
Floury-2 (fl2)	Semidominant	General reduction	Defect 22-kDa zein
Floury-3 (fl3)	Semidominant	General reduction	
Defective-endospermB30 (De*B30)	Dominant	General reduction	Defect 20-kDa zein
Mucronate (Mc1)	Dominant	General reduction	Abnormal 16-kDa γ-zein
Zpr10(22)	Recessive	10-kDa reduction	

Table 1. Some features of maize mutans affecting zein accumulation.

Genetics has played an important role in discovering a series of opaque endosperm mutants and demonstrating their effects on genes mediating zein deposition (Motto et al., 2009). For example, the recessive mutation *opaque-2* (*o2*) induce a specific decrease in the accumulation of 22-kDa α-zeins, while the *opaque-15* (*o15*) mutation exerts its effect primarily on the 27-kDa γ-zeins. The *floury1* (*fl1*) mutation is somewhat different, since it does not affect the amount or composition of zein proteins but rather results in the abnormal placement of α-zeins within the PB: *Fl1* encodes a transmembrane protein that is located in the protein body ER membrane. Similarly, Myers et al. (2011) have found that the *opaque5* (*o5*) mutant phenotype is caused by a reduction in the galactolipid content of the maize endosperm, with no change in zein proteins. Furthermore, these workers reported that *O5* locus encodes the *monogalactosyldiacylglycerol synthase* (*MGD1*) and specifically affects galactolipids necessary for amyloplast and chloroplast function. A further interesting maize opaque endosperm mutant, termed *mto140*, which also shows retarded vegetative growth has been studied by Holding et al., (2010). The seeds showed a general reduction in zein storage protein accumulation and an elevated lysine phenotype typical of other opaque endosperm mutants; however, it is distinct from the other opaque mutants because it does not result from quantitative or qualitative defects in the accumulation of specific zeins but rather from a disruption in amino acid biosynthesis. Because the opaque phenotype co-segregated with a Mutator transposon insertion in an *arogenate dehydrogenase* gene (*zmAroDH-1*), this has led the previous authors the characterization of the four-member family of maize *arogenate dehydrogenase* genes (*zmAroDH-1–4*) which share highly similar sequences. Their differential expression patterns, as well as subtle mutant effects on the accumulation of tyrosine and phenylalanine in endosperm, embryo, and leaf tissues, suggested that the functional redundancy of this gene family provides metabolic plasticity for the synthesis of these important amino acids.

The *o2* mutation has been widely studied at the genetic, biochemical and molecular levels. *O2* encodes a basic leucine zipper (*bZIP*) transcriptional regulator that is specifically expressed in the endosperm (reviewed in Motto et al., 2009). These studies showed that *O2* activates the expression of 22-kDa α-zein and 15-kDa β-zein genes by interacting with the TC-CACGT(a/c)R(a/t) and GATGYRRTGG sequences of their promoters, therefore displaying a broad binding specificity and recognizing a variety of target sites in several distinct genes. *O2* also regulates directly or indirectly a number of other non-storage protein genes, including *b-32*, encoding a type 1 ribosome-inactivating protein, one of the two cytosolic isoforms of the *pyruvate orthophosphate dikinase* gene (*cyPPDK1*), and *b-70*, encoding a heat shock protein 70 analogue, possibly acting as a chaperonin during PB formation. *O2* also regulates the levels of *lysine-ketoglutarate reductase* (Brochetto-Braga et al., 1992) and *aspartate kinase1* (Azevedo et al., 1997). These broad effects suggest that *O2* plays an important role in the developing grain as a coordinator of the expression of genes controlling storage protein, and nitrogen (N) and carbon (C) metabolism.

The *O7* gene was recently cloned by two different groups, using a combination of map-based cloning and transposon tagging and confirmed by transgenic functional complementation (Miclaus et al., 2011; Wang et al., 2011). Moreover, these last workers via sequence analysis indicated that the *O7* gene showed similarities with members of the larger family of *acyl-CoA synthetase*-like genes (*ACS*), although its exact enzymatic activity is uncertain. In particular, Miclaus et al (2011), have hypothesized a mechanism in which the *O7* protein functions in post-translational modification of zein proteins, thus contributing to membrane biogenesis and stability of PBs and conferring the normal vitreous phenotype of the kernel. Alternatively, Wang et al. (2011) have suggested, by analysis of amino acids and key metabolites, that O7 gene function might affect amino acid biosynthesis by affecting α-ketoglutaric acid and oxaloacetic acid phenotype, indicating a conserved biological function of O7 in cereal crops. In this respect, Hartings et al. (2011), in a study to clarify the role that O2 and O7 play in endosperm gene expression through transcriptomic analyses, indicated that the o2 and o7 mutants alter gene expression in a number of enzymatic steps in the tricarboxylic acid cycle (TCA) and glycolysis pathways that are of central importance for the amino acid metabolism in developing seeds. Although, a systematic characterization of such enzymes will be necessary before any inferences are warranted, the cloning of O7 revealed a novel regulatory mechanism for storage protein synthesis and highlighted an effective target for the genetic manipulation of storage protein contents in cereal seeds, maize included.

An alternative approach to understand the relationship between zein synthesis and the origin of the opaque endosperm phenotype is to perturb zein accumulation transgenically. In this respect, a number of laboratories have reported a reduction in 22-kDa (Segal et al., 2003) and 19-kDa α-zeins (Huang et al., 2004) by RNAi and by seed-specific expression of lysine rich protein (Rascon-Cruz et al., 2004; Yu et al., 2004).

3.1.2 Regulation of storage protein synthesis

The expression of zein genes is regulated coordinately and zein mRNAs accumulate at high concentrations during early stages of endosperm development (reviewed in Motto et al., 2009). From these studies it was also noted that the coordinate expression of zein genes in maize is controlled primarily at the level of transcription according to specific spatial/temporal patterns. Therefore, attention has turned to understanding the regulatory mechanisms responsible for zein gene expression. Highly conserved cis-regulatory sequences have been identified in the promoter of prolamine genes and corresponding trans-activity factors (cf Motto et al., 2009). Zein gene expression can also be affected by other regulatory mechanisms, such as methylation, aminoacid supply and phosphorylation. In this context, Locatelli et al. (2009) have provided evidence that O2-mediated transcriptional activation occurs in two-phases, first a potentiated and second a transcriptional activated phase, both characterized by a specific profile of chromatin modifications. The dependency on O2 activity in the establishment of these chromatin states was different for distinct sub-sets of O2 targets, indicating a gene-specific interaction of O2 with chromatin modifying mechanisms in driving transcription.

3.1.3 Practical applications and perspectives

Despite efforts to develop opaque mutations that are commercially useful, its inherent phenotypic deficiencies, such as soft endosperm texture, lower yield, increased seed susceptibility to pathogens and mechanical damages, have limited their use. To overcome these drawbacks Quality Protein Maize (QPM) strains were created by selecting genetic modifiers that convert the starchy endosperm of an o2 mutant to a hard, vitreous phenotype. Genetic studies have shown that there are multiple, unlinked o2 modifiers (Opm), (review in Gibbon and Larkins, 2005). Genetic analysis of o2 modifiers identified several disperse quantitative trait loci (QTLs). Although their molecular identities have remained unknown, QTLs could be correlated with observed increases in 27-kDa γ-zein transcript and protein in QPM (Holding et al. 2008, and references therein). Two different QTLs, which are candidates for *o2* modifier genes, affect 27-kDa γ-zein gene expression. The first of these is associated with increased expression and the other is linked to *o15*, a mutation at a different chromosome 7 location, which causes decreased 27-kDa γ-zein expression suggesting that the amount of γ-zeins would become critical to keep starch granules embedded in the vitreous area. To examine the role of γ-zeins in QPM, Wu et al., (2010) have used an RNAi construct, designed from the inverted coding sequences of the 27-kDa γ-zein gene, to knock down both 27- and 16-kDa γ-zeins by taking advantage of their DNA sequence conservation. Their findings reinforce the fact that different zeins have evolved to play distinct roles in the development of the endosperm.

Although maize endosperm storage protein genes have been studied for many years, many questions regarding their sequence relationships and expression levels have not been solved, such as structure, synthesis and assembly into protein bodies, and their genetic regulation (Holding and Larkins, 2009). The development of tools for genome-wide studies of gene families makes a comprehensive analysis of storage protein gene expression in maize endosperm possible with the identification of novel seed proteins that were not described previously (Woo et al., 2001). For example, to advance our understanding of the nature of the mutations associated with an opaque phenotype, Hunter et al. (2002) assayed the patterns of gene expression in a series of opaque endosperm mutants by profiling endosperm mRNA transcripts with an Affimetrix GeneChip containing approximately 1,400 selected maize gene sequences. Their results revealed distinct, as well as shared, gene expression patterns in these mutants. Similar research on the pattern of gene expression in *o2*, *o7*, and in the *o2o7* endosperm mutants was carried out by Hartings et al. (2011) by profiling endosperm mRNA transcripts at 14 DAP. Their results, based on a unigene set composed of 7,250 ESTs, allowed to identify a series of mutant related up-regulated (17.1%) and down-regulated (3.2%) transcripts. In addition, the same authors identified several differentially expressed ESTs, homologous to gene encoding enzymes involved in amino acid synthesis, C metabolism (TCA cycle and glycolysis), storage protein and starch metabolism, gene transcription and translation processes, signal transduction, and in protein, fatty acid, and lipid synthesis. Those analyses demonstrate that the mutants investigated are pleiotropic and play a critical role in several endosperm metabolic processes. Although, by necessity, these data are descriptive and more work is required to define gene functions and dissect the complex regulation of gene expression, the genes isolated and characterized to date give us an intriguing insight into the mechanisms underlying amino acid metabolim in the endosperm.

A useful strategy to develop more quickly new QPM varieties has been proposed by Wu and Messing (2011). In fact, conversion of QPM into local germplasm is a lengthy process that discourages the spread of the benefits of QPM because breeders have to monitor a high-lysine level, the recessive *o2* mutant allele, and the modifiers *o2*, (Mo2s). Accordingly, to overcome this problem these last authors presented a simpler and accelerated QPM selection. Instead of using the recessive *o2* mutation, they used an RNAi construct directed against both 22- and 19-kDa zeins, but linked to the visible green fluorescent protein (GFP) marker gene. Indeed, when such a green and nonvitreous phenotype was crossed with QPM lines, the Mo2s produced a vitreous green kernel, demonstrating that high lysine and kernel hardness can be selected in a dominant fashion.

3.2 Starch synthesis

Maize, like other cereals, accumulate starch in the seed endosperm as an energy reserve. Moreover, its starch is one of the most important plant products and has various direct and indirect applications in food, feed, and industries. For this reason attempts to increase starch accumulation have received a great deal of attention by plant breeders and plant scientists. Starch biosynthesis is a central function in plant metabolism that is accomplished by a multiplicity of conserved enzymatic activities (see Hannah & James 2008, for a review). Roughly three-quarters of the total starch is amylopectin, which consists of branched glucose chains that form insoluble, semi-crystalline granules. The remainder of the starch is amylose, which is composed of linear chains of glucose that adopt a helical configuration within the granule (Myers et al., 2000). Briefly starch synthesis has two fundamental activities represented by starch synthase, which catalyzes the polymerization of glucosyl units into $\alpha(1/4)$-linked "linear" chains, and starch-branching enzyme, which catalyzes the formation of $\alpha(1/6)$-glycoside bond branches that join linear chains. Acting together, the starch synthases and starch-branching enzymes assemble the relatively highly branched polymer amylopectin, with approximately 5% of the glucosyl residues participating in $\alpha(1/6)$-bonds, and the lightly branched molecule amylose. A third activity necessary for normal starch biosynthesis is provided by starch-debranching enzyme (DBE), which hydrolyzes $\alpha(1/6)$-linkages. Two DBE classes have been conserved separately in plants (Beatty et al., 1999). These are referred as pullulanase-type DBE (PUL) and isoamylase-type DBE (ISA), based on similarity to prokaryotic enzymes with particular substrate specificity. ISA functions in starch production are implied from genetic observations that mutations typically result in reduced starch content, abnormal amylopectin structure, altered granule morphology, and accumulation of abnormally highly branched polysaccharides similar to glycogen.

3.2.1 Genes affecting starch biosynthesis

Starch biosynthesis in seeds is dependent upon several environmental, physiological, and genetic factors (reviewed in Boyer and Hannah, 2001). Moreover, the maize kernel is a suitable system for studying the genetic control of starch biosynthesis. A large number of mutations that cause defects in various steps in the pathway of starch biosynthesis in the kernel have been described. Their analysis has contributed greatly to the understanding of starch synthesis (reviewed in Boyer and Hannah, 2001). In addition, these mutations have facilitated the identification of many genes involved in starch biosynthetic production. As there seems little point in reviewing these data, we will simply summarize in Table 2 cloned maize genes and their gross phenotypes. Although, the effects shown in this table may not necessarily be the primary effect of a mutant, these are the ones presently known. More recently, Kubo et al. (2010) have described novel mutations of *sugary1* (*su1*) and *isa2* loci, coding for isoamylase-type starch-DE enzyme (ISA) ISA1 and ISA2, respectively: Their data indicate that in maize endosperm these enzymes function to support starch synthesis either as a heteromeric multisubunit complex containing both ISA1 and the noncatalytic protein ISA2 or as a homomeric complex containing only ISA1. In particular, it was found that i) homomeric ISA has specific functions that determine amylopectin structure that are not provided by heteromeric ISA and ii) tissue-specific changes in relative levels of ISA1 and ISA2 transcripts, or functional changes in the ISA1 protein, could explain how maize endosperm acquired the homomeric enzyme.

Genotype	Mayor biochemical changes[a]	Enzyme affected
Shrunken-1 (sh1)	↑ Sugars, ↓ Starch	↓ Sucrose synthase
Shrunken-2 (sh2)	↑ Sugars, ↓ Starch	↓ ADPG-pyrophosphorylase, ↑ Hexokinase
Brittle-1 (bt1)	↑ Sugars, ↓ Starch	↓ Starch granule-bound phospho-oligosaccharide synthase
Brittle-2 (bt2)	↑ Sugars, ↓ Starch	↓ ADPG-pyrophosphorylase
Shrunken-4 (sh4)	↑ Sugars, ↓ Starch	↓ Pyridoxal phosphate
Sugary-1 (su)	↑ Sugars, ↓ Starch	↑ Phytoglycogen branching enzyme, ↓ Phytoglycogen debranching enzyme
Waxy (wx)	↑ 100% Amylopectin	↓ Starch-bound starch syntase, ↑ Phytoglycogen branching enzyme
Amylose-extender(ae)	↑ Apparent amylose, ↑ Loosely branched polysaccharide	↓ Branching enzyme IIb
Dull-1 (du1)	↑ Apparent amylase	↓ Starch synthase II, ↓ Branching enzyme IIa, ↑ Phytoglycogen branching enzyme

Table 2. Summary of mutant effects in maize where an associated enzyme lesion has been reported.[a] Changes relative to normal. ↑, ↓ = increase or decrease, respectively. Sugars = the alcohol-soluble sugars.

Many biochemical and molecular studies on starch synthesis have been also focused on identifying the rate limiting enzymes to control metabolism. In this context, ADP-glucose pyrophosphorylase (AGPase) plays a key role in regulating starch biosynthesis in cereal seeds. The AGPase in the maize endosperm is a heterotetramer of two small subunits encoded by Brittle2 (Bt2) gene, and two large subunits, encoded by the Shrunken2 (Sh2) gene. Transgenic approaches focused on allosteric regulation of AGPase, although studies of the kinetic mechanism of maize endosperm AGPase has uncovered complex regulatory properties (Kubo et al., 2010), increase starch content and caused an increased seed weight than lines expressing wild-types (Giroux et al., 1996; Wang et al., 2007). Additional research has been also devoted to the over-expression of the wide-type genes encoding maize AGPase. For example Li et al. (2011), have transferred the Bt2 and Sh2 genes from maize, with an endosperm-specific promoter from 27-kDa zein or an endosperm-specific promoter from 22-kDa zein, into elite inbred lines, solely and in tandem, by Agrobacterium tumefaciens-mediated transformation. They found that developing transgenic maize kernels exhibited higher Bt2 and Sh2 gene expression, higher AGPase activity, higher seed weight, and the kernels accumulated more starch compared with non-transgenic plants. The over-expression of either gene enhanced AGPase activity, seed weight (+15%) and starch content compared with the wild type, but the amounts were lower than plants with over-expression of both Bt2 and Sh2. Collectively, these results indicate that over-expression of those genes in transgenic maize plants could improve kernel traits and provide a feasible approach for enhancing starch content and seed weight in maize.

3.2.2 Regulation of starch synthesis

In spite of the above mentioned studies as a complex metabolic pathway, the regulation of starch biosynthesis is still poorly understood. This is surprising, considering the number and variety of starch mutations identified so far, which may indicate that nutrient flow is the key regulatory stimulus in carbohydrate interconversion. In this connection, it has been argued that glucose also serves as a signal molecule in regulating gene expression, in some cases, different sugars or sugar metabolites might act as the actual signal molecules (reviewed in Koch, 2004). There is evidence that regulation of major grain-filling pathway is highly integrated in endosperm. Gene responses to sugars and C/N balance have been implicated. For example, Sousa et al. (2008) have recently identified in maize a gene for *Sorbitol dehydrogenase1* (*Sdh1*). They showed that this gene is highly expressed early in seed development throughout the endosperm, with greatest levels in the basal region, compatible with SDH involvement in the initial metabolic steps of carbohydrate metabolism. The same authors also presented genetic, kinetic, and transient expression evidence for regulation at the transcriptional level by sugars and hypoxia. Moreover, many pleiotropic defective kernel (*dek*) mutations that fail to initiative or complete grain-filling have been identified, but not studied in detail. These are likely to include mutations in "housekeeping genes" as well as important developmental mutants or transcription factors. In this respect, a key challenge is to devise molecular and genetic strategies that can be used to effectively analyse this large, complex phenotypic class. As far as transcription factors are concerned, Fu & Xue (2010) have recently identified in rice candidate regulators for starch biosynthesis by gene coexpression analysis. Among these genes, *Rice Starch Regulator1* (*RSR1*), an APETALA2/ethylene-responsive element binding protein family transcription factor, was found to negatively regulate the expression of type I starch synthesis genes; moreover, *RSR1* deficiency results in the enhanced expression of starch synthesis genes in seeds. Collectively these results demonstrate the potential of co-expression analysis for studying rice starch biosynthesis and the regulation of a complex metabolic pathway and provide informative clues, including the characterization of *RSR1*, to facilitate the improvement of seed quality and nutrition. It is expected that similar orthologous loci will be soon identified in maize; this will allow us to deeper our knowledge on regulatory mechanisms affecting starch biosynthesis in maize.

Different approaches in this area are needed to identify direct interaction among starch biosynthetic enzymes, as well as modifying factors that regulate enzyme activity. In this respect, Wang et al. (2007) described a study in which a bacterial *glgC16* gene, which encodes a catalytically active allosteric-insensitive enzyme, was introduced into maize. The results of this study showed that developing transgenic maize seeds exhibited higher AGPase activity (a rate limiting step in glycogenesis and starch synthesis), in the presence of an inhibitory level of Pi in vitro, compared with the untransformed control. More interestingly, the same authors fuond the seed weight of transgenic plants was increased significantly. Furthermore, tools for genome-based analyses of starch biosynthesis pathway are now available for maize and other cereals. This may eventually help to explain species differences in starch granule shape and size, and thus provide the potential for agricultural advances. Recently, Prioul et al. (2008) have provided information on carbohydrate metabolism by comparing gene expression at three levels - transcripts, proteins and enzyme activities - in relation to substrate or product in developing kernels from 10 to 40 DAP. Their study have identified two distinct patterns: during endosperm development: invertases and hexoses are predominant at the beginning, whereas enzyme patterns in the starch pathway, at the three levels, anticipate and parallel starch accumulation, suggesting that, in most cases, transcriptional control is responsible for the regulation of starch biosynthesis.

3.3 Lipids

While intensive agricultural and industrial uses of the maize kernel is widely due to its high starch content, the oil stored in the maize kernel also has considerable importance. Moreover, its oil is the most valuable co-product from industrial processing of maize grain through wet milling or dry milling and is high-quality oil for human.

Research in this field (for review see Val et al., 2009) indicate that i) the mature embryo is approximately 33% lipid in standard hybrids and contains about 80% of the kernel lipids; ii) high-oil maize shows a greater feed efficiency than normal-oil maize in animal feed trials: the caloric content of oil is 2.25 times greater than that of starch on a weight basis and its fatty acid composition, mainly oleic and linoleic acids; iii) maize oil is highly regarded for its low level of saturated fatty acids, on average 11% palmitic acid and 2% stearic acid, and its relatively high levels of polyunsaturated fatty acids such as linoleic acid (24%); and iv) maize oil is relatively stable, since it contains only small amounts of linolenic acid (0.7%) and high levels of natural antioxidants. Additionally, it was found that oil and starch are accumulated in different compartments of the maize kernel: 85% of the oil is stored in the embryo, whereas 98% of the starch is located in the endosperm. Therefore, the relative amounts of oil and starch are correlated with the relative sizes of the embryo and endosperm and successful breeding for high oil content in the Illinois High Oil strains has mainly been achieved through an increase in embryo size (Moose et al., 2004). Whereas the embryo represents less than 10% of the kernel weight in normal or high-protein lines, it can contribute more than 20% in high-oil lines. However, genetic components may also modulate oil content in the embryo, independently of its size, as shown by the cloning of a high-oil QTL in maize that is caused by an amino acid insertion in an acyl-CoA:diacylglycerol acyltransferase catalyzing the last step of oil biosynthesis (Zheng et al., 2008).

3.3.1 Lipid biosynthetic pathway and genetic inheritance

The primary determinant of amount of lipids in maize kernels is the genetic makeup (Lambert, 2001). In maize studies through genetic mapping of oil traits reported that multiple (>50) QTLs are involved in lipid accumulation (Laurie et al., 2004), making yield improvement through conventional breeding difficulty. High-oil varieties of maize were developed at the University of Illinois through successive cycles of recurrent selection (Dudley and Lambert, 1992). Although these lines have an improved energy content for animal feeding applications, the poor agronomic characteristics, including disease susceptibility and poor standability. These deficiencies precluded their commercial introduction on broad hectarage.

In spite of a good understanding of the oil biosynthetic pathway in plants and of the many genes involved in oil pathway have been isolated, the molecular basis for oil QTL is largely unknown. However, Zheng et al. (2008) have recently found that a oil QTL ($qHO6$) affecting maize seed oil and oleic-acid content, encodes an acyl-CoA:diaglycerol acytransferase (DGAT1-2), which catalyze the final step of oil synthesis.

As far as the composition in concerned, maize oil is mainly composed of palmitic, stearic, oleic, linoleic, and linolenic fatty acids. Evidence has shown that genetic variation existed also for the fatty acid composition of the kernel (Lambert, 2001). In essentially all studies, researchers suggested that major gene effects were being modulated by modifier genes for oil composition. Although it seems that sources of major genes for composition of maize oil can be utilized, other studies indicate that the inheritance of oleic, linoleic, palmitic, and stearic acid content when considered together is complex and under multigenic control (Sun et al., 1978). Molecular characterization of fatty acid desaturase-2 (*fad2*) and fatty acid desaturase-6 (*fad6*) in this plant indicates that *fad2* and *fad6* clones are not associated with QTLs for the ratio of oleic/linoleic acid, suggesting that some of the QTLs for the oleic/linoleic acid ratio do not involved variants of *fad2* and *fad6*, but rather involved other gene that may influence flux via enzymes encoded by *fad2* or *fad6*. Additional studies are needed to more precisely identify the genes and enzymes involved in determining the composition of maize oil. Application of powerful new technologies, such as transcription profiling, metabolic profiling, and flux analyses, should prove valuable to achieving this scope. In addition, identification of transcription factors or other regulatory proteins that exert higher level control of oil biosynthesis or embryo development will be particularly attractive candidate for biotechnology approaches in the future.

In maize, Pouvreau et al. (2011) have recently identified orthologs related, respectively, to the master regulators *LEAFY COTYLEDON1-2 (i.e. ZmLEC1)*, that directly activate in Arabidopsis genes involved in TAG metabolism and storage, and to the transcription factor WRINKLED1 (i.e.. *ZmWRI1a* and *ZmWRI1b*), necessary to mediate the regulatory action of the master regulators towards late glycolytic and oil metabolism. In this crop, both genes are preferentially expressed in the embryo and exhibit a peak of expression at the onset of kernel maturation. *ZmWRI1a* is induced by *ZmLEC1* (Shen et al., 2010). Additionally, transcriptomic analyses carried out on *ZmWRI1a* over-expressing lines have allowed to the previous workers to identify putative target genes of *ZmWRI1a* involved in late glycolysis, fatty acid or oil metabolism. Though not fully overlapping, the sets of *AtWRI1* and *ZmWRI1a* target genes are very resembling. Exhaustive analyses relying on ChIP experiments would allow determining whether these sets are identical. Interestingly, the DNA AW-box proposed to be bound by *AtWRI1* (Maeo et al., 2009) was also identified in promoter sequences of putative target genes of *ZmWRIa*, suggesting that even the *cis*-regulatory element recognized by *WRI1* seems have been conserved between dicots and monocots. Additionally this study has shown that transgenic *ZmWri1a-OE* kernels did not only induce a significant increase in saturated and unsaturated fatty acids with 16 to 18 C atoms but also cause a significant increase for several free amino acids (Lys, Glu, Phe, Ala, Val), intermediates or cofactors of amino acid biosynthesis (pyro-Glu, aminoadipic acid, Orn, nor- Leu), and intermediates of the TCA cycle (citric acid, succinic acid). Since the transcriptome analysis suggests that *ZmWri1a* essentially activates genes coding for enzymes in late glycolysis, fatty acid, CoA, and TAG biosynthesis, and considering that no misregulated candidates participate in any additional pathways, the increase in amino acids and TCA intermediates probably reflects secondary adjustments of the C and N metabolism to the increased oil biosynthesis triggered by *ZmWri1a*. The three amino acids Phe, Ala, and

Val are derived from PEP or pyruvate, and their increase may simply be a byproduct of a strongly increased C flux through glycolysis.

3.4 Carotenoid pigments

Along with their essential role in photosynthesis, carotenoids are of significant economic interest as natural pigments and food additives (reviewed in Botella-Pavía & Rodríguez-Concepción, 2006). Their presence in the human diet provides health benefits as nontoxic precursors of vitamin A and antioxidants, including protection against cancer and other chronic diseases (review by Fraser & Bramley 2004). These motives have promoted scientists to explore ways to improve carotenoid content and composition in staple crops (reviewed in Sandmann et al. 2006; Zhu et al. 2009). Analyses of genotypes with yellow to dark orange kernels exhibits considerable natural variation for kernel carotenoids, with some lines accumulating as much as 66 µg/g (e.g. Harjes et al., 2008), with provitamin A activity (β-cryptoxanthin, a- and β-carotene is typically small (15% to 18% of the total carotenoids fraction) compared to lutein or zeaxanthin (45% and 35%, respectively; Kurlich & Juvik, 1999; Brenna and Berardo, 2004). Moreover, a moderate to high heritability estimates indicate that breeding for increased levels of both carotenes and xanthophylls should be feasible.

3.4.1 Carotenoid biosynthesis and genetic control

Carotenoids are derived from the isoprenoid biosynthetic pathway and are precursors of the plant hormone abscisic acid (ABA) and of other apocarotenoids (Matthews and Wurtzel, 2007). In maize characterization of the carotenoid biosynthetic pathway has been facilitated by the analysis of mutants associated with reduced levels of carotenoids. In fact, by using this approach in maize three genes controlling early steps in the carotenoid pathway have been cloned. The use of these cloned genes as probes on mapping populations will enable the candidate gene approach to be used for studying the genetic control of quantitative variation in carotenoids. Accordingly, Wurtzel et al.. (2004) detect major QTLs affecting accumulation of β-carotene and β-cryptoxanthin indicating that these QTLs could be selected to increase levels of pro-vitamin A structures. Chander et al. (2007), using a RIL population found 31 QTL including 23 for individual and 8 for total carotenoid accumulations. Moreover, Harjes et al. (2008), via association mapping, linkage mapping, expression analysis, and mutagenesis, showed that variation in *lycopene epsil cyclase* (*lcyE*) locus alters flux down a-carotene versus β-carotene branches of the carotenoid pathway. Additional experimental evidence obtained by Yan et al. (2010) have documented that also the gene encoding *β-carotene hydroxylase1* (*crtRB1*) underlies a principal QTL associated with β-carotene concentration and conversion in maize kernels. Moreover, the same workers noted that the *crtRB1* alleles associated with reduced transcript expression correlate with higher β-carotene concentrations. Genetic variation at *crtRB1* also affects hydroxylation efficiency among encoded allozymes, as observed by resultant carotenoid profiles in recombinant expression assays. Similarly, studies on natural maize genetic diversity carried out by Vallabhaneni et al (2009), have provided the identification of hydroxylation genes associated with reduced endosperm provitamin A content. In particular transcript profiling led to discovery of the *Hydroxylase3* locus that coincidently mapped to a carotene QTL, thereby prompting investigation of allelic variation in a broader collection. Vallabhaneni & Wurtzel (2009) have sampled a maize germplasm collection via statistical testing of the correlation between carotenoid content and candidate gene transcript levels. They observed multiple pathway bottlenecks for isoprenoid biosynthesis and carotenoid biosynthesis acting in specific temporal windows of endosperm development. Transcript levels of paralogs encoding *isoprenoid isopentenyl diphosphate* and *geranylgeranyl diphosphate*-producing enzymes, such as DXS3 (*1-deoxy-D-xylulose-5-phosphate synthase3*), DXR (*DXP reductoisomerase*), HDR (*4-hydroxy-3-methylbut-2-enyl diphosphate reductase*), and GGPPS1 (*geranylgeranyl pyrophosphate synthase1*), were found to positively correlate with endosperm carotenoid content. Toledo-Ortiz et al. (2010) have recently identified in *Arabidopsis* seedlings that *phytochrome-interacting factor1 (PIF1)* and other transcription factors of the *phytochrome-interacting factor (PIF)* family down-regulate the accumulation of carotenoids by specifically repressing the gene encoding *PSY*, the main rate-determining enzyme of the pathway. Their results also suggest a role for *PIF1* and other *PIFs* in transducing light signals to regulate *PSY* gene expression and carotenoid accumulation during daily cycles of light and dark in mature plants. In this context, manipulating the levels of *PIF* transcription factors by transgenic or marker-assisted breeding approaches might help improve carotenoid accumulation in plants for the production of varieties with enhanced agronomical, industrial, or nutritional value.

4. New strategies for creating variation

The use of molecular biology to isolate, characterize, and modify individual genes followed by plant transformation and trait analysis will introduce new traits and more diversity into maize database. For example, maize-based diets (animals or humans) require lysine and tryptophan supplementation for adequate protein synthesis. The development of high-lysine maize to use in improved animal feeds illustrates the challenges that continually interlace metabolic engineering projects. From a biochemical standpoint, the metabolic pathway for lysine biosynthesis in plants is very similar to that in many bacteria. The key enzymes in the biosynthetic pathway are aspartakinase (AK) and dihydrodipicolinic acid synthase (DHDPS), both of which are feedback inhibited by lysine (Galili, 2004). Falco et al. (1995) isolated bacterial genes encoding lysine-insensitive forms of AK and DHDPS from *Escherichia coli* and *Corynebacterium*, respectively. A deregulated form of the plant DHDPS was created by site-specific mutagenesis (Shaver et al., 1996). The expression of the bacterial DHPS in maize seeds overproduced lysine, but they also contained higher level of lysine catabolic products then their wild-type parents (Mazur et al., 1999), despite the fact that lysine catabolism was suggested to be minimal in this tissue (Arruda et al., 2000). Likewise, a gene corresponding to a feedback-resistant form of the enzyme anthranilate synthase (AS) has been cloned from maize and re-introduced via transformation under the control of seed-specific promoters. This altered AS has reduced sensitivity to feedback inhibition by tryptophan; thus, tryptophan is overproduced and accumulates to higher than normal levels in the grain. This strategy has been successful in reaching commercially valuable levels of tryptophan in the grain (Anderson et al., 1997). More recently, Houmard et al. (2007) reported the increase in maize grains by specific suppression of lysine catabolism via RNAi. An important observation from these studies was that the lysine content was increased in the transgenic lines by 15-20% to 54.8%. These experiments showed that transgenic approaches, in addition to investigating relationships between zein synthesis and opaque endosperm, could be useful to increase kernel lysine content. Similarly, Reyes et al. (2008), using RNAi, have produced transgenic maize lines that had LKR/SDH suppressed in the embryo, endosperm or both. These authors noted a synergist increase in free lysine content in the mature kernel when LKR/SDH was suppressed in both embryo and endosperm; these results have also suggested new insights into how free lysine level is regulated and distributed in developing grains.

A different approach to enhance the level of a given amino acid in kernels is to improve the protein sink for this amino acid (Kriz, 2009). This goal can be achieved by transforming plants with genes encoding stable proteins that are rich in the desired amino acid(s) and that can accumulate to high levels. Among a variety of natural, modified or synthetic genes that were tested, the most significant increases in seed lysine levels were obtained by expressing a genetically-engineered hordothionine (HT12) or a barley high-lysine protein 8 (BHL8), containing 28 and 24% lysine, respectively (Jung and Falco, 2000). These proteins accumulated in transgenic maize to 3-6% of total grain proteins and when introduced together with a bacterial DHPS, resulted in a very high elevation of a total lysine to over 0.7% of seed dry weight (Jung and Falco, 2000) compared to around 0.2% in wild-type maize. Similarly, Rascon-Cruz et al. (2004) have found that the introduction of a gene encoding amarantin-protein from *Amaranth* plants, which is known to be balanced in its amino acid content, increases from 8 to 44% essential amino acid content. Bicar et al. (2008) have developed transgenic maize lines that produce milk *a*-lactalbumin in the endosperm. They noted that the lysine content of the lines examined was 29-47% greater in endosperm from transgene positive kernels. Furthermore, Wu et al. (2007) provided a novel approach to enrich the lysine content (up to 26%) in the maize grain by endosperm-specific expression of an *Arabidopsis* lysyl tRNA synthate. Combining these traits with seed-specific reduction of lysine catabolism offers an optimistic future for commercial application of high-lysine maize.

Single mutations in starch biosynthesis have been commercially used for the production of some specialty maize. For example, specialty varieties such as waxy can result in 99% amylopectins, while the use of "amylomaize varieties" (*amylose extender* endosperm mutants) have kernels up to 20% amylopectin and 80% amylose. These varieties are of interest for commercial purposes in starch industry, such as food ingredients, sweeteners, adhesives, and for the development of thermoplastics and polyurhetanes. However, advances in understanding the starch biosynthetic pathway provide new ways to redesign starch for specific purposes, such for ethanol production. Alteration in starch structure can be achieved by modifying genes encoding the enzymes responsible for starch synthesis, many of which have more than one isoform (Boyer and Hannah, 2001). Transgenic lines with modified expression of specific starch synthases, starch branching enzymes or starch debranching enzymes are being generated in attempts to produce starch granules with increased or decreased crystallinity, and thus altered susceptibility to enzymatic digestion. Another strategy is to reduce the energy requirements for the starch to ethanol conversion process. For example, gelatinization is the first step in bioethanol production from starch. It is conceivable that a modified starch with decreased gelatinization temperature might require less energy for the conversion process. Research showed that expression of a recombinant amylopullulanase in rice resulted in starch that when heated to 85°C was completely converted into soluble sugars (Chiang, 2005). The expression of microbial genes in transgenic plants represents also an opportunity to produce renewable resources of fructans. Transgenic maize expressing the *Bacillus amyloliquefaciens SacB* gene accumulates high-molecular weight fructose in mature seed (Caimi et al., 1996). This could potentially be exploited within the high-fructose maize syrup market. Moreover, Zhang et al. (2007) have developed transgenic maize endosperm, via the introduction of a *Streptococcus mutans gtfD* gene, that accumulates novel glucan (oligo- and polysaccharides composed solely of glucose molecules) polymers at levels relevant to commercial production. The expression of that gene yielded fully functional GTF-D enzyme as shown by accumulation of novel soluble α-(1→6)-linked glucan at high levels in the mature maize kernels (up to 14% of their dry weight). These findings suggest that the introduction of greater diversity in linkages within α-glucan polymers will enable the generation of specialty glucans to replace modified starches used for several products (e.g. thickening reagents, adhesives, textile modification, and papermaking polymers with economical and environmental benefits).

Efforts to increase oil content and composition in maize kernels through breeding have considerable success, but high oil lines have significant reduced yield (cf Moose et al., 2004). Several and complementary approaches might be considered to try and enhance oil content in maize kernels. This goal may be achieved by increasing the relative proportion of the oil-rich embryonic tissue within the grain. It has been recently reported that embryo size and oil content could be increased in transgenic maize by ectopic expression of the wheat *Purindoline a* and *b* (*PINA* and *PINB*) genes (Zhang, et al., 2010). While total oil content of the kernel was increased by 25% in these transgenic lines, the molecular mechanism responsible for the increase remains to be clarified. If no modification of kernel size was observed in these transgenic lines, other agronomic characteristics remain to be studied to evaluate the economic potential of such material. Another strategy to increase oil accumulation in the grain may consist in improving both oil content of embryonic tissues. A close examination of C metabolism in maize embryos suggested that the flux of C through NADP-ME may constitute a metabolic bottleneck (Alonso et al., 2010). Accordingly, the oil content of the kernel was positively correlated with malic enzyme activities in maturing embryos (Doehlert & Lambert, 1991), which makes NADP-ME an attractive target for engineering high oil concentrations in embryos of maize. Furthermore, in oilseed species, numerous biotechnological approaches have been carried out that were aimed at maximizing the flow of C into oil by overexpression of enzymes of the TAG assembling network. For example in maize, several attempts have been made to over-express *diacylglycerol acyltransferases* (*DGAT*). *DGAT* catalyses the transfer of an acyl chain from the acyl-CoA pool to the *sn-3* position of a diacylglycerol molecule, resulting in the synthesis of TAG. The embryospecific over-expression of both maize *DGAT1-2* and of fungal *DGAT2* (Zheng et al., 2008; Oakes et al., 2011) resulted in limited (1.25 fold) but statistically significant increases in kernel oil content. Whereas it has been shown that grain yield was not affected by expression of fungal *DGAT2*, data concerning the putative incidence of the over-expression of maize *DGAT1-2* on yield and other agronomic characteristics of the modified lines are missing. Nevertheless these works provide insights into the molecular basis of natural variation of oil and oleic-acid contents in plants and highlight *DAGT* as a promising target for increasing oil and oleic-acid content in other crops.

The recent identification of transcriptional regulators of the oil biosynthetic network in maize has opened the way for designing and testing new original biotechnological strategies. A study has shown that seed-specific expression *ZmWRI1*, a WRI1-like gene of maize, enhanced oil accumulation in transgenic maize without detectable abnormalities. However, expression of *ZmLEC1* under similar conditions severely affected growth and development of the resulting transgenic maize plants (Shen, et al., 2010). Similar results were obtained by constitutive overexpression of the *ZmWRI1* gene in the transgenic maize plants (Pouvreau et al., 2011). It was also found that *ZmWri1* not only increases the fatty acid content of the mature maize grain but also the content of certain amino acids (Lys, Glu, Phe, Ala, Val) of several compounds involved in amino acid biosynthesis (pyro-Glu, aminoadipic acid, Orn, nor- Leu), and of two intermediates of the TCA cycle(citric acid, succinic acid) (Pouvreau et al., 2011). Finally, a third approach to increase oil content in maize grains may consist in diverting C flux from starch to oil in the endosperm. Considering both the elevated amounts of ATP consumed in futile cycling processes and the rates of reductant production in endosperm tissues of maize kernels, Alonso et al (2010) have speculated that increasing biomass synthesis and redirecting part of the C flux toward fatty acid production by metabolic engineering could theoretically be obtained. This would require inhibiting futile cycling whilst overexpressing the whole set of enzymes involved in TAG production. To date, no successful attempt has been reported. If the use of *ZmWRI1* as a biotechnological tool for improving oil content in embryos of maize seems promising (see above), over-expression of *ZmWRIa* in the starchy endosperm was not sufficient to trigger oil accumulation in this compartment (Shen et al., 2010). Since there is no evidence that WRI1 regulates TAG assembly, it is not surprising that over-expression of ZmWRI1a only proves to be efficient in tissues already accumulating oil, and thus already expressing the TAG biosynthetic machinery. What is more, the structure and size of maize kernels may impair large accumulation of oil in the endosperm.

The cloning of carotenogenic genes in maize and in other organisms have opening up the possibility of modifying and engineering the carotenoid biosynthetic pathways in plants, although question remains about the rate-controlling steps that limit the predictability of metabolic engineering in plants. Engineering high levels of specific carotenoid structures requires controlled enhancement of total carotenoid levels (enhancing pathway flux, minimizing degradation, and optimizing sequestration) plus controlled composition for specific pathway end products. While most of the nuclear genes for the plastid-localized pathway are available (Li et al. 2007) and/or can be identified, questions remain about the rate-controlling steps that limit the predictability of metabolic engineering in plants. Predictable manipulation of the seed carotenoid biosynthetic pathway in diverse maize genotypes necessitates the elucidation of biosynthetic step(s) that control carotenoid accumulation in endosperm tissue. Studies have implicated PSY, the first committed enzyme, as rate controlling for endosperm carotenoids (e.g. Pozniak et al., 2007; Li et al., 2008). However, upstream precursor pathways may also positively influence carotenoid accumulation (Matthews and Wurtzel, 2000; Mahmoud & Croteau, 2001), while downstream degradative pathways may deplete the carotenoid pool (Galpaz et al., 2008).

Transgenic strategies can also be used as tools to complement breeding techniques in meeting the estimated levels of provitamin A. In this respect, Aluru et al. (2008) reported that the overexpression of the bacterial genes *crtB* (for PS) and *crtI* (for the four desaturation steps of the carotenoid pathway catalyzed by PDS and β-carotene desaturase in plants), resulted in an increase of total carotenoids of up to 34-fold with a preferential accumulation of β-carotene in the maize endosperm. The levels attained approach those estimated to have a significant impact on the nutritional status of target populations in developing countries. Furthermore, the same authors, via gene expression analyses, suggested that increased accumulation of β-carotene is due to an up-regulation of the endogenous lycopene β-cylase.

These experiments set the stage for the design of transgenic approaches to generate provitamin A-rich maize that will help alleviate vitamin A deficiency in developing countries. Similarly, Naqvi et al. (2009) produced transgenic maize plants with significantly increased contents for β-carotene, ascorbate, and folate in the endosperm via that simultaneous modification of 3 separate metabolic pathways. The transgenic kernels contained 169-fold the normal amount of β-carotene, 6-fold, and 2-fold the normal amount of ascorbate and folate, respectively. This finding, which largely exceeds any realized thus far by conventional breeding alone, opens the way for the development of nutritional complete cereals to benefits the consumers in developing countries. Moreover, this is a very important proof of concept for genetic manipulation of distinct metabolic pathways.

There is evidence indicating that tocophenols, in particular γ-tocophenol the predominant form of vitamine E in plant seeds, are indispensable for protection of the polyunsaturated fatty acid in addition to have benefits to the meet industry (Rocheford et al., 2002). The same authors have also shown that considerable variation is present among different maize inbreds from tocophenol levels, as well as different ratios of α-tocophenol to γ-tocophenol. This result suggested that breeders can use natural varieties, molecular marker assisted selection strategies and transgenic technologies to alter overall level of tocophenols and ratio of α- to γ-tocophenol. However, current nutritional research on the relative and unique benefits of α- to γ-tocophenol should be considered in developing breeding strategies to alter levels of these vitamin E compounds.

Another area in which transgenic approaches may help solve an important problem with maize as a feed grain is in the reduction of phytic acid levels. In maize, 80% of the total phosphorous (P) is found as phytic acid, and most of that is in the germ (O'Dell et al., 1972). Phytate P is very poorly digested by non-ruminant animals, therefore inorganic supplementation is necessary. Phytate is also a strong chelator that reduces the bioavailability of several other essential minerals such as Ca, Zn, Cu, Mn, and Fe. In addition, since the phytate in the diet is poorly digested, the excrement of monogastric animals (e.g. poultry and pigs), is rich in P and this contributes significantly to environmental pollution. *Low phytic acid* mutants (*lpa*) of maize are available; these have received considerable attention by breeders in order to develop commercially acceptable hybrids with reduced levels of phytic acid (Raboy, 2009).

In maize, several mutants with low levels of phytate have been isolated and mapped; this includes *lpa 1-1*, *lpa 2-1*, and *lpa 241*, (Raboy, 2009). The *lpa1* mutant does not accumulate *myo*-inositol monophosphate or polyphosphate intermediates. It has been proposed that *lpa1* is a mutation in *myo*-inositol supply, the first part of the phytic acid biosynthesis pathway (Raboy et al., 2000). The *lpa2* mutant has reduced phytic acid content in seeds and accumulates *myo*-inositol phosphate intermediates. Maize *lpa2* gene encodes a *myo*-inositol phosphate kinase that belongs to the Ins(1,3,4)P3 5/6-kinase gene family (Shi et al., 2003). The *lpa3* mutant seeds have reduced phytic acid content and accumulate *myo*-inositol, but not *myo*-inositol phosphate intermediates was found to encode *myo*-inositol kinase (Shi et al., 2005).

Despite efforts to elucidate and manipulate phytic acid biosynthesis, *low phytic acid* mutants have limited value to breeders because of adverse effects on agronomic traits such as low germination rates, reduced seed weight (*lpa1-1*), stunted vegetative growth and impaired seed development (*lpa241*). However, Shi et al. (2007) have recently identified the gene disrupted in maize *lpa1* mutants as a multidrug resistance-associated protein (MRP) ATP-binding cassette (ABC) transporter. Silencing expression of this transporter using the embryo-specific globuline promoter produced low-phytic acid, high phosphate transgenic maize seeds that germinate normally and do not show any significant reduction in seed dry weight.

To increase the amount of bioavailable iron in maize, Drakakaki et al. (2005) have generated transgenic maize plants expressing aspergillus phytase and iron-binding protein ferritin. This strategy has proven effective for increasing iron availability and enhancing its absorption. However, much work is still to be done to transfer this technology to tropical and subtropical maize genotypes normally grown in the areas of greatest need for enhanced iron content maize.

A relatively new area in plant biotechnology is the use of genetically-engineered maize to produce high-value end products such as vaccines, therapeutic proteins, industrial enzymes and specialty chemicals (see Hood & Howard, 2009 for a review). The long-term commercial expectations for this use of "plants as factories", often also called "molecular farming", are large. Transgenic maize seed has many attractive features for this purpose, including: i) well-suited for the production and storage of recombinant proteins; ii) ease of scale-up to essentially an infinite capacity; iii) well-established infrastructure for producing, harvesting, transporting, storing, and processing; iv) low cost of production; v) freedom from animal pathogenic contaminants; vi) relative ease of producing transgenic plants which express foreign proteins of interest. However, there is a need, apart the public issues related with the acceptance of genetically-engineered maize, for continued efforts in increasing expression in order to reduce cost effectiveness for products at protein accumulation levels in transgenic plants to broaden this new uses.

5. Conclusion and future perspectives

Two prominent features of agriculture in the 20th century have been the use of breeding and genetics to boost crop productivity and the use of agricultural chemicals to protect crops and enhance plant growth. In the 21st century, crops must produce good yields while conserving land, water, and labor resources. At the same time, industries and consumers require plants with an improved and novel variation in grain composition. We expect that genomics will bolster plant biochemistry as researchers seek to understand the metabolic pathways for the synthesis of these compounds. Identifying rate-limiting steps in synthesis could provide targets for genetically engineering biochemical pathways to produce augmented amounts of compounds and new compounds. Targeted expression will be used to channel metabolic flow into new pathways, while gene-silencing tools will reduce or eliminate undesirable compounds or traits. Therefore, developing plants with improved grain quality traits involves overcoming a variety of technical challenges inherent in metabolic engineering programs.

Metabolism is one of the most important and best recognized networks within biological systems. However, advances in the understanding of metabolic regulation still suffer from insufficient research concerning the modular operation of such networks. Elucidation of metabolic regulation within the context of the entire system, including transcriptional, translational and posttranslational mechanisms, is rarely attempted (Sweetlove et al., 2008).

Instead, to date, studies on metabolic regulation have mostly been limited to regulatory interactions within the metabolic pathways themselves (Sweetlove & Fernie, 2005; Sweetlove et al., 2008). Strategies to detect intermediary metabolic fluxes can now be estimated by computer- aided modeling of the central metabolic network and by mapping the pattern of metabolic fluxes underlying, via the possibility of labeling data collected by NMR and GC-MS and the biomass composition. For example Alonso et al., (2011), to map the pattern of metabolic fluxes underlying this efficiency, have labeled maize embryos to isotopic steady state using a combination of labeled ^{13}C-substrates. The resultant flux map reveals that even though 36% of the entering C goes through the oxidative pentose-phosphate pathway; this does not fully meet the NADPH demands for fatty acid synthesis. Metabolic flux analysis and enzyme activities have highlighted the importance of plastidic NADP-dependent malic enzyme, which provides one-third of the C and NADPH required for fatty acid synthesis in developing maize embryos.

It should worth to be mentioned that metabolic engineering of maize has been relatively slow due to the difficulty of maize transformation. Maize transformation with *Agrobacterium* is now more efficient than currently used particle gun transformation (reviewed in Jones, 2009; Reyes et al., 2010). In addition, larger DNA fragments can be inserted with *Agrobacterium* than those previously reported by other methods. The ability to routinely insert metabolic pathway quantities of DNA into the maize genome will further speed up maize metabolic engineering. Furthermore, site-directed mutagenesis via gene targeting, based on homologous recombination such as the application of designed zinc finger nucleases that induce a double stranded break at their target locus, are promising tools for genetic applications (Shukla et al., 2009; Saika et al., 2011). The use of these technologies may lead to both targeted mutagenesis and targeted gene replacement at remarkably high frequencies and enable to modify useful information, acquired from structural- and computational-based protein engineering, to be applied directly to molecular breeding of crops, including metabolic engineering.

Advances in plant genetics and genomic technologies are also contributing to the acceleration of gene discovery for maize product development. In the past few years there has been much progress in the development of strategies to discover new plant genes. In large part, these developments derive from four experimental approaches: firstly, genetic and physical mapping in plants and the associated ability to use map-based gene isolation strategies; secondly, transposon tagging which allows the direct isolation of a gene via forward and reverse genetic strategies as well as the development of the Targeting Induced Local Lesions IN Genomes (TILLING) technique; thirdly, protein-protein interaction cloning, that permits the isolation of multiple genes contributing to a single pathway or metabolic process. Finally, through bioinformatics/genomics, the development and use of large ESTs databases (http://www.maizegdb.org) and, DNA microarray technology to investigate mRNA-level controls of complex pathways. Moreover, new technologies and information continue to increase our understanding of maize; for instance, the complete DNA sequence of the maize genome, along with comprehensive transcriptome, proteome, metabolome, and epigenome information, is also a key resource for advancing fundamental knowledge of the biology of development seed quality-related traits to be applied in molecular breeding and biotechnology. These additional layers of information should help to further unravel the complexities of how genes and gene networks function to give plants including quality-traits. This knowledge will drive to improved predictions and capacities to assemble gene variation through molecular breeding as well as more optimal gene selection and regulation in the development of future biotechnology products.

In conclusion, although, conventional breeding, molecular marker assisted breeding, and genetic engineering have already had, and will continue to have, important roles in maize improvement, the rapidly expanding information from genomics and genetics combined with improved genetic engineering technology offer a wide range of possibilities for the improvement of the maize grain.

6. Acknowledgements

We apologize to all those whose contributions we did not include in this review because of space constraints, personal preferences, or simple oversight. In addition, we would like to thank the members of the laboratory for their research contributions which are described here. The work was supported by Ministero per le Politiche Agricole, Alimentari e Forestali, Rome.

7. References

Alonso, P.A., Val, D.L . & Shachar-Hill, Y.. (2010) Understanding fatty acid synthesis in developing maize embryos using metabolic flux analysis *Metabolic Engineering* , Vol. 12 No. 5 , pp. 488-497

Alonso, A.P., Val, D.L. & Shachar-Hill, Y.. (2011) Central metabolic fluxes in the endosperm of developing maize seeds and their implications for metabolic engineering *Metabolic Engineering* , Vol. 13 No. 1 , pp. 96-107

Aluru, M., Xu, Y., Guo, R., Wang, Z., Li, S., White, W., Wang, K. & Rodermel, S.. (2008) Generation of transgenic maize with enhanced provitamin A content *Journal of Experimental Botany* , Vol. 59 No. 13 , pp. 3551 -3562

Anderson, P.C., Chomet, P.S., Griffor, M.C. & Kriz, A.L.. (1997) Anthranilate synthase gene and method of use thereof for conferring tryptophan overproduction

Arruda, P., Kemper, E.L., Papes, F. & Leite, A.. (2000) Regulation of lysine catabolism in higher plants *Trends in Plant Science* , Vol. 5 No. 8 , pp. 324-330

Azevedo, R.A., Arruda, P., Turner, W.L. & Lea, P.J.. (1997) The biosynthesis and metabolism of the aspartate derived amino acids in higher plants *Phytochemistry* , Vol. 46 No. 3 , pp. 395-419

Bicar, E.H., Woodman-Clikeman, W., Sangtong, V., Peterson, J.M., Yang, S.S., Lee, M. & Scott, M.P.. (2007) Transgenic maize endosperm containing a milk protein has improved amino acid balance *Transgenic Research* , Vol. 17 No. 1 , pp. 59-71

Botella-Pavía, P. & Rodríguez-Concepción, M.. (2006) Carotenoid biotechnology in plants for nutritionally improved foods *Physiologia Plantarum* , Vol. 126 No. 3 , pp. 369-381

Boyer, C. & Hannah, L.. (2001) Kernel mutants of Corn, In: *Specialty Corns. 2nd ed.* , A.R. Hallauer , pp. 1-32 , C.R.C. Press, Boca Raton.

Brenna, O.V. & Berardo, N.. (2004) Application of Near-Infrared Reflectance Spectroscopy (NIRS) to the Evaluation of Carotenoids Content in Maize *Journal of Agricultural and Food Chemistry* , Vol. 52 No. 18 , pp. 5577-5582

Brochetto-Braga, M.R., Leite, A. & Arruda, P.. (1992) Partial Purification and Characterization of Lysine-Ketoglutarate Reductase in Normal and Opaque-2 Maize Endosperms *Plant Physiology* , Vol. 98 No. 3 , pp. 1139 -1147

Caimi, P.G., McCole, L.M., Klein, T.M. & Kerr, P.S.. (1996) Fructan Accumulation and Sucrose Metabolism in Transgenic Maize Endosperm Expressing a Bacillus amyloliquefaciens SacB Gene *Plant Physiology* , Vol. 110 No. 2 , pp. 355 -363

Chander, S., Guo, Y.Q., Yang, X.H., Zhang, J., Lu, X.Q., Yan, J.B., Song, T.M., Rocheford, T.R. & Li, J.S.. (2007) Using molecular markers to identify two major loci controlling carotenoid contents in maize grain *Theor. Appl. Genet.* , Vol. 116 No. 2 , pp. 223-233

Chiang, Chih-Ming., Yeh, Feng-Shi., Huang, Li-Fen., Tseng, Tung-Hi., Chung, Mei-Chu., Wang, Chang-Sheng., Lur, Hu-Shen., Shaw, Jei-Fu. & Yu, Su-May.. (2005) Expression of a bi-functional and thermostable amylopullulanase in transgenic rice seeds leads to autohydrolysis and altered composition of starch *Molecular Breeding* , Vol. 15 No. 2 , pp. 125-143

Doehlert, D. & Lambert, R.. (1991) Metabolic characteristics associated with starch, protein, and oil deposition in developing maize kernels *Crop Sci.* , Vol. 31 , pp. 151-157

Drakakaki, G., Marcel, S., Glahn, R.P., Lund, E.K., Pariagh, S., Fischer, R., Christou, P. & Stoger, E.. (2005) Endosperm-Specific Co-Expression of Recombinant Soybean Ferritin and Aspergillus Phytase in Maize Results in Significant Increases in the Levels of Bioavailable Iron *Plant Molecular Biology* , Vol. 59 No. 6 , pp. 869-880

Dudley, J. & Lambert, R.. (1992) Ninety generations of selection for oil and protein in maize *Maydica* , Vol. 37 , pp. 1-7

Falco, S.C.., Guida, T., Locke, M., Mauvais, J., Sanders, C., Ward, R.T.. & Webber, P.. (1995) Transgenic Canola and Soybean Seeds with Increased Lysine *Nat Biotech* , Vol. 13 No. 6 , pp. 577-582

Fraser, P.D. & Bramley, P.M. (2004) The biosynthesis and nutritional uses of carotenoids *Progress in Lipid Research* , Vol. 43 No. 3 , pp. 228-265

Fu, F-F. & Xue, H-W. (2010) Coexpression Analysis Identifies Rice Starch Regulator1, a Rice AP2/EREBP Family Transcription Factor, as a Novel Rice Starch Biosynthesis Regulator *Plant Physiology* , Vol. 154 No. 2 , pp. 927 -938

Galili, G.. (2004) New insights into the regulation and functional significance of lysine metabolism in plants *Ann. Rev. Plant Biol.* , Vol. 53 No. 1 , pp. 27-43

Galpaz, N., Wang, Q., Menda, N., Zamir, D. & Hirschberg, J.. (2008) Abscisic acid deficiency in the tomato mutant high-pigment 3 leading to increased plastid number and higher fruit lycopene content *The Plant Journal* , Vol. 53 No. 5 , pp. 717-730

Gibbon, B.C. & Larkins, B.A.. (2005) Molecular genetic approaches to developing quality protein maize *Trends in Genetics* , Vol. 21 No. 4 , pp. 227-233

Giroux, M.J., Shaw, J., Barry, G., Cobb, B.G., Greene, T., Okita, T. & Hannah, L.C.. (1996) A single mutation that increases maize seed weight *Proceedings of the National Academy of Sciences* , Vol. 93 No. 12 , pp. 5824 -5829

Gygi, S., Rochon, Y., Franza, B. & Aebersold, R.. (1999) Correlation between protein and mRNA abundance in yeast *Molecular and cellular biology* , Vol. 19 No. 3 , pp. 1720

Hannah, L.C.. (2007) Starch Formation in the Cereal Endosperm, In: *Endosperm* , Olsen, Odd Arne , pp. 179-193 , Springer Berlin Heidelberg

Hannah, L. & James, M.. (2008) The complexities of starch biosynthesis in cereal endosperms *Current Opinion in Biotechnology* , Vol. 19 No. 2 , pp. 160-165

Harjes, C.E., Rocheford, T.R., Bai, L., Brutnell, T.P., Kandianis, C.B., Sowinski, S.G., Stapleton, A.E., Vallabhaneni, R., Williams, M., Wurtzel, E.T., Yan, J. & Buckler, E.S.. (2008) Natural Genetic Variation in Lycopene Epsilon Cyclase Tapped for Maize Biofortification *Science* , Vol. 319 No. 5861 , pp. 330 -333

Hartings, H., Lauria, M., Lazzaroni, N., Pirona, R. & Motto, M. (2011) The Zea mays mutants opaque-2 and opaque-7 disclose extensive changes in endosperm metabolism as revealed by protein, amino acid, and transcriptome-wide analyses *BMC genomics* , Vol. 12 No. 1 , pp. 41

Holding, D.R., Hunter, B.G., Chung, T., Gibbon, B.C., Ford, C.F., Bharti, A.K., Messing, J., Hamaker, B.R. & Larkins, B.A.. (2008) Genetic analysis of opaque2 modifier loci in quality protein maize *Theoretical and Applied Genetics* , Vol. 117 No. 2 , pp. 157-170

Holding, D.R. & Larkins, B.A.. (2009) Zein Storage Proteins, In: *Molecular Genetic Approaches to Maize Improvement* , Kriz, Alan L. and Larkins, Brian A. , pp. 269-286 , Springer Berlin Heidelberg

Holding, D.R., Meeley, R.B., Hazebroek, J., Selinger, D., Gruis, F., Jung, R. & Larkins, B.A.. (2010) Identification and characterization of the maize arogenate dehydrogenase gene family *Journal of Experimental Botany* , Vol. 61 No. 13 , pp. 3663 –3673

Hood, E. & Howard, J.. (2009) Over-expression of novel proteins in maize, In: *Molecular genetic approaches to maize improvement* , A.L. Kriz, B.A. Larkins , pp. 91-105 , Springer Verlag, Berlin Heidelberg.

Houmard, N.M., Mainville, J.L., Bonin, C.P., Huang, S., Luethy, M.H. & Malvar, T.M.. (2007) High-lysine corn generated by endosperm-specific suppression of lysine catabolism using RNAi *Plant Biotechnology Journal* , Vol. 5 No. 5 , pp. 605-614

Huang, S., Adams, W.R., Zhou, Q., Malloy, K.P., Voyles, D.A., Anthony, J., Kriz, A.L. & Luethy, M.H.. (2004) Improving Nutritional Quality of Maize Proteins by Expressing Sense and Antisense Zein Genes *Journal of Agricultural and Food Chemistry* , Vol. 52 No. 7 , pp. 1958-1964

Hunter, B.G., Beatty, M.K., Singletary, G.W., Hamaker, B.R., Dilkes, B.P., Larkins, B.A. & Jung, R.. (2002) Maize Opaque Endosperm Mutations Create Extensive Changes in Patterns of Gene Expression *The Plant Cell Online* , Vol. 14 No. 10 , pp. 2591 -2612

John, P.C.L. (2007) Hormonal Regulation of Cell Cycle Progression and its Role in Development, In: *Cell cycle control and plant development* , Inze D. , pp. 311–334 , Blackwell, Oxford

Jung, R. & Falco, S.. (2000) Transgenic corn with an improved amino acid composition *8th Intl. Sump. on Plant Seeds, Gatersleben, Germany*

Koch, K. (2004) Sucrose metabolism: regulatory mechanisms and pivotal roles in sugar sensing and plant development *Current Opinion in Plant Biology* , Vol. 7 No. 3 , pp. 235-246

Kriz, Alan, L.. (2009) Enhancement of Amino Acid Availability in Corn Grain, In: *Molecular Genetic Approaches to Maize Improvement* , Kriz, Alan L. and Larkins, Brian A. , pp. 79-89 , Springer Berlin Heidelberg

Kubo, A., Colleoni, C., Dinges, J.R., Lin, Q., Lappe, R.R., Rivenbark, J.G., Meyer, A.J., Ball, S.G., James, M.G., Hennen-Bierwagen, T.A. & Myers, A.M.. (2010) Functions of Heteromeric and Homomeric Isoamylase-Type Starch-Debranching Enzymes in Developing Maize Endosperm *Plant Physiology* , Vol. 153 No. 3 , pp. 956 –969

Kurilich, A.C. & Juvik, J.A.. (1999) Quantification of Carotenoid and Tocopherol Antioxidants in Zea mays *Journal of Agricultural and Food Chemistry* , Vol. 47 No. 5 , pp. 1948-1955

Lai, J., Dey, N., Kim, Cheol-Soo., Bharti, A.K., Rudd, S., Mayer, K. F.X.., Larkins, B.A., Becraft, P. & Messing, J.. (2004) Characterization of the Maize Endosperm

Transcriptome and Its Comparison to the Rice Genome *Genome Research* , Vol. 14 No. 10a , pp. 1932 -1937

Lambert, R.. (2001) High-oil corn hybrids, In: *Specialty Corns. 2nd ed.* , Hallauer, A.R. , pp. pp 131-154 , C.R.C. Press, Boca Raton.

Laurie, C.C., Chasalow, S.D., LeDeaux, J.R., McCarroll, R., Bush, D., Hauge, B., Lai, C., Clark, D., Rocheford, T.R. & Dudley, J.W.. (2004) The Genetic Architecture of Response to Long-Term Artificial Selection for Oil Concentration in the Maize Kernel *Genetics* , Vol. 168 No. 4 , pp. 2141 -2155

Lawton, J.W.. (2002) Zein: A History of Processing and Use *Cereal Chemistry* , Vol. 79 No. 1 , pp. 1-18

Lending, C. & Larkins, B.. (1989) Changes in the zein composition of protein bodies during maize endosperm development *The Plant Cell Online* , Vol. 1 No. 10 , pp. 1011

Li, F., Murillo, C. & Wurtzel, E.T.. (2007) Maize Y9 Encodes a Product Essential for 15-cis-ζ-Carotene Isomerization *Plant Physiology* , Vol. 144 No. 2 , pp. 1181 -1189

Li, F., Vallabhaneni, R., Yu, J., Rocheford, T. & Wurtzel, E.. (2008) The maize phytoene synthase gene family: overlapping roles for carotenogenesis in endosperm, photomorphogenesis, and thermal stress tolerance *Plant physiology* , Vol. 147 No. 3 , pp. 1334

Liu, X., Fu, J., Gu, D., Liu, W., Liu, T., Peng, Y., Wang, J. & Wang, G.. (2008) Genome-wide analysis of gene expression profiles during the kernel development of maize (Zea mays L.) *Genomics* , Vol. 91 No. 4 , pp. 378-387

Llop-Tous, I., Madurga, S., Giralt, E., Marzabal, P., Torrent, M. & Ludevid, M.D.. (2010) Relevant Elements of a Maize γ-Zein Domain Involved in Protein Body Biogenesis *Journal of Biological Chemistry* , Vol. 285 No. 46 , pp. 35633 –35644

Locatelli, S., Piatti, P., Motto, M. & Rossi, V.. (2009) Chromatin and DNA Modifications in the Opaque2-Mediated Regulation of Gene Transcription during Maize Endosperm Development *The Plant Cell Online* , Vol. 21 No. 5 , pp. 1410 -1427

Maeo, K., Tokuda, T., Ayame, A., Mitsui, N., Kawai, T., Tsukagoshi, H., Ishiguro, S. & Nakamura, K.. (2009) An AP2-type transcription factor, WRINKLED1, of Arabidopsis thaliana binds to the AW-box sequence conserved among proximal upstream regions of genes involved in fatty acid synthesis *The Plant Journal* , Vol. 60 No. 3 , pp. 476-487

Mahmoud, S. & Croteau, R.. (2001) Metabolic engineering of essential oil yield and composition in mint by altering expression of deoxyxylulose phosphate reductoisomerase and menthofuran synthase *Proceedings of the National Academy of Sciences* , Vol. 98 No. 15 , pp. 8915

Matthews, P. & Wurtzel, E.. (2000) Metabolic engineering of carotenoid accumulation in Escherichia coli by modulation of the isoprenoid precursor pool with expression of deoxyxylulose phosphate synthase *Applied microbiology and biotechnology* , Vol. 53 No. 4 , pp. 396-400

Matthews, P. & Wurtzel, E.. (2007) Biotechnology of food colorant production., In: *Food colorants: chemical and functional properties* , C. Socaciu , pp. 347-398 , CRC Press: Boca Raton, FL

Mazur, B., Krebbers, E. & Tingey, S.. (1999) Gene discovery and product development for grain quality traits *Science* , Vol. 285 No. 5426 , pp. 372 -375

Miclaus, M., Xu, J. & Messing, J.. (2011) Differential gene expression and epiregulation of alpha zein gene copies in maize haplotypes. *PLoS Genetics* , Vol. 7 No. 6 , pp. e1002131

Miclaus, M., Wu, Y., Xu, Y., Dooner, H. & Messing, J.. (2011) *Opaque7* encodes an acyl-CoA synthetase-like protein in maize. *Genetics (in press)*

Moose, S.P., Dudley, J.W. & Rocheford, T.R.. (2004) Maize selection passes the century mark: a unique resource for 21st century genomics *Trends in Plant Science* , Vol. 9 No. 7 , pp. 358-364

Motto, M., Balconi, C., Hartings, H., Lauria, M. & Rossi, V.. (2009) Improvment of quality-related traits in maize grain: gene identification and exploitation *Maydica* , Vol. 54 No. 2/3 , pp. 321-342

Myers, A., James, M., Lin, Q., Yi, G., Stinard, P., Hennen-Bierwagen, T. & Becraft, P.. (2011) Maize *opaque5* encodes monogalactosyldiacylglycerol synthase and specifically affects galactolipids necessary for amyloplast and chloroplast function *Plant Cell,* ,Vol. 23 No 6, pp: 2331-2347

Myers, A.M., Morell, M.K., James, M.G. & Ball, S.G.. (2000) Recent Progress toward Understanding Biosynthesis of the Amylopectin Crystal *Plant Physiology* , Vol. 122 No. 4 , pp. 989 -998

Méchin, V., Balliau, T., Château-Joubert, S., Davanture, M., Langella, O., Négroni, L., Prioul, Jean-Louis., Thévenot, C., Zivy, M. & Damerval, C.. (2004) A two-dimensional proteome map of maize endosperm *Phytochemistry* , Vol. 65 No. 11 , pp. 1609-1618

Naqvi, S., Zhu, C., Farre, G., Ramessar, K., Bassie, L., Breitenbach, J., Perez Conesa, D., Ros, G., Sandmann, G., Capell, T. & Christou, P.. (2009) Transgenic multivitamin corn through biofortification of endosperm with three vitamins representing three distinct metabolic pathways *Proceedings of the National Academy of Sciences* , Vol. 106 No. 19 , pp. 7762 -7767

O'Dell, B.L., De Boland, A.R. & Koirtyohann, S.R.. (1972) Distribution of phytate and nutritionally important elements among the morphological components of cereal grains *Journal of Agricultural and Food Chemistry* , Vol. 20 No. 3 , pp. 718-723

Oakes, J., Brackenridge, D., Colletti, R., Daley, M., Hawkins, D.J., Xiong, H., Mai, J., Screen, S.E., Val, D., Lardizabal, K., Gruys, K. & Deikman, J.. (2011) Expression of Fungal diacylglycerol acyltransferase2 Genes to Increase Kernel Oil in Maize *Plant Physiology* , Vol. 155 No. 3 , pp. 1146 -1157

Pouvreau, B., Baud, S., Vernoud, V., Morin, V., Py, C., Gendrot, G., Pichon, Jean-Philippe., Rouster, J., Paul, W. & Rogowsky, P.M.. (2011) Duplicate Maize Wrinkled1 Transcription Factors Activate Target Genes Involved in Seed Oil Biosynthesis *Plant Physiology* , Vol. 156 No. 2 , pp. 674 -686

Pozniak, C., Knox, R., Clarke, F. & Clarke, J.. (2007) Identification of QTL and association of a phytoene synthase gene with endosperm colour in durum wheat *TAG Theoretical and Applied Genetics* , Vol. 114 No. 3 , pp. 525-537

Prioul, J.L., Méchin, V., Lessard, P., Thévenot, C., Grimmer, M., Chateau-Joubert, S., Coates, S., Hartings, H., Kloiber-Maitz, M., Murigneux, A., Sarda, X., Damerval, C. & Edwards, K.J.. (2008) A joint transcriptomic, proteomic and metabolic analysis of maize endosperm development and starch filling *Plant Biotechnology Journal* , Vol. 6 No. 9 , pp. 855-869

Raboy, V., Gerbasi, P.F., Young, K.A., Stoneberg, S.D., Pickett, S.G., Bauman, A.T., Murthy, P. P.N.., Sheridan, W.F. & Ertl, D.S.. (2000) Origin and Seed Phenotype of Maize low phytic acid 1-1 and low phytic acid 2-1 *Plant Physiology* , Vol. 124 No. 1 , pp. 355 -368

Raboy, V.. (2009) Seed Total Phosphate and Phytic Acid, In: *Molecular Genetic Approaches to Maize Improvement* , Kriz, Alan L. and Larkins, Brian A. , pp. 41-53 , Springer Berlin Heidelberg

Rascon-Cruz, Q., Sinagawa-Garcia, S., Osuna-Castro, J., Bohorova, N. & Paredes-López, O.. (2004) Accumulation, assembly, and digestibility of amarantin expressed in transgenic tropical maize *TAG Theoretical and Applied Genetics* , Vol. 108 No. 2 , pp. 335-342

Reyes, A.R., Bonin, C.P., Houmard, N.M., Huang, S. & Malvar, T.M.. (2008) Genetic manipulation of lysine catabolism in maize kernels *Plant Molecular Biology* , Vol. 69 No. 1-2 , pp. 81-89

Reyes, F., Sun, B., Guo, H., Gruis, D. & Otegui, M.. (2010) Agrobacterium tumefaciens-mediated transformation of maize endosperm as a tool to study endosperm cell biology *Plant physiology* , Vol. 153 No. 2 , pp. 624

Rocheford, T., Wong, J., Egesel, C. & Lambert, R.. (2002) Enhancement of Vitamin E Levels in Corn *J. Am. College Nutr.* , Vol. 21 , pp. 191-198

Saika, H., Oikawa, A., Matsuda, F., Onodera, H., Saito, K. & Toki, S.. (2011) Application of gene targeting to designed mutation breeding of high-tryptophan rice *Plant Physiology* , Vol. 156 No. 3 , pp. 1269

Sandmann, G., Römer, S. & Fraser, P.D.. (2006) Understanding carotenoid metabolism as a necessity for genetic engineering of crop plants *Metabolic Engineering* , Vol. 8 No. 4 , pp. 291-302

Segal, G., Song, R. & Messing, J.. (2003) A New Opaque Variant of Maize by a Single Dominant RNA-Interference-Inducing Transgene *Genetics* , Vol. 165 No. 1 , pp. 387 -397

Shaver, J.M., Bittel, D.C., Sellner, J.M., Frisch, D.A., Somers, D.A. & Gengenbach, B.G.. (1996) Single-amino acid substitutions eliminate lysine inhibition of maize dihydrodipicolinate synthase. *Proceedings of the National Academy of Sciences of the United States of America* , Vol. 93 No. 5 , pp. 1962-1966

Shen, B., Allen, W.B., Zheng, P., Li, C., Glassman, K., Ranch, J., Nubel, D. & Tarczynski, M.C.. (2010) Expression of ZmLEC1 and ZmWRI1 Increases Seed Oil Production in Maize *Plant Physiology* , Vol. 153 No. 3 , pp. 980 -987

Shi, J., Wang, H., Schellin, K., Li, B., Faller, M., Stoop, J.M., Meeley, R.B., Ertl, D.S., Ranch, J.P. & Glassman, K.. (2007) Embryo-specific silencing of a transporter reduces phytic acid content of maize and soybean seeds *Nat Biotech* , Vol. 25 No. 8 , pp. 930-937

Shukla, VK., Doyon, Y., Miller, JC., DeKelver, RC., Moehle, EA., Worden, SE., Mitchell, JC., Arnold, NL., Gopalan, S., Meng, X., Choi, VM., Rock, JM., Wu, YY., Katibah, GE., Zhifang, G., McCaskill, D., Simpson, MA., Blakeslee, B., Greenwalt, SA., Butler, HJ., Hinkley, SJ., Zhang, L., Rebar, EJ., Gregory, PD. & Urnov, FD.. (2009) Precise genome modification in the crop species *Zea mays* using zinc-finger nucleases *Nature* , Vol. 459 No. 7245 , pp. 437-41

Sousa, S.M., Paniago, M. d.G., Arruda, P. & Yunes, J.A.. (2008) Sugar levels modulate sorbitol dehydrogenase expression in maize *Plant Molecular Biology* , Vol. 68 No. 3 , pp. 203-213

Sun, D., Gregory, P. & Grogan, C.O.. (1978) Inheritance of saturated fatty acids in maize *Journal of Heredity* , Vol. 69 No. 5 , pp. 341 -342

Sweetlove, L., Fell, D. & Fernie, A.. (2008) Getting to grips with the plant metabolic network *Biochem. J* , Vol. 409 , pp. 27-41

Sweetlove, L. & Fernie, A.. (2005) Regulation of metabolic networks: understanding metabolic complexity in the systems biology era *New Phytologist* , Vol. 168 No. 1 , pp. 9-24

Val, L.D., Schwartz, S.H., Kerns, M.R. & Deikman, J.. (2009) Development of a High Oil Trait for Maize, In: *Molecular Genetic Approaches to Maize Improvement* , Kriz, Alan L. and Larkins, Brian A. , pp. 303-323 , Springer Berlin Heidelberg

Vallabhaneni, R., Gallagher, C.E., Licciardello, N., Cuttriss, A.J., Quinlan, R.F. & Wurtzel, E.T.. (2009) Metabolite Sorting of a Germplasm Collection Reveals the Hydroxylase3 Locus as a New Target for Maize Provitamin A Biofortification *Plant Physiology* , Vol. 151 No. 3 , pp. 1635 -1645

Vallabhaneni, R. & Wurtzel, E.T.. (2009) Timing and Biosynthetic Potential for Carotenoid Accumulation in Genetically Diverse Germplasm of Maize *Plant Physiology* , Vol. 150 No. 2 , pp. 562 -572

Verza, N.C., Silva, T.R., Neto, G.C., Nogueira, F.T.S., Fisch, P.H., Rosa, V.E., Rebello, M.M., Vettore, A.L., Silva, F.R. & Arruda, P.. (2005) Endosperm-preferred Expression of Maize Genes as Revealed by Transcriptome-wide Analysis of Expressed Sequence Tags *Plant Molecular Biology* , Vol. 59 No. 2 , pp. 363-374

Wan, Y., Poole, R., Huttly, A., Toscano-Underwood, C., Feeney, K., Welham, S., Gooding, M., Mills, C., Edwards, K., Shewry, P. & others. (2008) Transcriptome analysis of grain development in hexaploid wheat *BMC genomics* , Vol. 9 No. 1 , pp. 121

Wang, G., Sun, X., Wang, G., Wang, F., Gao, Q., Sun, X., Tang, Y., Chang, C., Lai§, J., Zhu, L., Xu, Z. & Song, R.. (2011) Opaque7 encodes an AAE3-like protein that affects storage protein synthesis in maize endosperm *Genetics (in press)*

Wang, Z., Chen, X., Wang, J., Liu, T., Liu, Y., Zhao, L. & Wang, G.. (2007) Increasing maize seed weight by enhancing the cytoplasmic ADP-glucose pyrophosphorylase activity in transgenic maize plants *Plant Cell, Tissue and Organ Culture* , Vol. 88 No. 1 , pp. 83-92

Washida, H., Sugino, A., Kaneko, S., Crofts, N., Sakulsingharoj, C., Kim, D., Choi, Sang-Bong., Hamada, S., Ogawa, M., Wang, C., Esen, A., Higgins, T.J.V. & Okita, T.W.. (2009) Identification of cis-localization elements of the maize 10-kDa delta-zein and their use in targeting RNAs to specific cortical endoplasmic reticulum subdomains *The Plant Journal: For Cell and Molecular Biology* , Vol. 60 No. 1 , pp. 146-155

White, P.J. & Johnson, L.A.. (2003) *Corn: chemistry and technology.* , American Association of Cereal Chemists

Wu, X.R., Kenzior, A., Willmot, D., Scanlon, S., Chen, Z., Topin, A., He, S.H., Acevedo, A. & Folk, W.R.. (2007) Altered expression of plant lysyl tRNA synthetase promotes tRNA misacylation and translational recoding of lysine *The Plant Journal* , Vol. 50 No. 4 , pp. 627-636

Wu, Y., Holding, D. & Messing, J.. (2010) $-Zeins are essential for endosperm modification in quality protein maize *Proceedings of the National Academy of Sciences* , Vol. 107 No. 29 , pp. 12810

Wu, Y. & Messing, J.. (2011) Novel Genetic Selection System for Quantitative Trait Loci of Quality Protein Maize *Genetics*

Wu, Y. & Messing, J.. (2010) RNA Interference-Mediated Change in Protein Body Morphology and Seed Opacity through Loss of Different Zein Proteins *Plant Physiology*, Vol. 153 No. 1 , pp. 337 –347

Wurtzel, E.T., Rocheford, T.R., Wong, J.C. & Lambert, R.J.. (2004) QTL and candidate genes phytoene synthase and ?-carotene desaturase associated with the accumulation of carotenoids in maize *TAG Theoretical and Applied Genetics*, Vol. 108 No. 2 , pp. 349-359

Yan, J., Kandianis, C.B., Harjes, C.E., Bai, L., Kim, Eun-Ha., Yang, X., Skinner, D.J., Fu, Z., Mitchell, S., Li, Q., Fernandez, M.G.S., Zaharieva, M., Babu, R., Fu, Y., Palacios, N., Li, J., DellaPenna, D., Brutnell, T., Buckler, E.S., Warburton, M.L. & Rocheford, T.. (2010) Rare genetic variation at Zea mays crtRB1 increases [beta]-carotene in maize grain *Nat Genet*, Vol. 42 No. 4 , pp. 322-327

Yu, J., Peng, P., Zhang, X., Zhao, Q., Zhy, D., Sun, X., Liu, J. & Ao, G.. (2004) Seed-specific expression of a lysine rich protein sb401 gene significantly increases both lysine and total protein content in maize seeds *Molecular Breeding*, Vol. 14 No. 1 , pp. 1-7

Zhang, J., Martin, J.M., Beecher, B., Lu, C., Hannah, L.C., Wall, M.L., Altosaar, I. & Giroux, M.J.. (2010) The ectopic expression of the wheat Puroindoline genes increase germ size and seed oil content in transgenic corn *Plant Molecular Biology*, Vol. 74 No. 4-5 , pp. 353-365

Zhang, S., Dong, J., Wang, T., Guo, S., Glassman, K., Ranch, J. & Nichols, S.. (2007) High level accumulation of $-glucan in maize kernels by expressing the gtfD gene from Streptococcus mutans *Transgenic research*, Vol. 16 No. 4 , pp. 467-478

Zheng, P., Allen, W.B., Roesler, K., Williams, M.E., Zhang, S., Li, J., Glassman, K., Ranch, J., Nubel, D., Solawetz, W., Bhattramakki, D., Llaca, V., Deschamps, S., Zhong, Gan-Yuan., Tarczynski, M.C. & Shen, B.. (2008) A phenylalanine in DGAT is a key determinant of oil content and composition in maize *Nat Genet*, Vol. 40 No. 3 , pp. 367-372

Zhu, C., Naqvi, S., Capell, T. & Christou, P.. (2009) Metabolic engineering of ketocarotenoid biosynthesis in higher plants *Archives of Biochemistry and Biophysics*, Vol. 483 No. 2 , pp. 182-190

12

Expression of Sweet Potato Senescence-Associated Cysteine Proteases Affect Seed and Silique Development and Stress Tolerance in Transgenic *Arabidopsis*

Hsien-Jung Chen[1], Guan-Jhong Huang[2], Chia-Hung Lin[1],
Yi-Jing Tsai[1], Zhe-Wei Lin[1], Shu-Hao Liang[1] and Yaw-Huei Lin[3]
[1]*Department of Biological Sciences, National Sun Yat-sen University, Kaohsiung*
[2]*School of Chinese Pharmaceutical Sciences and Chinese Medicine Resources,*
College of Pharmacy, China Medical University, Taichung
[3]*Institute of Plant and Microbial Biology, Academia Sinica, Nankang, Taipei*
Taiwan

1. Introduction

Leaf is in general the main site of photosynthesis and acts as a carbohydrate source for nutrients to support the growth in sink organs of plants. Therefore, its longevity and senescent level may affect the photosynthesis efficiency and thus crop yield. There are endogenous and exogenous factors affecting leaf senescence, including plant growth regulators, sucrose starvation, dark, cold, drought, salt, wound, pathogen infection and insect attack (Yoshida, 2003; Lim et al., 2007). Leaf senescence is the final developmental stage of leaves and has been considered as a type of programmed cell death. During leaf senescence, it is not only a degradative process but also a recycling one. Therefore, macromolecules and organelles can be degraded into small molecules, salvaged and mobilized from the senescent cells to other sinks, such as young leaves, developing seeds, or storage tissues (Buchanan-Wollaston, 1997; Quirino et al., 2000).

In sweet potato, several morphological, biochemical and physiological changes have also be observed during leaf senescence, including leaf yellowing, decrease of chlorophyll contents, reduction of photochemical Fv/Fm, elevation of H_2O_2 amount, increase of plastoglobuli number in chloroplast, activation of senescence-associated gene expression, and finally cell death (Chen et al., 2000; Chen et al., 2003; Chen et al. 2010a). Several full-length cDNAs encoding putative isocitrate lyase, papain-like cysteine proteases and asparaginyl endopeptidase, have been cloned from senescent leaves (Chen et al., 2000, 2004, 2006, 2008, 2009, 2010b), which likely play roles in association with lipid degradation and gluconeogenesis, and protein degradation and re-mobilization. These data support the occurrence of macromolecule and organelle degradation into small molecules for recycling and re-mobilization during sweet potato leaf senescence.

During senescence, breakdown of leaf proteins by proteases provides a large pool of cellular nitrogen for recycling (Makino & Osmond, 1991). In plants, different degradation pathways have been described and the vacuolar degradation pathway is assumed to be involved in bulk protein degradation by virtue of the resident proteases in the vacuole (Vierstra, 1996). There are two types of vacuoles described in plants: the storage vacuole and the lytic vacuole (Marty, 1999). Protein storage vacuoles are found in seed tissues and accumulate proteins that are re-mobilized and used as the main nutrient resource for germination (Senyuk et al., 1998; Schlereth et al., 2001). Most cells in vegetative tissues have lytic vacuoles, containing a wide range of proteases in an acidic environment. Substrate proteins must be transported and sequestered into these lytic vacuoles before degradation. Therefore, senescence-associated vacuoles are lytic vacuoles and involved in the degradation of imported chloroplast proteins in tobacco leaves (Martı́nez et al., 2008).

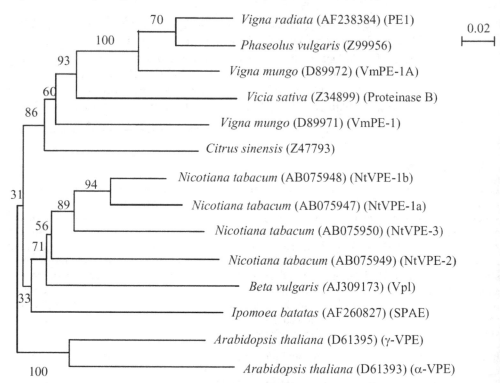

Fig. 1. Phylogenetic tree analysis of plant asparaginyl endopeptidases (Adapted and Modified from Chen et al., 2004).

The molecular mechanisms for vacuolar protein degradation and nutrient recycling pathway in senescent leaves are generally not clear. Phylogenetic tree analysis indicated that sweet potato asparaginyl endopeptidase (SPAE) exhibited high amino acid sequence identities and closely-related association with plant vacuolar processing enzymes (VPEs) or legumains, including legumain-like protease LLP of kidney bean (*Phaseolus vulgaris*), legumain-like protease VsPB2 of vetch (*Vicia sativa*), vacuolar processing enzymes of *Arabidopsis thaliana*, and asparaginyl endopeptidases VmPE-1 of *Vigna mungo* (Fig. 1). Sweet potato papain-like cysteine protease (SPCP2) also showed high amino acid sequence identities and closely-related association with a subgroup of cysteine proteases, including *Actinidia deliciosa* CP3, *Arabidopsis thaliana* RD19, *Brassica oleracea* BoCP4, *Phaseolus vulgaris* CP2, *Solanum melongena* SmCP, *Vicia sativa* CPR2, and *Vigna mungo* SH-EP (Fig. 2). These data suggest the possible physiological roles and functions for *SPAE* and *SPCP2* related to these mentioned vacuolar processing enzymes and papain-like cysteine proteases, respectively.

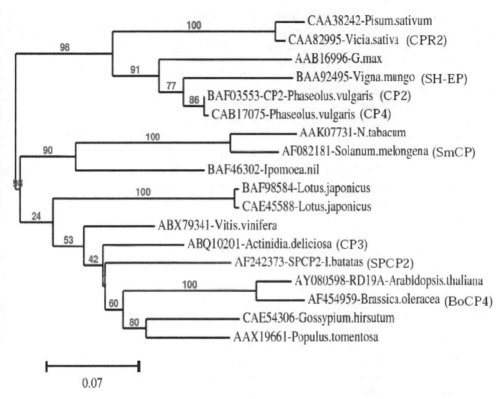

Fig. 2. Phylogenetic tree analysis of plant papain-like cysteine proteases (Adapted from Lin, 2010).

2. Association of vacuolar processing enzyme and papain-like cysteine protease with seed storage globulin protein degradation

Vacuolar processing enzyme is a novel group of cysteine endopeptidase and has recently been found in seeds. The enzyme exhibits strict cleavage specificity for the peptide bonds of seed globulin storage proteins with asparagines at the P1 position, and is called asparaginyl endopeptidase (Ishii, 1994). The substrate specificity was observed with purified asparaginyl endopeptidases from developing seeds of castor bean (Hara-Nishimura et al., 1991) and soybean (Scott et al., 1992; Hara-Nishimura et al., 1995), from mature seeds of jack bean (Abe et al., 1993), and from germinating seeds of vetch (Becker et al., 1995). Many seeds accumulate protein reserves in the storage vacuoles, and a number of these proteins undergo proteolytic cleavage, including the 7S and 11S seed storage globulins (Müntz & Shutov, 2002). The 11S seed globulin storage proteins are synthesized as precursors and are cleaved post-translationally in storage vacuoles by an asparaginyl endopeptidase during seed development (Ishii, 1994). In castor bean and soybean seeds, vacuolar processing enzymes were found in the protein bodies and likely associated with the conversion of proproteins into their corresponding mature forms in vacuoles (Hara-Nishimura et al., 1991; Shimada et al., 1994).

Asparaginyl endopeptidases also play a role with bulk degradation and mobilization of storage proteins during seed germination and seedling growth. For example, the asparaginyl endopeptidase, which was also called "legumain-like proteinase" (LLP), was isolated from cotyledons of kidney bean (*Phaseolus vulgaris*) seedlings. It was the first proteinase ever known which *in vitro* extensively degraded native phaseolin, the major storage globulin of this grain legume (Senyuk et al., 1998). In vetch (*Vicia sativa*) seeds, the legumain-like VsPB2 and proteinase B together with additional papain-like cysteine proteinases were responsible for the bulk breakdown and mobilization of storage globulins during seed germination (Schlereth et al., 2000). In *Arabidopsis*, the seed protein profiles were compared between the wild type and a seed-type vacuolar processing enzyme βVPE mutant using a two dimensional gel/mass spectrometric analysis. A significant increase in accumulation of several legumin-type globulin propolypeptides was found in βVPE mutant seeds (Gruis et al., 2002).

For papain-like cysteine protease, the vacuolar SH-EP is synthesized in cotyledons of germinated *Vigna mungo* seeds and is responsible for the degradation of seed globulin storage proteins accumulated in protein bodies. In *Vicia faba* (vetch), globulins such as legumin and vicillin are major seed storage proteins present in the protein bodies of cotyledon, radicle, axis, and shoots. Papain-like cysteine protease such as CPR2 and CPR4 are found in cotyledon and axis of dry and imbibed seeds. Gene expression studies concluded that storage globulin mobilization in germinating vetch seeds is started by the stored cysteine proteases (CPRs), however, the bulk globulin mobilization is mediated by *de novo* synthesized CPRs (Schlereth et al., 2000; Schlereth et al., 2001; Tiedemann et al., 2001). These data suggest that papain-like cysteine proteases may play physiological roles and functions in association with seed storage globulin protein degradation and mobilization during seed germination and seedling growth. In addition to the possible physiological function with seed storage globulin protein degradation, papain-like cysteine proteases have also been implied to play a role in cope with environmental cues. For example, a dehydration-responsive papain-like cysteine protease RD19 was cloned and results showed that its expression was strongly induced under high-salt and osmotic stress conditions, which suggests a possible physiological role of RD19 in association with the regulation of plant cell osmotic potential in *Arabidopsis thaliana* (Koizumi et al., 1993; Xiong et al., 2002). In broccoli, the florets showed water loss during post-harvest storage. Gene expression of papain-like cysteine proteases BoCP4, which exhibited high amino acid sequence identity with *Arabidopsis* RD19, was also found to be dehydration-responsive and was repressed by water and sucrose (Coupe et al., 2003).

Many vacuolar enzymes are synthesized as pro-proteins and become active after proteolytically processed. In seed storage tissues, specific endoplasmic reticulum (ER)-derived compartments containing precursors of cysteine proteases have been described

(Chrispeels & Herman, 2000; Toyooka et al., 2000; Hayashi et al., 2001; Schmid et al., 2001). Germination of the seeds induces the expression and processing of those proteases into the mature active forms, which in turn participate in the degradation of cellular materials in storage tissues and provide nutrients to the growing embryo. The mechanism of asparaginyl endopeptidases (VmPE-1) and papain-like cysteine protease (SH-EP) associated with bulk seed storage globulin protein degradation has been studied in *Vigna mungo*. The vacuolar cysteine protease SH-EP is synthesized in cotyledons of germinated *Vigna mungo* seeds with an N-terminal and a C-terminal prosegments (Okamoto & Minamikawa, 1999; Okamoto et al., 1999). Okamoto & Minamikawa (1995) isolated a processing enzyme, designated VmPE-1. VmPE-1 is a member of the asparaginyl endopeptidases and is involved in the post-translational processing of SH-EP. In addition, the cleavage sites of the *in vitro* processed intermediates and the mature form of SH-EP were identical to those of SH-EP purified from germinated cotyledons of *V. mungo*. Therefore, it is proposed that the asparaginyl endopeptidase (VmPE-1)-mediated processing functions mainly in the activation of proSH-EP during seed germination (Okamoto et al., 1999). The activated SH-EP plays a major role in the degradation of seed storage proteins accumulated in cotyledonary vacuoles of *Vigna mungo* seedlings (Mitsuhashi et al., 1986). These reports demonstrate a role of asparaginyl endopeptidase associated with papain-like cysteine protease in the bulk breakdown and mobilization of storage globulins during seed germination.

3. Characterization of sweet potato asparaginyl endopeptidase *SPAE* and papain-like cysteine protease *SPCP2*

Recently, similar compartments have also been described in vegetative tissues of *Arabidopsis* (Hayashi et al., 2001). These precursor protease vesicles derived from ER are plant specific compartments and contain vesicle-localized vacuolar processing enzyme (γVPE) precursor, which is critical for maturation of the vacuolar protease AtCPY. The vacuolar protease AtCPY in turn participates in the degradation of cellular components including vacuolar invertase AtFruct4 and various proteins in organs undergoing senescence in *Arabidopsis* (Rojo et al., 2003). A mechanism of senescence-induced activation of vesicle-localized vacuolar processing enzyme precursor by releasing its inactive form from the precursor protease vesicle into the acidic lumen of the vacuole is suggested. This activation triggers the processing of downstream proteases for protein degradation and recycling in senescing tissues (Rojo et al., 2003). These data suggest sweet potato asparaginyl endopeptidase *SPAE* and papain-like cysteine protease *SPCP2* may also play roles with functions related to protein degradation for nutrient remobilization during leaf senescence.

3.1 *SPAE*

SPAE had been cloned from senescent leaves with PCR-selective subtractive hybridization and exhibited high amino acid sequence homologies to seed vacuolar legumains/asparaginyl endopeptidases of kidney bean (*Phaseolus vulgaris*), spring vetch (*Vicia sativa*) and jack bean (*Canavalia ensiformis*) (Chen et al., 2004). The conserved catalytic residues (His and Cys) and central β-strands that supported the catalytic residues of human and mouse legumains (Chen et al., 1998) were also found in SPAE, plant legumain/asparaginyl endopeptidase, vacuolar processing enzymes, and the other cysteine proteases (Chen et al., 2004).

Asparaginyl endopeptidase *SPAE* encoded a pre-proprotein precursor, which contained a putative mature protein (325 amino acid residues) and an N-glycosylation site at its C-terminus. The deduced molecular mass of mature SPAE protein was, thus, likely between 33 and 36 kDa that detected by protein gel blot with polyclonal antibody against putative SPAE protein (Chen et al., 2004). Asparaginyl endopeptidase is an atypical cysteine endopeptidase with a reported insensitivity to the inhibitor L-3-carboxy-2,3-trans-epoxypropionyl-leucyl-amino(4-guanidino)butane (E-64) (Okamoto & Minamikawa, 1999). A cysteine protease activity band with a molecular mass near 36 kDa similar to the protein gel blot results was also detected and exhibited insensitivity to E-64 inhibitor (Chen et al., 2004). These data provide indirect evidence to support the existence of asparaginyl endopeptidase in senescent leaves.

In sweet potato, *SPAE* gene expression level is higher in dark- or ethephon-treated leaves similar to that in natural senescent leaves. Hormones such as jasmonic acid (JA) and abscisic acid (ABA) also caused the decrease of chlorophyll contents in treated leaves; whereas, did not significantly alter *SPAE* gene expression level compared to that of untreated dark control in mature green leaves within a 3-day period (Chen et al., 2004). These data suggest that *SPAE* is a senescence-associated gene and its expression in natural or induced senescent leaves is likely controlled by ethylene, but not by JA and ABA.

3.2 *SPCP2*

SPCP2 had been cloned from senescent leaves with PCR-selective subtractive hybridization. The open reading frame of *SPCP2* contained 1101 nucleotides (366 amino acids) and exhibited high amino acid sequence identities with a subgroup of vacuolar cysteine proteases including *Actinidia deliciosa* CP3, *Arabidopsis thaliana* RD19, *Brassica oleracea* BoCP4, *Phaseolus vulgaris* CP2, *Vicia sativa* CPR2, and *Vigna mungo* SH-EP (Chen et al., 2010). These data suggest an intracellular localization of SPCP2 in the vacuole. For SH-EP, a C-terminal KDEL sequence (endoplasmatic retention signal) was proved to be associated with its vacuole-targeting (Okamoto et al., 2003). However, no significant C-terminal KDEL sequence was found for SPCP2. For RD-19, a vacuolar localization was also suggested. However, it can be re-localized to nucleus in the presence of PopP2, an avirulent gene product of *R. solanacearum* (Bernoux et al., 2008; Poueymiro & Genin, 2009). Therefore, it is possible to assume that different vacuolar targeting mechanisms and signal peptides are involved and associated with different related cysteine protease genes.

SPCP2 gene expression was enhanced in natural senescent leaves and can be induced by dark, ethephon, ABA and JA (Chen et al., 2010). Buchanan-Wollaston et al. (2005) analyzed gene expression patterns and signal transduction pathways of senescence in *Arabidopsis* induced by different factors. Transcriptome analysis demonstrated that pathways such as dark, ethylene, and JA are all required for gene expression during developmental senescence. Genes associated with essential metabolic processes such as degradation of proteins and peptides and nitrogen mobilization can utilize alternative pathways for induction (Buchanan-Wallaston et al., 2005). Therefore, a possible explanation which is likely associated with multiple signal transduction pathways is suggested for the induction of sweet potato *SPCP2* gene expression by different factors, including development, dark, ABA, ethephon and JA.

4. Ectopic expression of asparaginyl endopeptidase *SPAE* and papain-like cysteine protease *SPCP2* in transgenic *Arabidopsis*

Sweet potato full-length cDNAs of asparaginyl endopeptidase *SPAE* and papain-like cysteine protease *SPCP2* were individually constructed in the T-DNA portion of recombinant pBI121 vector under the control of *CaMV 35S* promoter for transformation of *Arabidopsis* with *Agrobacterium*–mediated floral dip transformation method (Clough & Bent, 1998). Transgenic *Arabidopsis* plants ectopically expressing sweet potato asparaginyl endopeptidase *SPAE* (Chen et al., 2008) or papain-like cysteine protease *SPCP2* (Chen et al., 2010) were produced, identified and characterized.

4.1 Expression of sweet potato asparaginyl endopeptidase *SPAE* altered seed and silique development in transgenic *Arabidopsis*

Three transgenic *Arabidopsis* plants were isolated and identified with floral dip transformation method (Clough & Bent, 1998). Genomic PCR and protein gel blot analysis confirmed that these *Arabidopsis* plants (YP1, YP2 and YP3) were transgenic and sweet potato *SPAE* gene was expressed and properly processed into mature form with a predicted molecular mass near 36 kDa (Chen et al., 2008). Similar results have also been observed and reported for various plant vacuolar processing enzymes, including *Vigna mungo* VmPE-1 (Okamoto et al., 1999), *Arabidopsis* βVPE (Gruis et al., 2002), *Arabidopsis* γVPE (Kuroyanagi et al., 2002; Rojo et al., 2003). These data suggest that transgenic *Arabidopsis* plants may use similar mechanisms for sweet potato SPAE processing, and thus can produce mature sweet potato SPAE protein products.

Transgenic *Arabidopsis* plants exhibited earlier floral transition from vegetative growth and leaf senescence (Chen et al., 2008). Early transition of vegetative phase to reproductive phase has been considered as a type of senescence. The reasons and mechanisms that sweet potato *SPAE* gene expression can promote earlier floral transition and enhance senescence in transgenic *Arabidopsis* plants are not clear. However, Raper et al. (1988) and Rideout et al. (1992) hypothesized that floral transition is stimulated by an imbalance in the relative availability of carbohydrate and nitrogen in the shoot apical meristem. Barth et al. (2006) suggest that the flowering phenotype is likely linked to the endogenous ascorbic acid content. Degradation and removal of flowering repressor(s) by ectopic SPAE expression in transgenic *Arabidopsis* plants provides another possibility.

Expression of sweet potato *SPAE* in transgenic *Arabidopsis* plants caused altered development of seed and silique, elevated percentage of incompletely-developed siliques, and fewer silique numbers per plant than that of control (Figs. 3 and 4). The reasons for altered phenotypic characteristics in transgenic *Arabidopsis* by sweet potato SPAE expression are not clear. However, sweet potato SPAE is in close association with plant vacuolar processing enzymes of seeds from phylogenetic analysis (Chen et al., 2004). Vacuolar processing enzymes have been reported to be in association with the degradation and mobilization of globulin storage proteins during seed germination and seedling growth in *Phaseolus vulgaris* (Senyuk et al., 1998), *Vigna mungo* (Okamoto et al., 1999), *Vicia sativa* (Schlereth et al., 2000; Schlereth et al., 2001), and *Arabidopsis thaliana* (Gruis et al., 2002). In *Vigna mungo*, VmPE-1 has been demonstrated to increase in the cotyledons of germinating seeds and was involved in the post-translational processing of a vacuolar cysteine endopeptidase, designated SH-EP, which degraded seed storage proteins (Okamoto &

Minamikawa, 1999). A possible explanation that inappropriate pre-degradation of globulin-type storage protein during seed development and maturation by constitutively expressed sweet potato SPAE in transgenic *Arabidopsis* is suggested. The inappropriate pre-degradation of globulin-type storage protein may result in partial repression of seed and silique development which in turn leads to higher incompletely-developed silique percentage and lower silique numbers per plant. These data also suggest that sweet potato asparaginyl endopeptidase SPAE may have enzymatic function similar to seed vacuolar processing enzymes for protein degradation and nutrient recycling during leaf senescence.

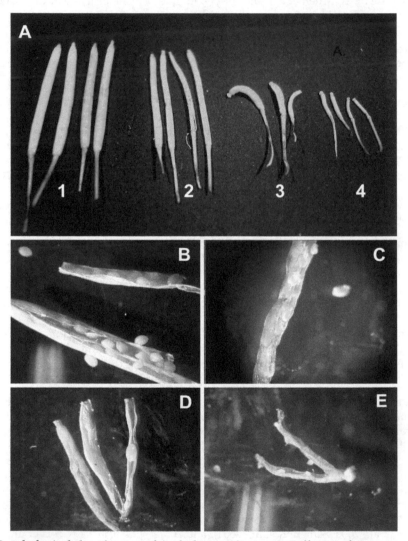

Fig. 3. Morphological classification of Arabidopsis siliques. A. Different silique types (types 1, 2, 3 and 4) classified. **B.** Dissection of type 1 silique; **C.** Dissection of type 2 silique; **D.** Dissection of type 3 silique; **E.** Dissection of type 4 silique (Adapted from Chen et al., 2008).

Expression of Sweet Potato Senescence-Associated Cysteine Proteases Affect Seed and
Silique Development and Stress Tolerance in Transgenic Arabidopsis

245

A. Number of siliques per plant

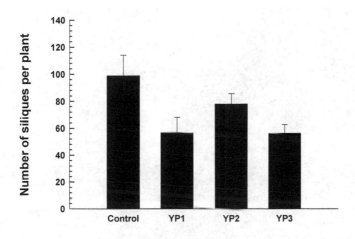

B. Percentage of incomplete silique development

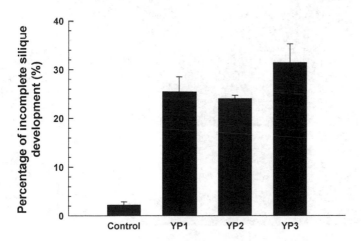

Fig. 4. Comparison of silique number per plant and incompletely-developed silique percentage among control and transgenic T1 plants ectopically expressing sweet potato *SPAE*. **A.** Comparison of silique number per plant. **B.** Comparison of incompletely-developed silique percentages. C and YP1/YP2/YP3 denote non-transformant control and transgenic *Arabidopsis* plants, respectively. The data are from the average of 5 plants per treatment and shown as mean ± S.E. (Adapted from Chen et al., 2008).

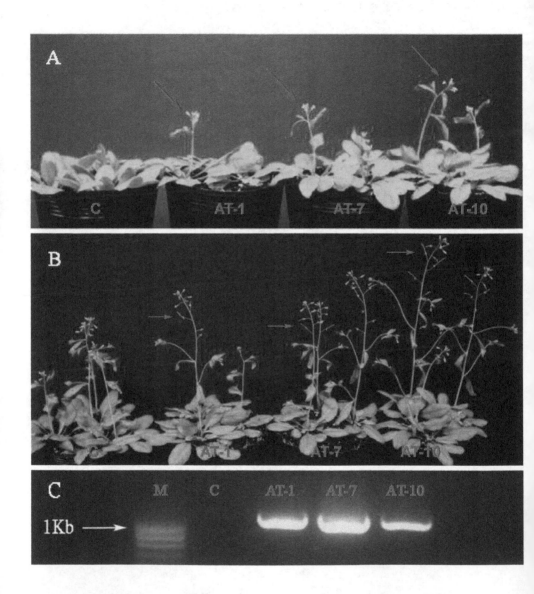

Fig. 5. Comparison of growth patterns among control and transgenic T1 Arabidopsis plants ectopically expressing sweet potato *SPCP2*. **A.** Transition from vegetative growth to flowering was observed and compared 30 days after seed germination. **B.** The appearance and size of inflorescences and siliques were observed and compared 35 days after seed germination. **C.** RT-PCR analysis of *SPCP2*. C and AT denote control and transgenic T1 *Arabidopsis* plants, respectively.

Expression of Sweet Potato Senescence-Associated Cysteine Proteases Affect Seed and
Silique Development and Stress Tolerance in Transgenic Arabidopsis

247

4.2 Expression of sweet potato papain-like cysteine protease *SPCP2* altered seed and silique development and enhanced stress tolerance in transgenic *Arabidopsis*

Transgenic *Arabidopsis* plants were isolated and identified with floral dip transformation method (Clough & Bent, 1998). Genomic PCR and RT-PCR analysis confirmed that the presence and expression of sweet potato papain-like cysteine protease *SPCP2* in transgenic *Arabidopsis* plants (Chen et al., 2010). Transgenic *Arabidopsis* plants also exhibited slightly earlier transition from vegetative to reproductive growth (Fig. 5). The reasons and mechanisms are not clear. However, an imbalance in the relative availability of carbohydrate and nitrogen in the shoot apical meristem (Raper et al., 1988; Rideout et al., 1992), the change of endogenous ascorbic acid content (Barth et al., 2006), and possible non-specific degradation and removal of flowering repressor(s) by ectopic SPCP2 expression are suggested.

Expression of sweet potato SPCP2 in transgenic *Arabidopsis* plants also caused elevated number of incompletely-developed silique (Fig. 6), and reduced average fresh weight per seed and lower germination percentage (Chen et al., 2010). The reasons for the altered phenotypic characteristics in transgenic *Arabidopsis* by ectopic *SPCP2* gene expression are not clear. However, SPCP2 exhibited high amino acid sequence identities with plant papain-like cysteine proteases, such as *Phaseolus vulgaris* CP2, *Vicia sativa* CPR2, and *Vigna mungo* SH-EP. These papain-like cysteine proteases together with vacuolar processing enzymes have been implicated in association with the degradation and mobilization of globulin storage proteins during seed germination and seedling growth in *Phaseolus vulgaris* (Senyuk et al., 1998), *Vigna mungo* (Okamoto et al., 1999), and *Vicia sativa* (Schlereth et al., 2000; Schlereth et al., 2001). These reports provide a possible explanation for the altered phenotypic characteristics observed in transgenic *Arabidopsis* plants, and suggest that sweet potato SPCP2 may have an enzymatic function similar to papain-like cysteine proteases, including *Vigna mungo* SH-EP and *Vicia sativa* CPR2 for protein degradation and nutrient recycling during leaf senescence.

Expression of sweet potato *SPCP2* in transgenic *Arabidopsis* plants exhibited higher salt and drought stress tolerance (Fig. 7), and contained higher relative water content than control (Fig. 8). The reasons for the altered stress responses in transgenic *Arabidopsis* by ectopic SPCP2 gene expression are not clear. However, SPCP2 exhibited high amino acid sequence identities with plant cysteine proteases, such as *Arabidopsis* RD19 and broccoli Bocp4. *Arabidopsis* RD19 was a drought-inducible cysteine protease (Koizumi et al., 1993), and belonged to osmotic stress-responsive genes (Xiong et al., 2002). Under osmotic stress such as drought, high salinity (NaCl or PEG) and ABA treatments, RD19 mRNA transcript was significantly enhanced compared to untreated control (Xiong et al., 2002). Broccoli Bocp4 exhibited high sequence identity to dehydration-responsive *Arabidopsis* RD19, and was also significantly induced in broccoli florets, which were kept in air (dry situation) but not in water or 2% sucrose solution 12 h post harvest (Coupe et al., 2003). Sweet potato cysteine protease *SPCP2* was also inducible by salt and drought stresses in detached leaves (Fig. 9), and its ectopic expression in transgenic *Arabidopsis* caused higher salt and drought resistances (Figs. 7 and 8). Our results agree with these reports and suggest a possible role of sweet potato cysteine protease *SPCP2* in osmotic stress regulation and salt/drought stress tolerance.

A.

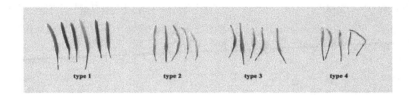

B.

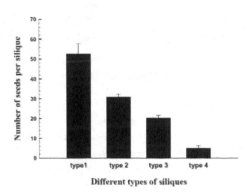

C.

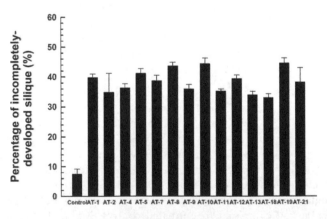

Fig. 6. Comparison of incompletely-developed silique percentages among control and transgenic T1 plants ectopically expressing sweet potato *SPCP2*. **A.** The appearance and size of different silique types (types 1, 2, 3 and 4) were observed and compared 35 days after seed germination. **B.** The average seed number of different silique type. **C.** Comparison of incompletely-developed silique percentage among control and transgenic T1 plants. C and AT denote control and transgenic T1 *Arabidopsis* plants,respectively (Adapted from Chen et al., 2010).

A. NaCl

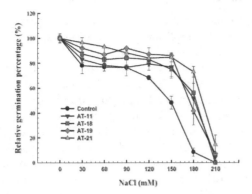

B. Drought

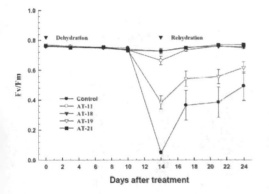

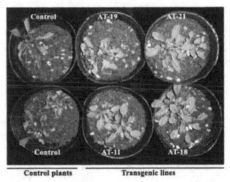

Drought stress at day 14

Fig. 7. Comparison of salt and drought stress tolerance among control and transgenic T1 *Arabidopsis* plants ectopically expressing sweet potato *SPCP2*. **A.** Salt. For salt treatment, seeds were germinated on half strength MS medium plus 3% sucrose and different concentrations of NaCl for ca. 2 weeks, and the relative germination percentages were recorded and compared. **B.** Drought. For drought treatment, upper panel of B is the photochemical Fv/Fm comparison among control and transgenic T1 *Arabidopsis* plants during dehydration and rehydration treatment. Lower panel of B is the morphological comparison among control and transgenic T1 *Arabidopsis* plant at day 14 after drought treatment. The data were the average of total 5 petri dishes for A or 5 seedlings per transgenic line for B, and shown as mean ± S.E. Control and AT-11/AT-18/AT-19/AT-21 denote wild type and transgenic T1 *Arabidopsis* plants, respectively. ▲ indicates the time points of dehydration and rehydration (Adapted from Chen et al., 2010).

5. Correlation of cysteine protease expression with storage protein degradation in sweet potato storage root during sprouting

In sweet potato storage root, trypsin protease inhibitors (TIs) are the main storage proteins and composed of a multiple gene family. It has been implicated that cysteine proteases are likely associated with the degradation of storage root trypsin inhibitors during sprouting (Huang et al., 2005). Therefore, expression and correlation of sweet potato asparaginyl endopeptidase SPAE and papain-like cysteine protease SPCP2 with the degradation and mobilization of the two major storage root trypsin inhibitor bands during sprouting were studied. The sprouts appeared and were visible within the first week of incubation of storage root at room temperature, whereas, degradation of trypsin inhibitors became significant in the later incubation. RT-PCR analysis of *SPAE* and *SPCP2* also demonstrated that their gene expression was significantly higher in the sprout and flesh of sprouting storage root than that of non-sprouting storage root (Fig. 10), and correlated well with the time course of degradation of the two major trypsin inhibitor bands (unpublished data). These data suggest that the asparaginyl endopeptidase SPAE and papain-like cysteine protease SPCP2 may play roles in association with storage root major trypsin inhibitor degradation during sprouting.

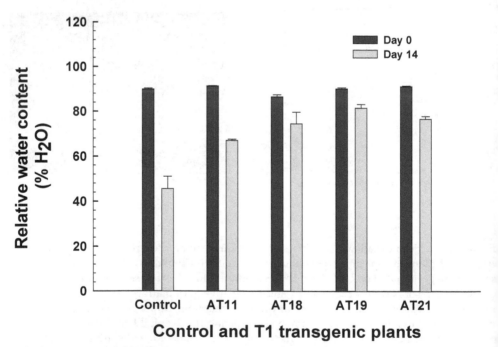

Fig. 8. Comparison of the relative water content (H2O%) between control and transgenic T1 *Arabidopsis* plants ectopically expressing sweet potato *SPCP2* at day 14 after drought treatment. The data were the average of total 5 seedlings per transgenic plants, and shown as mean ± S.E. Control and AT-11/AT-18/AT-19/AT-21 denote wild type and transgenic T1 *Arabidopsis* plants, respectively.

A. NaCl

B. Drought

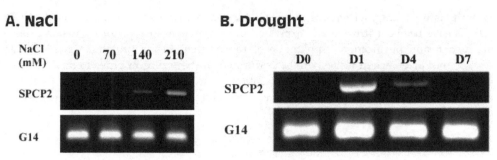

Fig. 9. Induction of sweet potato papain-like cysteine protease SPCP2 gene expression by salt and drought treatments. **A.** Salt treatment. Sweet potato detached leaves were treated with different salt concentrations (0, 70, 140 and 210 mM, respectively,) for 9 days and collected individually for RT-PCR analysis. **B.** Drought treatment. Detached sweet potato leaves were placed on dry paper tower in the dark for 0, 1, 4 and 7 days, and then collected individually for RT-PCR analysis. Sweet potato *G14* encoded a constitutively expressed metallothionein-like protein and was used as a control.

A.

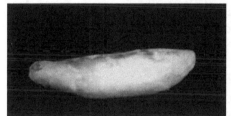
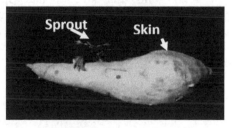

Non-sprouting storage root   **Sprouting storage root**

B.

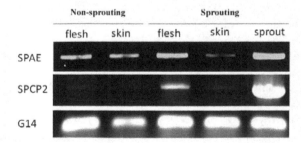

Fig. 10. Expression patterns of sweet potato asparaginyl endopeptidase SPAE and papain-like cysteine protease *SPCP2* in non-sprouting and sprouting st The orage roots. **A.** Storage root morphology. **B.** RT-PCR analysis of *SPAE* and *SPCP2*. Sweet potato *G14* encoded a constitutively expressed metallothionein-like protein and used as a control.

Sweet potato asparaginyl endopeptidase *SPAE* and papain-like cysteine protease *SPCP2* cDNAs have been constructed and expressed in recombinant PET vector for fusion protein production and purification. Application of the purified fusion protein to sweet potato storage root or detached leaves will be performed in the future in order to study further whether they can promote (1) the degradation of storage root major trypsin inhibitors during sprouting, (2) protein degradation and recycling during leaf senescence, or (3) stress tolerance.

6. Conclusion

Sweet potato asparaginyl endopeptidase *SPAE* and papain-like cysteine protease *SPCP2* are senescence-associated genes and significantly enhanced their expression in senescent leaves. Phylogenetic tree analysis shows that *SPAE* and *SPCP2* exhibit close association with vacuolar processing enzyme and papain-like cysteine protease, respectively, which are involved in seed globulin storage protein degradation. Ectopic expression of sweet potato *SPAE* and *SPCP2* in transgenic *Arabidopsis* plants also caused altered phenotypic characteristics, including abnormal seed and silique development, elevated incompletely-developed silique percentage, fewer silique numbers per plant, reduced seed germination percentage, and enhanced tolerance to drought and salt stresses. Based on these data, it can be concluded that sweet potato asparaginyl endopeptidase *SPAE* and papain-like cysteine protease *SPCP2* may play physiological roles in association with protein degradation and nutrient recycling during leaf senescence with enzymatic functions similar to seed globulin storage protein degradation and re-mobilization during germination and seedling growth.

7. References

Abe, Y., Shirane, K., Yokosawa, H., Matsushita, H., Mitta, M., Kato, I. & Ishii, S. (1993). Asparaginyl endopeptidases of jack bean seeds. *J. Biol. Chem.*, 268, 3523-3529

Barth, C., De Tullio, M. & Conklin, P.L. (2006). The role of ascorbic acid in the control of flowering time and the onset of senescence. *J. Exp. Bot.*, 57, 1657-1665

Bernoux, M., Timmers, T., Jauneau, A., Brie, C., de Wit, P.J.G.M., Marco, Y. & Deslandesa, L. (2008). RD19, an *Arabidopsis* cysteine protease required for RRS1-R–mediated resistance, is relocalized to the nucleus by the *Ralstonia solanacearum* PopP2 effector. *Plant Cell*, 20, 2252-2264

Buchanan-Wollaston, V. (1997). The molecular biology of leaf senescence. *J. Exp. Bot.*, 48, 181-199

Buchanan-Wollaston, V., Page, T., Harrison, E., Breeze, E. & Lim, P.O. (2005). Comparative transcriptome analysis reveals significant differences in gene expression and signaling pathways between developmental and dark/starvation-induced senescence in *Arabidopsis*. *Plant J.* 42, 567-585

Chen, H.J., Hou, W.C., Jane, W.N. & Lin, Y.H. (2000). Isolation and characterization of an isocitrate lyase gene from senescent leaves of sweet potato (*Ipomoea batatas* cv. Tainong 57). *J. Plant Physiol.*, 157, 669-676

Chen, H.J., Hou, W.C., Yang, C.Y., Huang, D.J., Liu, J.S. & Lin, Y.H. (2003). Molecular
    cloning of two metallothionein-like protein genes with differential expression
    patterns from sweet potato (*Ipomoea batatas* (L.) Lam.) leaves. *J. Plant Physiol.*, 160,
    547-555
Chen, H.J., Hou, W.C., Liu, J.S., Yang, C.Y., Huang, D.J. & Lin, Y.H. (2004). Molecular
    cloning and characterization of a cDNA encoding asparaginyl endopeptidase
    from sweet potato (*Ipomoea batatas* (L.) Lam) senescent leaves. *J. Exp. Bot.*, 55, 825-
    835
Chen, H.J., Huang, G.J., Hou, W.C., Liu, J.S. & Lin, Y.H. (2006). Molecular cloning and
    characterization of a granulin-containing cysteine protease SPCP3 from sweet
    potato (*Ipomoea batatas*) senescent leaves. *J. Plant Physiol.*, 163, 863-876
Chen, H.J., Wen, I.C., Huang, G.J., Hou, W.C. & Lin, Y.H. (2008). Expression of sweet
    potato asparaginyl endopeptidase caused altered phenotypic characteristics in
    transgenic *Arabidopsis*. *Bot. Stud.*, 49, 109-117
Chen, H.J., Huang, G.J., Chen, W.S., Su, C.T., Hou W.C. & Lin, Y.H. (2009). Molecular
    cloning and expression of a sweet potato cysteine protease SPCP1 from senescent
    leaves. *Bot. Stud.*, 50, 159-170.
Chen, H.J., Tsai, Y.J., Chen, W.S., Huang, G.J., Huang, S.S. & Lin, Y.H. (2010a). Ethephon-
    mediated effects on leaf senescence are affected by reduced glutathione and
    EGTA in sweet potato detached leaves. *Bot. Stud.*, 51, 171-181
Chen, H.J., Su, C.T., Lin, C.H., Huang, G.J. & Lin, Y.H. (2010b). Expression of sweet potato
    cysteine protease SPCP2 altered developmental characteristics and stress
    responses in transgenic *Arabidopsis* plants. *J. Plant Physiol.*, 167, 838-847
Chen, J.M., Rawlings, N.D., Stevens, R.A.E. & Barrett, A.J. (1998). Identification of the
    active site of legumain links it to caspases, clostripain and gingipains in a new
    clad of cysteine endopeptidases. *FEBS Lett.*, 441, 361-365
Chrispeels, M.J. & Herman, E.M. (2000). Endoplasmic reticulum-derived compartments
    function in storage and as mediators of vacuolar remodeling via a new type of
    organelle, precursor protease vesicles. *Plant Physiol.*, 123, 1227-1233.
Clough, S.J. & Bent, A.F. (1998). Floral dip: a simplified method for *Agrobacterium*-
    mediated transformation of *Arabidopsis thaliana*. *Plant J.*, 16, 735-743
Coupe, S.A., Sinclair, B.K., Watson, L.M., Heyes, J.A. & Eason. J.R. (2003). Identification of
    dehydration-responsive cysteine proteases during post-harvest senescence of
    broccoli florets. *J. Exp. Bot.*, 54, 1045 - 1056
Gruis, D.F., Selinger, D.A., Curran, J.M. & Jung, R. (2002). Redundant proteolytic
    mechanisms process seed storage proteins in the absence of seed-type members
    of the vacuolar processing enzyme family of cysteine proteases. *Plant Cell*, 14,
    2863-2882
Hara-Nishimura, I., Inoue, K. & Nishimura, M. (1991). A unique vacuolar processing
    enzyme responsible for conversion of several proprotein precursors into the
    mature forms. *FEBS Lett.*, 294, 89-93
Hara-Nishimura, I., Shimada, T., Hiraiwa, N. & Nishimura, M. (1995). Vacuolar
    processing enzyme responsible for maturation of seed proteins. *J. Plant Physiol.*,
    145, 632-640

Huang, D.J., Chen, H.J., Hou, W.C., Chen, T.E., Hsu, W.Y. & Lin, Y.H. (2005). Expression and function of a cysteine proteinase cDNA from sweet potato (*Ipomoea batatas* [L.] Lam 'Tainong 57') storage roots. *Plant Sci.*, 169, 423–431

Koizumi, M., Yamaguchi-Shinozaki, K. & Shinozaki, K. (1993). Structure and expression of two genes that encode distinct drought-inducible cysteine proteinases in *Arabidopsis thaliana*. *Gene*, 129, 175-182

Kuroyanagi, M., Nishimura, M. & Hara-Nishimura, I. (2002). Activation of *Arabidopsis* vacuolar processing enzyme by self-catalytic removal of an auto-inhibitory domain of the C-terminal propeptide. *Plant Cell Physiol.*, 43, 143-151

Lim, P.O., Kim, H.J. & Nam, H.G. (2007). Leaf senescence. *Annu. Rev. Plant Biol.*, 58, 115-136

Lin, C.H. (2010). Ectopic expression of sweet potato cysteine protease *SPCP2* promotes earlier flowering and enhances drought stress tolerance. Master thesis, Department of Biological Sciences, National Sun Yat-sen University, Kaohsiung, Taiwan

Makino, A. & Osmond, B. (1991). Effects of nitrogen nutrition on nitrogen partitioning between chloroplasts and mitochondria in pea and wheat. *Plant Physiol.*, 96, 355-362

Martı́nez, D.E., Costa, M.L., Gomez, F.M., Otegui, M.S. & Guiamet, J.J. (2008). 'Senescence-associated vacuoles' are involved in the degradation of chloroplast proteins in tobacco leaves. *Plant J.*, 56, 196-206

Marty, F. (1999). Plant vacuoles. *Plant Cell*, 11, 587-599

Mitsuhashi, W., Koshiba, T. & Minamikawa, T. (1986). Separation and characterization of two endopeptidases from cotyledons of germinating *Vigna mungo* seeds. *Plant Physiol.*, 80, 628-634

Müntz, K. & Shutov, A.D. (2002). Legumains and their functions in plants. *Trends in Plant Sci.*, 7, 340-344

Okamoto, T. & Minamikawa, T. (1995). Purification of a processing enzyme (VmPE-1) that is involved in post-translational processing of a plant cysteine endopeptidase (SH-EP). *Eur. J. Biochem.*, 231, 300-305

Okamoto, T. & Minamikawa, T. (1999). Molecular cloning and characterization of *Vigna mungo* processing enzyme 1 (VmPE-1), an asparaginyl endopeptidase possibly involved in post-translational processing of a vacuolar cysteine endopeptidase (SH-EP). *Plant Mol. Biol.*, 39, 63-73

Okamoto, T., Yuki, A., Mitsuhashi, N. & Mimamikawa, T. (1999). Asparaginyl endopeptidase (VmPE-1) and autocatalytic processing synergistically activate the vacuolar cysteine proteinase (SH-EP). *Eur. J. Biochem.*, 264, 223-232

Okamoto, T., Shimada, T., Hara-Nishimura, I., Nishimura, M. & Minamikawa, T. (2003). C-terminal KDEL sequence of A KDEL-tailed cysteine proteinase (sulfhydryl-endopeptidase) is involved in formation of KDEL vesicle and in efficient vacuolar transport of sulfhydryl-endopeptidase1. *Plant Physiol.*, 132, 1892–1900

Poueymiro, M. & Genin, S.P. (2009). Secreted proteins from *Ralstonia solanacearum*: a hundred tricks to kill a plant. *Curr. Opin. Microbiol.*, 12, 44–52

Quirino, B.F., Noh, Y.S., Himelblau, E. & Amasino, R.M. (2000). Molecular aspects of leaf
    senescence. *Trends Plant Sci.*, 5, 278-282

Raper, C.D.J., Thomas, J.F., Tolley-Henry, L. & Rideout, J.W. (1988). Assessment of an
    apparent relationship between availability of soluble carbohydrates and reduced
    nitrogen during floral initiation in tobacco. *Bot. Gaz.*, 149, 289-294

Rideout, J.W., Raper, C.D. & Miner, G.S. (1992). Changes in ratio of soluble sugars and
    free amino nitrogen in the apical meristem during floral transition of tobacco. *Int.
    J. Plant Sci.*, 153, 78-88

Rojo, E., Zouhar, J., Carter, C., Kovaleva, V. & Raikhel, N.V. (2003). A unique mechanism
    for protein processing and degradation in *Arabidopsis thaliana*. *Proc. Natl. Acad.
    Sci. USA*, 100, 7389-7394

Schlereth, A., Becker, C., Horstmann, C., Tiedemann, J. & Muntz, K. (2000). Comparison of
    globulin mobilization and cysteine proteinases in embryonic axes and cotyledons
    during germination and seedling growth of vetch (*Vicia sativa* L.). *J. Exp. Bot.*, 51,
    1423-1433

Schlereth, A., Standhardt, D., Mock, H.P. & Muntz, K. (2001). Stored cysteine proteinases
    start globulin mobilization in protein bodies of embryonic axes and cotyledons
    during vetch (*Vicia sativa* L.) seed germination. *Planta*, 212, 718-727

Scott, M.P., Jung, R., Muntz, K. & Nielsen, N.C. (1992). A protease responsible for post-
    translational cleavage of a conserved Asn-Gly linkage in glycinin, the major seed
    storage protein of soybean. *Proc. Natl. Acad. Sci. USA*, 89, 658-662

Senyuk, V., Rotari, V., Becker, C., Zakharov, A., Horstmann, C., Muntz, K. & Vaintraub,
    L. (1998). Does an asparaginyl-specific cysteine endopeptidase trigger phaseolin
    degradation in cotyledons of kidney bean seedlings? *Eur. J. Biochem.*, 258, 546-
    558

Shimada, T., Hiraiwa, N., Nishimura, M. & Hara-Nishimura, I. (1994). Vacuolar
    processing enzyme of soybean that converts proproteins to the corresponding
    mature forms. *Plant Cell Physiol.*, 35, 713-718

Shimada, T., Yamada, K. & Kataoka, M. (2003). Vacuolar processing enzymes are essential
    for proper processing of seed storage proteins in *Arabidopsis thaliana*. *J. Biol.
    Chem.*, 278, 32292-32299

Tiedemann, J., Schlereth, A. & Müntz, K. (2001). Differential tissue-specific expression
    of cysteine proteinases forms the basis for the fine-tuned mobilization of
    storage globulin during and after germination in legume seeds. *Planta*, 212, 728-
    738

Toyooka, K., Okamoto, T. & Minamikawa, T. (2000). Mass transport of proform of a
    FDEL-tailed cysteine proteinase (SH-EP) to protein storage vacuoles by
    endoplasmic reticulum-derived vesicle is involved in protein mobilization in
    germinating seeds. *J. Cell Biol.*, 148, 453-463

Vierstra, R.D. (1996). Proteolysis in plants: mechanisms and functions. *Plant Mol. Biol.*, 32,
    275-302

Xiong, L., Lee, H., Ishitani, M. & Zhu, J.K. (2002). Regulation of osmotic stress-responsive
    gene expression by the LOS6/ABA1 locus in *Arabidopsis*. *J. Biol. Chem.*, 277, 8588-
    8596

Yoshida, S. (2003). Molecular regulation of leaf senescence. *Curr. Opin. Plant Biol.*, 6, 79-84

Permissions

The contributors of this book come from diverse backgrounds, making this book a truly international effort. This book will bring forth new frontiers with its revolutionizing research information and detailed analysis of the nascent developments around the world.

We would like to thank Assoc. Prof. Yelda Özden Çiftçi, for lending her expertise to make the book truly unique. She has played a crucial role in the development of this book. Without her invaluable contribution this book wouldn't have been possible. She has made vital efforts to compile up to date information on the varied aspects of this subject to make this book a valuable addition to the collection of many professionals and students.

This book was conceptualized with the vision of imparting up-to-date information and advanced data in this field. To ensure the same, a matchless editorial board was set up. Every individual on the board went through rigorous rounds of assessment to prove their worth. After which they invested a large part of their time researching and compiling the most relevant data for our readers. Conferences and sessions were held from time to time between the editorial board and the contributing authors to present the data in the most comprehensible form. The editorial team has worked tirelessly to provide valuable and valid information to help people across the globe.

Every chapter published in this book has been scrutinized by our experts. Their significance has been extensively debated. The topics covered herein carry significant findings which will fuel the growth of the discipline. They may even be implemented as practical applications or may be referred to as a beginning point for another development. Chapters in this book were first published by InTech; hereby published with permission under the Creative Commons Attribution License or equivalent.

The editorial board has been involved in producing this book since its inception. They have spent rigorous hours researching and exploring the diverse topics which have resulted in the successful publishing of this book. They have passed on their knowledge of decades through this book. To expedite this challenging task, the publisher supported the team at every step. A small team of assistant editors was also appointed to further simplify the editing procedure and attain best results for the readers.

Our editorial team has been hand-picked from every corner of the world. Their multi-ethnicity adds dynamic inputs to the discussions which result in innovative outcomes. These outcomes are then further discussed with the researchers and contributors who give their valuable feedback and opinion regarding the same. The feedback is then collaborated with the researches and they are edited in a comprehensive manner to aid the understanding of the subject.

Apart from the editorial board, the designing team has also invested a significant amount of their time in understanding the subject and creating the most relevant covers. They scrutinized every image to scout for the most suitable representation of the subject and create an appropriate cover for the book.

The publishing team has been involved in this book since its early stages. They were actively engaged in every process, be it collecting the data, connecting with the contributors or procuring relevant information. The team has been an ardent support to the editorial, designing and production team. Their endless efforts to recruit the best for this project, has resulted in the accomplishment of this book. They are a veteran in the field of academics and their pool of knowledge is as vast as their experience in printing. Their expertise and guidance has proved useful at every step. Their uncompromising quality standards have made this book an exceptional effort. Their encouragement from time to time has been an inspiration for everyone.

The publisher and the editorial board hope that this book will prove to be a valuable piece of knowledge for researchers, students, practitioners and scholars across the globe.

List of Contributors

Hülya Akdemir and Yelda Ozden Çiftçi
Gebze Institute of Technology, Department of Molecular Biology and Genetics, Plant Biotechnology Laboratory, Kocaeli, Turkey

Jorge Gago and Pedro Pablo Gallego
Applied Plant and Soil Biology, Faculty of Biology, University of Vigo, Vigo, Spain

Aneta Wiktorek-Smagur
Nofer Institute of Occupational Medicine, Poland

Katarzyna Hnatuszko-Konka, Aneta Gerszberg, Tomasz Kowalczyk, Piotr Luchniak and Andrzej K. Kononowicz
Department of Genetics Plant Molecular Biology and Biotechnology University of Lodz, Poland

Pelayo Pérez-Piñeiro, Jorge Gago and Pedro P.Gallego
Applied Plant and Soil Biology, Dpt. Plant Biology and Soil Science, Faculty of Biology, University of Vigo,Vigo, Spain

Mariana Landín
Dpt. Pharmacy and Pharmaceutical Technology, Faculty of Pharmacy, University of Santiago, Santiago de Compostela, Spain

Naruemon Khemkladngoen, Naoki Wada, Suguru Tsuchimoto, Joyce A. Cartagena, Shin-ichiro Kajiyama and Kiichi Fukui
Osaka University, Japan

Wataru Takahashi and Tadashi Takamizo
Forage Crop Research Division, NARO Institute of Livestock and Grassland Science, Japan

Rosangela L. Brandão, Newton Portilho Carneiro, Antônio C. de Oliveira, Gracielle T. C. P. Coelho and Andréa Almeida Carneiro
Embrapa Maize and Sorghum, Brazil

Bimal Kumar Ghimire
Department of Ethnobotany and Social Medicine Studies, Sikkim University, Sikkim, India

Jung Dae Lim
Department of Herbal Medicine Resource, Kangwon National University, Samcheok, South Korea

Chang Yeon Yu
Division of Applied Plant Science, Kangwon National University Chunchon, Kangwondo, South Korea

Tomonori Sonoki
Faculty of Agriculture and Life Science, Hirosaki University, Japan

Yosuke Iimura
National Institute of Advanced Industrial Science and Technology, Japan

Shinya Kajita
Graduate School of Bio-Applications and Systems Engineering, Tokyo University of Agriculture and Technology, Japan

Yoshihiro Narusaka, Mari Narusaka, Satoshi Yamasaki and Masaki Iwabuchi
Research Institute for Biological Sciences (RIBS), Okayama, Japan

Nikon Vassilakos
Benaki Phytopathological Institute, Greece

H. Hartings, M. Fracassetti and M. Motto
CRA-Unità di Ricerca per la Maiscoltura, Bergamo, Italy

Hsien-Jung Chen, Chia-Hung Lin, Yi-Jing Tsai, Zhe-Wei Lin and Shu-Hao Liang
Department of Biological Sciences, National Sun Yat-sen University, Kaohsiung, Taiwan

Guan-Jhong Huang
School of Chinese Pharmaceutical Sciences and Chinese Medicine Resources, College of Pharmacy, China Medical University, Taichung, Taiwan

Yaw-Huei Lin
Institute of Plant and Microbial Biology, Academia Sinica, Nankang, Taipei, Taiwan

Printed in the USA
CPSIA information can be obtained
at www.ICGtesting.com
JSHW011444221024
72173JS00004B/937

9 781632 395986